Impacts of Climate Change on Rainfall Extremes and Urban Drainage Systems

Impacts of Climate Change on Rainfall Extremes and Urban Drainage Systems

Patrick Willems, Jonas Olsson, Karsten Arnbjerg-Nielsen, Simon Beecham, Assela Pathirana, Ida Bülow Gregersen, Henrik Madsen and Van-Thanh-Van Nguyen

IWA Publishing
London • New York

Published by **IWA Publishing**
Alliance House
12 Caxton Street
London SW1H 0QS, UK
Telephone: +44 (0)20 7654 5500
Fax: +44 (0)20 7654 5555
Email: publications@iwap.co.uk
Web: www.iwapublishing.com

First published 2012

British Library Cataloguing in Publication Data
A CIP catalogue record for this book is available from the British Library

ISBN 9781780401256 (Paperback)
ISBN 9781780401263 (eBook)

Contents

About the IGUR

The International Working Group on Urban Rainfall (IGUR) of the International Water Association (IWA) and the International Association of Hydraulic Engineering and Research (IAHR) operates under the umbrella of the IWA/IAHR Joint Committee on Urban Drainage and aims to disseminate knowledge, technology and existing procedures relating to rainfall measurement, analysis and modelling for urban stormwater management. The IGUR coordinates and pursues activities related to all aspects of rainfall related phenomena in urban hydrology. It has members from different climatic regions in both developing and developed countries. IGUR members are actively working in areas related directly to precipitation data and are fully conversant with precipitation studies in their countries. More information can be found on the IGUR website: http://www.kuleuven.be/hydr/gurweb

People interested in the activities of the IGUR or interested in joining the working group can contact the

–IGUR Chair, Prof. Patrick Willems (Patrick.Willems@bwk.kuleuven.be)
–IGUR Secretary, Dr. Thomas Einfalt (einfalt@hydrometeo.de)

About the Authors

Patrick Willems is Professor in water engineering at University of Leuven (Hydraulics Division, Department of Civil Engineering, and Leuven Sustainable Earth Research Center) and part-time lecturer at the Vrije Universiteit Brussel. He is author of more than 200 publications, about 45 in peer-reviewed international journals. He is promoter of 15 PhD researchers, many of them focusing on climate change impact estimates on hydrological extremes (floods, droughts and extreme surface water pollution levels, in and outside Europe). He is currently chairman of the IWA/IAHR International Working Group on Urban Rainfall.

University of Leuven, Hydraulics division, Kasteelpark Arenberg 40, BE-3001 Leuven, Belgium (Patrick.Willems@bwk.kuleuven.be)

Jonas Olsson is senior researcher in hydrology at the Swedish Meteorological and Hydrological Institute (SMHI). His main fields of research are climate hydrology, statistical hydrology and dynamical runoff modelling and forecasting. His main work task is project management and supervision and he has functioned as WP-leader in several EU-projects. He has published some 45 peer-reviewed papers, is a frequent reviewer of papers and applications and has had a range of other scientific assignments. In terms of education he lectures at university courses as well as co-supervises Ph.D. and M.Sc. students.

Swedish Meteorological and Hydrological Institute, Research & Development (hydrology), SE-601 76 Norrkoping, Sweden (jonas.olsson@smhi.se)

Karsten Arnbjerg-Nielsen is Associate Professor at the Department of Environmental Engineering at the Technical University of Denmark. His main fields of research are urban hydrology, rainfall modelling, and adaptation to climate change impacts. He worked as a consultant from 1998 to 2008 as project manager and Head of Innovation before returning to academia. He has published more than 20 ISI publications and is Editor of Water Science and Technology responsible for the topic climate change. His contributions to this book were supported by **Ida Bülow Gregersen**, who is currently a PhD fellow at the Department of Environmental Engineering at the Technical University of Denmark, supervised by Karsten Arnbjerg-Nielsen, Henrik Madsen and Dan Rosbjerg. Her research area is statistical modeling of rainfall extremes, with a focus on trend identification and urban hydrology.

Technical University of Denmark, Department of Environmental Engineering, DK-2800 Lyngby, Denmark (kan@env.dtu.dk and idbg@env.dtu.dk)

Simon Beecham is currently Professor of Sustainable Water Resources Engineering and Head of the School of Natural and Built Environments at the University of South Australia. He is also a Board Director for Water Quality Research Australia Limited (WQRA), which is a national research centre established by the Australian water industry in 2008. Simon's research interests centre around the effects of climate change on total water cycle management and in particular on water sensitive urban design. He is also the author of the Syfon software program which has been used to design the roofwater harvesting systems for Sydney's Stadium Australia and Melbourne's MCG, as well as the siphonic roof drainage systems at the Norman Foster designed Chek Lap Kok airport in Hong Kong and the new International Terminal Buildings at Adelaide, Sydney and Kuala Lumpur airports.

University of South Australia, School of Natural and Built Environments, North Terrace, Adelaide, South Australia 5000, Australia, (simon.beecham@unisa.edu.au)

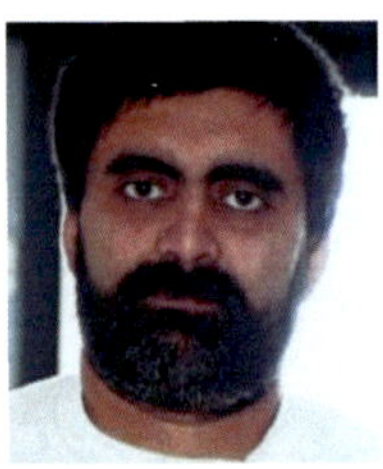

Assela Pathirana works since 2006 as a senior lecturer on Urban Drainage and Sewerage at UNESCO-IHE in Delft Netherlands. His main research interests are focused on the interactions between urban environment, atmosphere and hydrological cycle, which include: climate change, scaling issues with special emphasis on translating climatic forcing to drivers of changes in urban-scale hydrology, single and multi-objective optimization in urban water applications. He teaches graduate courses in Urban Drainage and Sewerage, Asset Management of Urban Water Infrastructure, Water transport and distribution, Water and Climate and Computer Programming. During his post-doctoral research (2001–2003) at Chuo University, in Tokyo, Japan, he explored dynamic downscaling of rainfall with regional atmospheric models.

UNESCO-IHE Institute for Water Education, Westvest 7, 2611 AX Delft, The Netherlands (a. pathirana@unesco-ihe.org)

Henrik Madsen is head of innovation at DHI responsible for the research and development activities within climate change adaptation. His main fields of research are hydrological modelling and forecasting, water resources management, extreme value analysis, stochastic modelling, and climate change impact and adaptation analysis. He has more than 50 peer-reviewed international journal publications. He is member of the editorial board of Hydrology and Earth System Sciences and Journal of Hydroinformatics.

DHI Water & Environment, Water Resources Department, DK-2970 Hørsholm, Denmark (hem@dhigroup.com)

Van-Thanh-Van Nguyen is Endowed Brace Professor and Chair of Civil Engineering Department at McGill University. He is also Director of the Brace Centre for Water Resources Management and Associate Director of the Global Environmental and Climate Change Centre. His professional contributions over the past 25 years have been mostly in Hydrology and Water Resources Management. He has served as expert consultant to various organizations in Canada and in other countries, and has been invited professors at universities in Canada, Japan, Singapore, and Malaysia. He was an active member in several professional associations (President, Hydrological Science Section, Asia-Oceania Geosciences Society, 2006–2008).

Mc Gill University, Department of Civil Engineering and Applied Mechanics, Montreal, Quebec, Canada H3A 2K6, (van.tv.nguyen@mcgill.ca)

Acknowledgements

The authors would like to thank the anonymous reviewers for their initial assessment of this book and for their supportive suggestions, which helped improving this book in a significant way.

The contributions by Patrick Willems to this book were made possible through various research projects funded by the Belgian Science Policy Office, the Flemish Environment Agency and Flanders Hydraulics Research since 2005, and a travel grant by the Fund for Scientific Research – Flanders for a 6 months research stay at the Laboratoire des Sciences du Climat et de l'Environnement at Gif-sur-Yvette, France.

The contributions by Jonas Olsson were mainly funded by the Swedish Research Council FORMAS, through the project HYDROIMPACTS 2.0 (contract no. 2009-525). Additional funding was provided by the Swedish Environment Protection Agency, through project CLEO (contract no. 802-0115-09), and by EU FP7, through project SUDPLAN (contract no. ICT-2009-6.4).

The contributions by Karsten Arnbjerg-Nielsen and Ida Bülow Gregersen were partly funded by the Danish Strategic Research Council as part of the project "Center for Regional Change in the Earth System", contract no. 09-066868.

The contributions by Van-Thanh-Van Nguyen and his research team to this book were made possible through various research projects funded by the Natural Sciences and Engineering Research Council of Canada over the past years.

These financial supports are gratefully acknowledged.

The following persons have made contributions to the text: Thomas Einfalt (hydro & meteo GmbH & Co. KG, Germany), Johan Södling and Wei Yang (from the Swedish Meteorological and Hydrological Institute).

Acronyms

AIC	Akaike Information Criterion
AMO	Atlantic Multidecadal Oscillation
AMS	Annual Maxima Series
ANN	Artificial Neural Network
AOGCM	Atmosphere-Ocean Global Circulation Model
AR4	Fourth Assessment Report (2007) of the IPCC
AR5	Fifth Assessment Report (2007) of the IPCC
ARF	Areal Reduction Factor
ARI	Average Recurrence Interval
BLRP	Bartlett-Lewis Rectangular Pulse
C-C	Clausius-Clapeyron
CDF	Cumulative Distribution Function
CFL	Courant–Friedrichs–Lewy
CMIP	Coupled Model Intercomparison Project of the World Climate Research Programme (CMIP3: Phase 3, CMIP4: Phase 4, CMIP5: Phase 5)
CORDEX	COordinated Regional climate Downscaling Experiment of the World Climate Research Programme
CSIRO	Commonwealth Scientific and Industrial Research Organisation, Australia
CSO	Combined Sewer Overflow
CUSUM	Cumulative Sum
CV	Coefficient of Variation
DC	Delta Change
DDF	Depth–Duration–Frequency
DBS	Distribution-Based Scaling
DKK	Danish Kroner
ECMWF	European Centre for Medium-Range Weather Forecasts
EEA	European Environment Agency

ENSEMBLES	Research project European-scale ensemble RCM prediction supported by the European Commission's 6th Framework Programme (http://www.ensembles-eu.org/)
EUR	Euro
FFT	Fast Fourier Transform
FNL	NCEP Final Analysis (reanalysis)
GCM	General Circulation Model or Global Climate Model
GCOS	Global Climate Observing System
GFS	Global Forecast System
GHCN	Global Historical Climatology Network
GHG	Green House Gasses
GLM	Generalized Linear Model
GMT	Greenwich Mean Time
GPD	Generalized Pareto Distribution
GTS	Global Telecommunications System
HCN	Historical Climatology Network of the USA
IDF	Intensity-Duration-Frequency
IPCC	Intergovernmental Panel on Climate Change
LAM	Limited Area Model
LARS-WG	a stochastic weather generator
LID	Low Impact Development
LIUD	Low Impact Urban Design
LSM	Land Surface Model
MAP	Mean Annual Precipitation
ML	Maximum Likelihood
MMM	Mesoscale Meteorological Modelling
MODIS	Moderate Resolution Imaging Spectroradiometer
MOM	Methods of Moments
MOS	Model Output Statistics
NASA	National Aeronautics and Space Administration of the USA
NCEP	National Centers For Environmental Prediction of NOAA
NSRP	Neyman-Scott Rectangular Pulses
NOAA	US National Oceanic and Atmospheric Administration
PBL	Planetary Boundary Layer
PC	Personal Computer
PDO	Pacific Decadal Oscillation
PDS	Partial Duration Series
PMP	Probable Maximum Precipitation
POT	Peak-Over-Threshold
PP	Perfect Prognosis
PRUDENCE	Research project on European-scale RCM projections supported by the European Commission's 5th Framework Programme (http://prudence.dmi.dk)
PWM	Probability Weighted Moments
Q-Q	Quantile-Quantile

RCM	Regional Climate Model
RSM	Regional Spectral Model
RTD	Real-Time Control
SDS	Statistical Down Scaling
SDSM	Statistical Down Scaling Model
SEK	Swedish Kronor
SEA	South Eastern Australia
SEACI	South Eastern Australian Climate Initiative
SOI	Southern Oscillation Index
SRES	Special Report on Emissions Scenarios of the IPCC
STR	Sub-Tropical Ridge
STRC-EMS	Science and Training Resource Center – Environmental Model System of the U.S. National Weather Service
SuDS	Sustainable Drainage Systems
SWMM	Storm Water Management Model
UCAR	University Corporation for Atmospheric Research
UHI	Urban Heat Island
UKWIR	UK Water Industry Research
USD	U.S. Dollar
USGS	United States Geological Survey
WCRP	World Climate Research Programme
WFT	Windowed Fourier Transform
WRF	Weather Research and Forecasting (model)
WSUD	Water Sensitive Urban Design
WT	Weather Type
WMO	World Meteorological Organization

Executive Summary

Cities are becoming increasingly vulnerable to flooding because of rapid urbanization, installation of complex infrastructure, and changes in the precipitation patterns caused by anthropogenic climate change. While there are many previous works that have dealt with the effects of urbanization, the number of quantitative assessment studies related to the impacts of climate change on urban drainage remains, however, rather limited. This is partly due to the particular difficulties of dealing with this type of climate impact assessment for urban catchments. In particular, downscaling of outputs from global circulation models or regional climate models to urban catchment scales are needed because these models do not contain an adequate description of the rainfall governing processes at relevant high temporal and spatial resolutions. More specifically, for urban catchments these resolutions could be as small as a few kilometres spatially and as small as a few minutes temporally. Consequently, the expected results from such impact studies can be highly uncertain and dependent on the feasibility and reliability of the downscaling process. This problem becomes more challenging when dealing with the extreme runoff events since the properties of such extreme events do not necessarily reflect those of the average precipitation.

This book provides a state-of-the-art review of existing methodologies and results that are relevant to the assessment of climate change impacts on urban rainfall extremes as well as on urban hydrology and hydraulics. In particular, this overview focuses on the limitations and pitfalls of current methods, which are important for every user of results from urban impact studies. Further, the various issues and challenges facing the research community in dealing with the assessment and adaptation of climate change impacts for urban drainage infrastructure design and management are discussed.

This book has been prepared by the International Working Group on Urban Rainfall (IGUR) of the International Water Association (IWA) and the International Association of Hydraulic Engineering and Research (IAHR). The IGUR operates under the umbrella of the IWA/IAHR Joint Committee on Urban Drainage. The reader has to be aware that this book is extensive but it has not been possible to cover all investigations and research papers. We would therefore appreciate any feedback regarding significant omissions for potential inclusion in future editions.

Electronic supplement:

This book is accompanied by an electronic supplement provided on the following IWA Water Wiki website: http://www.iwawaterwiki.org/xwiki/bin/view/Articles/ICCREUDS

Some training material is uploaded on restricted pages. To access those pages, you need to create first your own IWA Water Wiki user account. Once this is done, you can email your Wiki username together with the password "CCIGUR" to the Water Wiki Community Manager (WaterWiki@iwap.co.uk), who will grant you access to be restricted pages.

The electronic supplement contains the electronic training material referred to in the Appendices (scripts in R, Matlab, Python; and example datasets). This supplement will also contain future text updates.

Chapter 1

Introduction

1.1 NEED FOR ASSESSING CLIMATE CHANGE IMPACTS ON URBAN DRAINAGE

For more than a century, large-scale separate and combined sewer systems have been constructed across many cities worldwide. As the name suggests, combined sewer systems convey both urban runoff and sewage in the same (combined) pipe drainage system. This is the most common type of urban drainage system in Western Europe and North American regions. The alternative solution is a separate system, which consists of parallel sewers for storm and waste water (e.g. Burian *et al.* 1999; Butler & Davies, 2010). Separate systems are widely used in many countries in Asia, Australia, Europe and North America for newly developed urban areas. In separate sytems, sewage is conveyed in smaller diameter pipe systems while urban runoff is conveyed separately, usually in either open channels or street pipe drainage systems. They are built to reduce the pollution effect of urban drainage on receiving waters, and to enhance the efficiency of the wastewater treatment plant (less diluted wastewater). For instance, in Japan separate systems are only constructed since the 1980s and currently about 20% of the sewer systems are of the combined type. This percentage of combined systems is much higher in Europe, for example about 70% in the UK (Butler & Davies, 2010). For clarity in this book, both combined and separate systems will henceforth be referred to as *urban drainage systems*.

In general, these urban drainage systems have reduced the vulnerability of the cities to the health risks since they are often built as part of municipal sanitation programs. However, the installation of these systems could make them more vulnerable to rainfall extremes, partly due to the lack of consideration to what occurs when the design criteria are exceeded. In particular, urban land use is constantly changing in response to the continuous changes in demographic and socio-economic conditions of the population (O'Loughlin *et al.* 1995). As a consequence of these environmental changes, designers and managers must now cope with the increase in surface imperviousness and the shorter response time of urban catchments, which boost stormwater runoff volumes and velocities beyond the capacity of existing drainage systems.

For most cities, it is expected that these trends will continue over the coming decades. At the same time, many highly developed regions already realise that their urban design and planning processes urgently need to incorporate more sustainable approaches. Many urban water systems are particularly vulnerable to rapid population growth and climate change (Semadeni-Davies *et al.* 2008). In the presence of climate change induced uncertainty, urban water systems need to be more resilient and multi-sourced. This is partly because of decreasing volumetric rainfall trends in many parts of the world, which might have severe effects on reservoir yields and operational practices. In addition, severe intensity rainfall events can cause failure of drainage system capacity and subsequent urban flood inundation problems (Beecham & Chowdhury, 2012).

Besides this increased vulnerability, there is also strong evidence that the probabilities and risks of urban flooding and sewer surcharge are changing due to the increasing trends of some climatic parameters such as precipitation and temperature extremes (Stone *et al.* 2000; Alexander *et al.* 2006; Allan & Soden, 2008). In particular, in their Fourth Assessment Report (AR4) the Intergovernmental Panel on Climate Change (IPCC) of the World Meteorological Organization and the United Nations Environment Program reports for the late 20th century a worldwide increase in the frequency of extreme rain storms as most likely a result of global warming (IPCC, 2007a; WMO, 2009a; Giorgi *et al.* 2011). Extremes were by the IPCC (2007a) defined as events that are relevant from a disaster risk management perspective, for example urban flood disasters. The increase in rainfall extremes is most pronounced in the period of anthropogenic greenhouse gas (GHG) induced twentieth-century warming (approximately 0.5 deg. C worldwide in the period 1976–2000) after the so-called climate shift (IPCC, 2007a). The study by Min *et al.* (2011) revealed that human-induced increases in GHG have contributed to the observed intensification of heavy rainfall events over approximately two-thirds of the data-covered parts of the Northern Hemisphere land areas. Based on climate model simulations with different future GHG emission scenarios, IPCC (2007a) furthermore concluded that it is very likely that this trend will continue in the 21st century. The consequences of these changes have to be assessed in a perspective of sustainable development. Water managers have to anticipate these changes in order to limit flood risks for communities. Also the insurance industry, as well as the various water users and policy makers, need quantification of these risks so as to develop and adapt policies.

Consequently, the number of hydrological impact studies of climate change has increased greatly in recent years. These studies, however, most often focus on river discharge extremes and low flow risks. The number of climate change studies dealing with urban drainage impacts is still rather limited, partly because they require a specific focus on small urban catchment scales (normally on a scale of 1–10 km^2) and short duration precipitation extremes (normally less than 1 hour). This is because of the small characteristic time scales of the processes involved in the hydrological cycle within urban areas. These processes react very quickly to rainfall.

Despite a significant increase in computational power in recent years, the spatial resolution of climate models still remains relatively coarse and they are therefore unable to resolve significant climate features relevant at the fine scales of urban drainage systems. They also have limitations in the accuracy with which they describe precipitation extremes (e.g. high-intensity convective storms leading to urban flooding). This is due to an incomplete knowledge and inadequate description of the complex nonlinear and dynamical phenomena during a convective storm leading to the most extreme events on a local scale. As such, the climate model results cannot be used directly for providing an adequate assessment of the impacts of future climate change on urban hydrological processes, which is usually undertaken through simulation with urban hydrological and sewer system models. This poses strong challenges to the urban drainage impact modeller.

1.2 OVERVIEW OF CLIMATE CHANGE IMPACT ASSESSMENT FOR URBAN DRAINAGE

Evaluating regional impacts on urban drainage from possible future climate change requires a methodology to estimate extreme and short-duration rainfall statistics for the time period and the geographical region of interest. In general, two physical systems are involved: the climate system and the urban drainage system (Figure 1.1). Climate models can simulate the effects of climate forcing scenarios such as changes in GHG emissions or GHG concentrations in the atmosphere, for example due to anthropogenic activities, on the climate system. Various types of climate models – global (GCM) and/or regional (RCM) – can be used, providing climate system outputs (climatic variables including extreme rainfall). As GCMs and RCMs are effectively deterministic models of atmospheric processes, they calculate a single value of climate variables at each time step and for each grid cell.

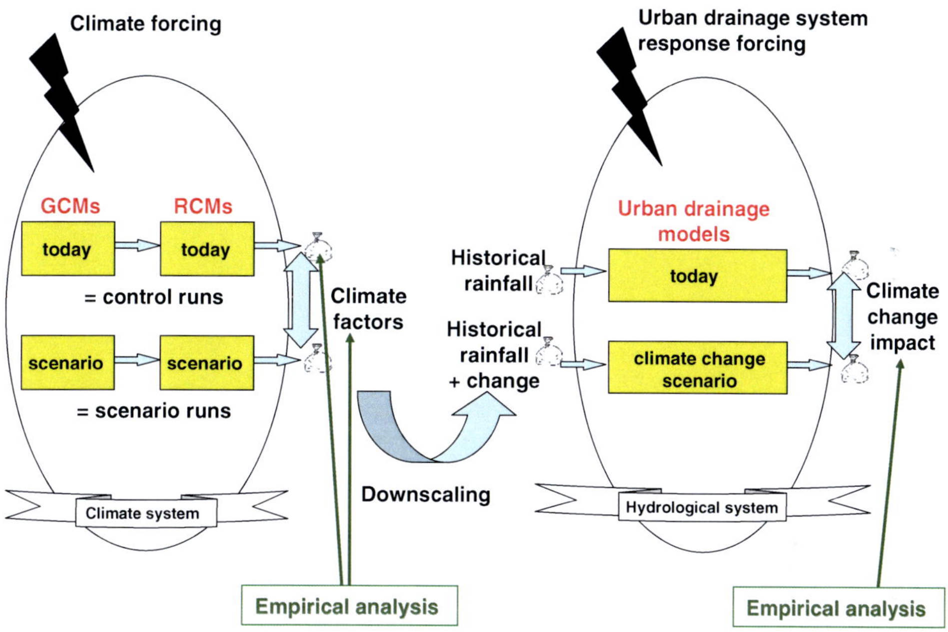

Figure 1.1 Different aspects involved in urban hydrological impact analysis of climate change.

Also urban drainage impact models can take several forms and these can require different rainfall inputs. Most common is the use of simulation models in which rainfall inputs are translated into discharge (either long-term rainfall series or design storms) (Butler & Davies, 2010). Other models are (semi-)probabilistic where probability distributions of urban runoff discharge are calculated based on the rainfall input distribution (e.g. Bacchi *et al.* 2008). Some urban drainage models account for evapotranspiration (mainly important where vegetated areas are considered) and temperature (mainly for snow melt calculations), but these inputs are generally of secondary importance in comparison with the rainfall input.

For historical periods, the results of the climate models can be validated based on historical observations. Also based on historical records, climate change effects can be investigated by analysing trends in available series (i.e. long-term rainfall series). This is termed "empirical analysis" in Figure 1.1. For future conditions, simulation models for both the climate system and the urban drainage system are needed. Changes simulated in the climate system output (rainfall) due to (anthropogenic GHG) climate forcing need to be transferred to changes in the urban drainage model inputs.

The changes imposed by the climate forcing should be compared to the inherent natural variation of precipitation. The rainfall generating processes occur over temporal scales ranging from multi-decadal to sub-minute resolutions with corresponding changes in spatial scales. Therefore great care should be taken when analysing outcomes of historical and simulated precipitation series, as trends in time series of 20 to 40 years can be due to natural variation rather than a change in precipitation patterns. As such the assumption of inter-annual independence is clearly violated.

Another important feature of GCMs and RCMs is their spatial and temporal resolution. Section 1.1 already highlighted that this resolution is too coarse for urban drainage applications. Consequently all applications of impacts of climate change in urban drainage must make assumptions about how anticipated future precipitation patterns will impact at the urban catchment temporal and spatial scales. These scales are related to the area of the urban catchment (which is typically limited to the size of a town, city or district). Due to the limited area, the relevant temporal scales are generally short and are controlled by the concentration time of the urban drainage system (the time the rainwater needs to move from the most remote location in the urban catchment to the impact location of interest; Chow, 1964; Chow et al. 1988). This means that rainfall information is needed with time steps smaller or equal to the smallest concentration time in the system. To bridge the gap between the climate model scales and the local urban drainage scales and to account for the inaccuracies in describing precipitation extremes, downscaling techniques and bias correction methods are required. GCM projections can be downscaled by using a higher resolution RCM nested within a GCM, called *dynamical downscaling*. *Statistical downscaling* relates large-scale climate variables to local scale climate using empirical-statistical relationships. Traditionally, statistical downscaling of GCM projections has been considered, but in recent years statistical downscaling methods that optimally combine dynamic and statistical downscaling have been developed.

The changes in downscaled local short-duration rainfall extremes then can be assessed and transferred to changes in the inputs for urban hydrological impact models. The models were calibrated based on historical rainfall data, which usually take the form of design rain storms or full rainfall time series. These rainfall data will be changed according to the results obtained from the downscaled climate model projections. Finally, the changes in impact results between the today's climate and the climate change scenarios are to be assessed. If long time series of observations are available for the impact variables (i.e. sewer runoff flows, flood frequencies, sewer overflow frequencies), impact assessment can also be done after trend analysis on the series ("empirical analysis" in Figure 1.1). However, this trend analysis cannot go beyond the period covered by the historical observations. While the model-based impact assessment can look into future trends.

It is important to be aware of the uncertainties introduced at each stage of the process. When attempting to make a future projection, as opposed to a hypothetical scenario or numerical experiment, the uncertainty begins with the need to arbitrarily choose a climate forcing scenario, and this initial uncertainty is then compounded by further modelling variability all the way to the final (urban) catchment-scale projection. The uncertainty introduced at each step comes from several sources, such as natural variability, physical parameterisations of the models, and the lack of process descriptions (known or unknown) that are important for modelling climate change.

The above methodology outlines the impact assessment of climate change to urban drainage, focusing on its main driver, namely the changes in short-duration rainfall statistics. However, many other drivers affect the performance of an urban drainage system, particularly urbanization and changes in urban drainage management and planning. This book focuses on estimation of climate change impacts on urban drainage, but the reader has to be aware that these other drivers might be as important. For example, urbanization and associated increasing population can lead to a significant increase in water use and increased impermeable areas. Urban areas might also be affected by other types of climate change impacts such as sea level rise and increase in river flood frequency. When combined, these changes could have impacts that are more significant than those caused by changes in short-duration rainfall extremes only and/or due to climate change only. Also note that urban water management practices are likely to improve into the future, and this might potentially offset some or all of the negative impacts.

1.3 SCOPE AND LIMITATIONS

This book aims to present a state-of-the-art review of climate change impact assessment in the field of urban drainage. More specifically, the objectives of the book are:

- To give an overview of current practices with respect to rainfall analysis and modelling for urban drainage simulations;
- To review trend analyses in historical urban rainfall extremes;
- To introduce the basic concepts of atmospheric modelling;
- To describe the fundamentals of dynamical and statistical downscaling of rainfall;
- To review evaluations of downscaled rainfall;
- To review expected future changes in urban rainfall extremes and the corresponding impacts on urban drainage;
- To give an overview of adaptation issues, principles and methods; and
- To provide practical tools and instructions.

The book provides on the one hand a review of methods and difficulties concerning the assessment of climate change impacts on urban rainfall extremes and urban drainage systems. On the other hand, it provides a practical and useful guide on these methods. The audience of the book is therefore not only scientists, but also practitioners (urban drainage engineers, urban planners) and students.

While the book aims to give a representative overview of current knowledge, practices and challenges associated with climate change impact investigations in the field of urban drainage and rainfall extremes, the authors are aware of some limitations:

- Because of its focus on urban drainage, this book mainly focuses on extreme rainfall at small (sub-daily) time scales. However, some references to investigations using daily time scales are also provided.
- The book presents many case studies based mainly on European conditions, but also from other continents, such as North America (USA and Canada), Asia and Australia.
- While the scope of this book is extensive, it has not been possible to cover all investigations and research papers. We have relied on what we view as important scientific contributions but we would appreciate any feedback regarding significant omissions for potential inclusion in future editions.
- The authors are aware that climate science evolves very rapidly, which means that new knowledge and methodologies might have become available after the date that the book manuscript was delivered to the publisher. The authors therefore recognise that future updates will be required. Readers are invited to send their additions and comments to the authors. Text updates will be provided together with the electronic supplement through the IWA Water Wiki that accompanies this book.

1.4 BOOK OUTLINE

The nine chapters of this book discuss the various aspects and steps involved in climate change impact investigations in the field of urban hydrology, as outlined in Section 1.2.

Chapter 2 presents first the techniques commonly applied when modelling and analysing rainfall extremes, particularly in a stationary context. This involves the stochastic generation of rainfall series, based on multifractal, cascade or other processes, at various time and space scales. In cases when available measured or simulated rainfall have a relatively coarse temporal resolution (e.g. daily, grid averaged), sub-daily and fine-scale (e.g. point) rainfall series can be generated by rainfall disaggregation methods. The statistical description of the probability distribution of the rainfall extremes is required for most rainfall generation or disaggregation methods, but also for urban drainage impact or design investigations. Intensity-Duration-Frequency (IDF) relationships describe the results of the extreme rainfall distributions for a range of scales, and can be applied for the construction of design storms.

Statistical methods for analysing trends and non-stationary properties of rainfall series and urban impacts is the main subject of *Chapter 3*. This chapter also discusses decadal and multi-decadal climate oscillations and how the impact of climate change can be separated from natural climate oscillations or variability. While trends in rainfall are mainly climate driven, this is generally different for urban drainage situations, which are also affected by urbanization or other types of land use trends or urban design and management practices.

Chapter 4 introduces climate science. It discusses atmospheric modelling, the difference between weather and climate (modelling), reliability issues and uses this knowledge to explain the features of GCMs. *Chapter 5* builds further on that knowledge to explain dynamical downscaling and RCMs as limited area models. It is explained how these models need to be nested in GCMs and how these models can be applied for various types of sensitivity and scenario analysis. Before results from RCMs can be used for climate change impact studies, they need to be assessed. How this assessment can be done and what results are typically obtained is the topic of *Chapter 6*.

The use of statistical downscaling is detailed in *Chapter 7* for all the main types of most existing methods. Also the most recent developments are discussed, together with the assessment of their accuracy and relability.

From the downscaled climate model results, future changes in rainfall extremes can be assessed, and this is described in *Chapter 8*. These changes can be presented in the form of climate factors, which depend on time scale, return period and region. Methods as well as regional results are included. How the climate factors can be used to generate hypothetical future rainfall series as input for urban drainage simulation models is the main topic of *Chapter 9*. This chapter also discusses methods used and results obtained in the assessment of climate change impacts on urban drainage hydrologic parameters such as storm runoffs, sewer floods, surcharges, overflows and other types of relevant variables. The uncertainty in the impact assessment might be high and this needs to be addressed and quantified for the decision-making process in urban water management. This process is described in more detail in *Chapter 10*, where the need for climate change adaptation, flexible designs and other new design philosophies are presented. In particular, information is provided to explain how the uncertainty in the climate change impact assessment could be managed, and how adaptive management may also involve active social learning and better integration between various aspects of urban management, such as spatial planning and urban design.

Concluding remarks are provided in *Chapter 11*.

This book also aims to provide practical tools and instructions and therefore technical details on some state-of-the-art methods are presented in *Appendices A and B* for methods on statistical analysis of rainfall extremes and statistical downscaling. *Appendix C* focuses on dynamical downscaling and explains how a local area atmospheric model can be simulated on a personal computer. This is demonstrated using a popular state-of-the-art model. Scripts are provided in R, Matlab and Python, together with example datasets.

Chapter 2

Modelling and analysis of rainfall extremes in a stationary context

As described in Chapter 1, high-intensity short-duration rainfall extremes are the main driver of urban flooding as well as sewer overflows. Urban drainage impact studies therefore often focus on modelling of these rainfall extremes and their probability of occurrence. This involves statistical techniques of extreme value analysis, applied to rainfall series. The series can be historical observations or generated by a stochastic rainfall model. Two main approaches exist to model rainfall series stochastically: by point process theory (Section 2.1) and by multifractals and cascade processes (Section 2.2). Multifractals and cascade-based methods are also used to disaggregate rainfall from coarse (i.e. daily) to fine (i.e. sub-daily) temporal scales (Section 2.3).

Through statistical extreme value analysis, the frequencies or return periods of different intensity levels may be studied (Section 2.4). This analysis can be carried out for various aggregation levels (durations or time spans over which the rainfall intensities are averaged), covering the range of concentration times occurring in the urban drainage catchments under study (typically between 5 minutes and 1 hour). Results are typically summarized in the form of extreme rainfall intensity – duration – frequency (IDF) curves or relationships (Section 2.5). From the IDF relationships, synthetic design storms can be derived, which are used for the design of urban drainage systems, limiting the frequency of flooding or surcharge of the pipe systems to acceptable levels (Section 2.6).

In addition to the dependence of the rainfall statistics on time scales, the dependence on spatial scales is also important. The most accurate rainfall data are obtained from rain gauges. However, spatial rainfall over the urban catchment area is required for urban drainage impact modelling. Section 2.7 discusses how the difference between point rainfall and areal catchment rainfall can be addressed. The latter is also of relevance to overcome the mismatch in spatial scale discussed earlier between the spatially averaged GCM/RCM results and the point rainfall data available in historical time series.

This chapter mainly deals with techniques that are commonly used for modelling rainfall extremes in a stationary context. Many of these techniques are still applicable in the non-stationary context of climate change, after some adjustments (as shown in the next chapter).

2.1 STOCHASTIC RAINFALL GENERATION BY POINT PROCESS THEORY

Rainfall series at high resolution that are required for urban drainage impact studies can be produced by stochastic rainfall generators. Different types of stochastic models exist to generate point rainfall. One set of methods is based on *Markov chain modelling* (e.g. Richardson, 1981; Stern & Coe, 1984; Woolhiser, 1992; Srikanthan & McMahon, 2002). They consider that rainfall intensities have a mixed probability distribution, composed of a discrete component at zero (probability of zero rainfall) and a continuous component (probability distribution of non-zero rainfall). Through a two-state Markov chain model, temporal sequences of wet and dry states are modelled considering a matrix of transition probabilities. These probabilities describe the transition from wet or dry (on the previous time step, i.e. previous day) to wet or dry on the next time step: p_{01} the transition probability from dry to wet, p_{11} from wet to wet, $1 - p_{01}$ from dry to dry, and $1 - p_{11}$ from wet to wet. The rainfall generation procedure is summarized in Figure 2.1. First a random number (u) is generated between 0 and 1. When u is lower than the transition probability (p_c) based on the state of the previous day, a wet state rainfall intensity is generated. In the other case, a dry day is generated.

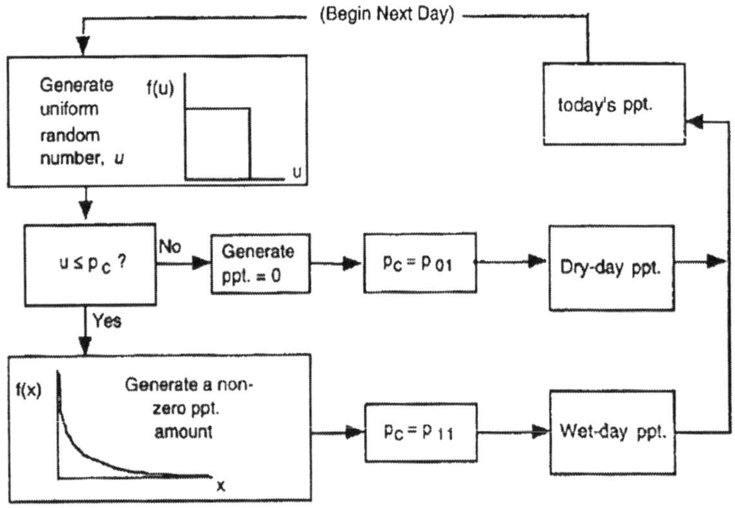

Figure 2.1 Two-state Markov chain based stochastic rainfall generation process.

For the wet state rainfall intensity generation, rainfall probability distributions are randomly sampled. Common distributions are the exponential, mixed exponential, gamma, Weibull, lognormal distributions and the Generalized Pareto Distribution (GPD). In order to describe the full range of rainfall intensities from low to medium and high intensities, hybrid distributions might be needed, for example the gamma distribution for the low and moderate intensities and the GPD for the extremes (Furrer & Katz, 2008). The wet state rainfall intensities are generated from the rainfall intensity distributions independent or dependent on the rainfall intensity of the previous time step (Fahrmeir & Tutz, 1994).

The distribution parameters and transition probabilities can vary seasonally or in time depending on atmospheric indices (e.g. Hyndman & Grunwald, 2000; Wheather *et al.* 2005; Furrer & Katz, 2008). Examples of such atmospheric indices are the El Nino-Southern Oscillation (ENSO), the North Atlantic Oscillation (NAO), atmospheric pressure, weather type or type of atmospheric circulation pattern, temperature, wind speed, humidity or a slowly varying trend function. Considering dependency of the model parameters with these indices, trends or other non-stationarities not accurately described by periodic seasonal patterns can be captured. The Markov chain model is in that case called "non-homogenous" as the parameters are conditional on one or more of the above-mentioned indices. The dependency relationship is nowadays most often described with *Generalized Linear Models* (GLMs) (see McCullagh & Nelder, 1989, and Box 2.1). These have the advantage that the explanatory indices can be treated as continuous variables (Katz & Parlange, 1996). It avoids that classes have to be defined for these indices and different stochastic models or parameters have to be defined for each discrete class. Chandler and Wheather (2002) made use of GLMs for daily rainfall modelling conditioned upon large-scale atmospheric predictors such as sea level pressure, temperature and relative humidity as predictors. Temporal dependence was incorporated by including rainfall values from the previous days as predictors. Seasonal variation was represented parsimoniously using Fourier series and two-way interactions between predictors in the GLM were used to recover the autoregressive correlation structure (see GLIMCLIM software by Chandler, 2011).

Box 2.1 Generalized Linear Models – GLMs

A GLM is a natural extension of the simple linear regression model, by allowing that the dependent variables have other than a normal distribution. The GLM generalizes linear regression by allowing an arbitrary function of the dependent variables (called link function) to vary linearly with the independent variables (rather than assuming that the dependent variables themselves vary linearly). In this way, they also allow the magnitude of the variance of the dependent variable to be a function of the dependent variable.

In a GLM, each outcome of the (vector of) dependent variables Y is assumed to be generated from a particular distribution. In applications of rainfall generation, the exponential family of distributions is most useful. This family covers a large range of probability distributions that includes the normal, exponential, gamma, binomial and Poisson distributions, among others. The mean of the distribution of Y depends in a GLM on the (vector of) independent variables X through:

$$E(Y) = g^{-1}(X\beta)$$

where $E(Y)$ is the expected value of Y; $X\beta$ is the linear predictor, a linear combination of unknown parameters β of the GLM and g is the link function, which is a monotonic function that depends on the distribution of Y.

In this framework, the variance is typically a function V of the mean:

$$V(Y) = V(g^{-1}(X\beta))$$

The unknown GLM parameters β are typically estimated with maximum likelihood or Bayesian methods (McCullagh & Nelder, 1989).

In the approach by Chandler and Wheather (2002), GLMs are used to describe wet day occurrence using a Markov based approach and the rainfall amount using a gamma distribution. They used as independent variables X large-scale atmospheric predictors such as sea level pressure, temperature and relative humidity as predictors. In the rainfall occurrence model, the probability of a wet day (p) as the dependent variable was modelled using logistic regression:

$$\ln\left(\frac{p}{1-p}\right) = X\beta$$

The link function thus is based on the logit of the probability: $\text{logit}(p) = \ln\left(\frac{p}{1-p}\right)$, assuming the typical binomial distribution for the wet day probability.

For the mean rainfall intensity of a wet day ($\bar{R}$), a natural logarithmic link function is used by Chandler and Wheather (2002) assuming a gamma distribution for the wet day rainfall intensity: $\ln(E(\bar{R})) = X\beta$.

The at-site rainfall generators have been extended to multi-site models by accounting for the spatial dependencies between stations (Wilks, 1998; Wheater *et al.* 2005; Yang *et al.* 2005; Chandler & Bate, 2007; Brissette *et al.* 2007; Apipattanavis *et al.* 2007; Maraun *et al.* 2010). As for the at-site approach, the multi-site rainfall generation can be done conditional on weather types. These types can be defined a priori based on distinct patterns in atmospheric circulation (see Section 7.4). Another approach is to relate atmospheric circulation variables through a finite number of hidden (unobserved) rainfall patterns (states) to multi-site rainfall intensities. This is what is done in a Non-homogeneous Hidden Markov Model (NHMM). This NHMM determines the most distinct patterns in multi-site rainfall records rather than patterns in atmospheric circulation. These patterns (rainfall states) are then defined as conditionally dependent on a set of atmospheric predictor variables (Vrac & Naveau, 2007; Vrac *et al.* 2007c; Gelati *et al.* 2010).

Another type of generator that defines rainfall occurrence and amount separately, as the Markov chain approach does, are *spell length based generators*. One of the well-known generators of this type is the LARS Weather Generator (LARS-WG; Rascko *et al.* 1991; Semenov & Barrow, 1997). It randomly generates lengths of alternate wet and dry spells from probability distributions. For the wet spells, the precipitation amount is randomly defined, also by a probability distribution. All distributions are specified as histograms derived from observed series. LARS-WG therefore is often called a *semi-empirical generator* (Semenov *et al.* 1998; Semenov & Stratonovitch, 2010). The spell length based generator was developed after it was noticed that the Markov chain based generators fail to describe adequately the length of dry and wet series (i.e. persistent events such as drought and prolonged rainfall). These can be very important in some applications (e.g. agricultural impacts). Because they are less important for urban flood related applications, this type of generator is less commonly used in urban drainage. LARS-WG, however, became popular in climate change impact studies (see Section 7.5). The generator also might have an added value when studying the impact of long dry spells on sedimentation or other types of impacts in sewers (see Section 9.3).

Markov chain based stochastic models and the LARS-WG were mainly developed for daily or coarser time scales of the rainfall series. The condition limits their use in urban drainage, unless they are combined with a stochastic disaggregation method (Section 2.3). In such combined approach, the stochastic model can be used to generate daily rainfall intensities, followed by a disaggregation step to generate sub-daily intensities conditional on the daily intensity (e.g. Chun *et al.* 2009).

Another set of methods for stochastic point rainfall generation make use of the stochastic representation of rain cells in time and/or space. The best known stochastic rainfall generator of this type for point rainfall is the *rectangular pulse model*. Originally developed for the spatial distribution of galaxies, two versions exist for modelling of rainfall, namely the Neyman-Scott (Kavvas & Delleur, 1981; Cowpertwait *et al.* 1996; Kilsby *et al.* 2007) and Bartlett-Lewis (Rodriguez-Iturbe *et al.* 1987a,b; Verhoest *et al.* 1997; Onof *et al.* 2000; Vandenberghe *et al.* 2011) models. Both of these schematize rain storms as a cluster of rain cells by means of rectangular pulses (Figure 2.2). They are hereafter referred to as the Neyman-Scott Rectangular Pulses (NSRP) and Bartlett-Lewis Rectangular Pulses (BLRP) models. To describe the rain storm occurrences in time, or the rain cells within a rain storm, Poisson processes are assumed. This means that the rain storm or cell arrivals are random in time with exponential interarrival times, which are independent from each other.

The models use in the order of 5 to 8 parameters, describing the probability distributions of pseudo-physical rain storm properties such as the mean rain cell intensity (I in Figure 2.2) or cell volume (R), the mean cell duration (D), the dry spell lengths or inter-arrival times of storms (Q), the waiting times from the origin to the rain cell origins (B), and so on. Although these properties can be interpreted physically, their distributions are commonly not estimated directly. Distribution functions are assumed and their parameters optimized by means of an objective function defined based on a number of rainfall statistics such as the mean, variance, skewness, autocorrelation, dry period probability, probability of a dry-dry sequence and probability of a wet-wet sequence (Onof *et al.* 2000; Vanhaute *et al.* 2012). Another approach makes use of triangular pulses to model rain cells (Mariën & Vandewiele, 1986). Applications exist for daily as well as sub-daily intensities.

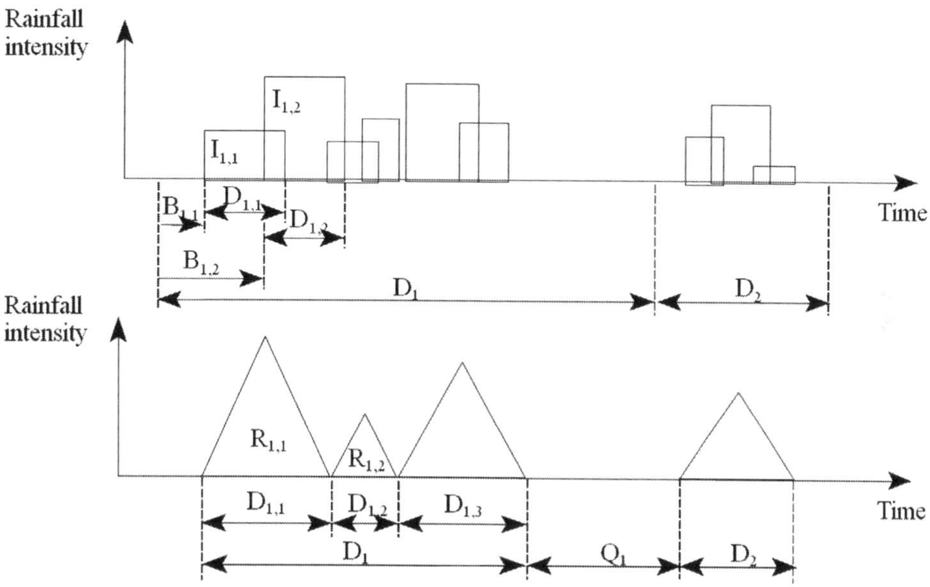

Figure 2.2 Schematic representation of rain storms in point rainfall generators (top: based on rectangular pulses; bottom: based on triangular pulses).

However, several authors have reported that these generators tend to underestimate the rainfall extremes for short durations (e.g. Verhoest *et al.* 1997; Back *et al.* 2011). Cowpertwait and O'Connell (1997) made a modified version of the NSRP model by allowing for two types of rain cells: "heavy" short-duration convective cells and "light" long-duration stratiform cells. Modifications to the BLRP model were proposed by Cowpertwait (2004) and Cowpertwait *et al.* (2007). They obtained a model with 12 parameters that allows better estimates of short duration rainfall extremes. This was done by considering an extra Poisson process at the scale of an individual rain cell, and by allowing different storm types, hence superposing several processes. They concluded that the model may simulate extreme rainfall well for durations of 1 and 24 hours, but that it underestimates the extreme rainfall for 5-minute durations. Verhoest *et al.* (2010) reported that the BLRP generator occasionally creates unrealistic rainfall cells and proposed a truncation of the distribution from which cell durations are drawn. Furthermore, rainfall models based on the Neyman-Scott stochastic point process could be found physically inconsistent over different time scales since they may not be able to preserve the rainfall characteristics over these different time resolutions (Foufoula-Georgiou & Guttorp, 1987).

To allow the NSRP model to be conditioned on atmospheric indices (as can be done with Makov chain models), Fowler *et al.* (2000, 2005) and Burton *et al.* (2008) coupled a semi-Markov chain based generator to the NSRP model of Cowpertwait and O'Connell (1997). The semi-Markov chain model is used to model the temporal occurrence and persistence of atmospheric states. Parameters sets of the NSRP model are then conditioned on these states.

Spatial versions of the NSRP and BLRP rainfall generators exist as well. In the most simple formulation the spatial structure is assumed to be circular discs moving over the catchment with uniform velocity with cells being generated according to a two-dimensional Poisson process (Cowpertwait, 1995; Cowpertwait *et al.* 2002; Fowler *et al.* 2005; Burton *et al.* 2008).

Willems (2001) developed a spatial analogy of the NSRP and BLRP generators, based on the spatial rain storm structure such as that illustrated in the radar image of Figure 2.3. The smallest building blocks of this spatial rainfall model are the rain cells, which are assumed to have Gaussian shapes (e.g. Jinno *et al.* 1993). These rain cells, which are a few kilometers in diameter, are embedded in a clustered way within small mesoscale areas of size 100 to 1000 km^2. In their simplest form, the cells inside such areas are assumed to move with nearly identical velocity. At larger scales, small mesoscale areas occur in a clustered way within lower intensity "large mesoscale areas", which in turn are embedded within some synoptic scale rainfall field of even lower intensity (e.g. Austin & Houze, 1972). The precipitation associated with stratiform rainfall is spatially related to these large mesoscale areas. Although rain cells and cell clusters most often appear in large and small mesoscale areas, they can also occur in isolation outside such regions (as is the case for convective rain storms). Willems (2001) calibrated the probability distributions of the rain storm and cell properties directly based on statistical analysis of a large number of storms observed by a dense network of rain gauges and radar images.

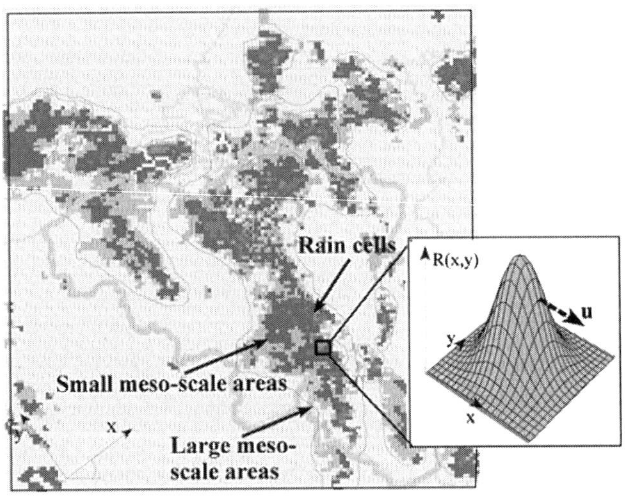

Figure 2.3 Representation of spatial rainfall structure, based on a radar image (after Willems, 2001).

2.2 MULTIFRACTAL AND CASCADE PROCESSES

Under the hypothesis of scale-invariant statistical properties of the rainfall process, the rainfall variability at different scales can be explained using concepts originating from fractal theory (Mandelbrot, 1982; Schertzer & Lovejoy, 1987, 1991; Gupta & Waymire, 1990, 1993). This approach is based on scaling laws, which describe the scale-invariant properties or relationships that connect the statistical properties of rainfall for different scales. Based on the scaling properties, the variability of rainfall at different temporal and spatial scales can be described using a small number of parameters.

By definition, a function $f(t)$ possesses scaling properties if $f(t)$ is proportional to the scaled function $f(\lambda t)$ for all positive values of the scale factor λ, where the proportionality factor depends on λ as well. When applied to the moments of the rainfall distribution or its parameter values, the scaling concept uses the moments or parameters known at one specific scale (temporal or spatial scale) to estimate the moments or parameters at all other scales through application of a scaling factor.

In the case of simple scaling, a power relation exists for each of the rainfall distribution parameters β depending on the scale (e.g. time scale D):

$$\beta_D = a\ D^b$$

or:

$$\beta_{\lambda D} = \lambda^b\ \beta_D$$

where λ is the scale factor and b the scaling exponent. The scaling exponent equals the slope of the linear relationship between β and D in a double logarithmic plot. It is constant in the simple scaling case, which means that the variability in the rainfall process does not change with time scale. It also means that for a time scale $\lambda\ D$ the same distribution holds as for time scale D if the rainfall intensities x are scaled with a factor λ^b (Gupta & Waymire, 1990; Burlando & Rosso, 1996):

$$F_{\lambda D}(\lambda^b x) = F_D(x)$$

The scale invariance then also holds for the distribution moments:

$$E\left[X^l_{\lambda D}\right] = \lambda^{lD} E\left[X^l_D\right]$$

where $E\left[X^l_D\right]$ denotes the non-central moment of order l for the rainfall distribution at time scale D.

When the slope of the relation between the moments $E\left[X^l_D\right]$ and the time scale D is plotted against the moment order l, a linear increase is found in the simple scaling case. When the relation is non-linear but concave, extreme rainfall statistics are called multi-scaling (Gupta & Waymire, 1990).

Scaling laws of precipitation have been found and applied by several authors. Various techniques have been used, including analyses of power spectra, empirical probability distribution functions and statistical moments. Early analyses of temporal and spatial rainfall observations were undertaken by Fraedrich and Larnder (1993), Tessier *et al.* (1993) and Olsson *et al.* (1993). Several later investigations were aimed at establishing functional relationships between the scaling parameters and physical or geographical characteristics such as storm type, altitude and climate region (e.g. Harris *et al.* 1996; Olsson & Niemczynowicz, 1996; Perica & Foufoula-Georgiou, 1996; Svensson *et al.* 1996).

The scaling functions can be used to derive probability distributions of sub-daily rainfall intensities from the distribution of daily rainfall intensities (Willems, 2000). The scaling properties of rainfall are also commonly applied in stochastic rainfall generation techniques based on fractal models that generally employ a random cascade process as the generating mechanism.

A wide range of scaling-based modelling approaches have been developed, focusing on the temporal structure (e.g. Menabde *et al.* 1997a; Cârsteanu *et al.* 1999; Veneziano & Iacobellis, 2002), the spatial structure (e.g. Menabde *et al.* 1997a,b, 1999a,b; Deidda, 1999) or the full space-time structure (e.g. Lovejoy & Schertzer, 1990; Over & Gupta, 1996; Seed *et al.* 1999; Venugopal *et al.* 1999; Deidda, 2000; Jothiyangkoon *et al.* 2000).

Some work has focused explicitly on extreme rainfall. Nguyen *et al.* (2002) have obtained scaling functions for the first three non-central moments of annual maximum point rainfall data between

5 minutes and 1 hour, and between 1 hour and 1 day for Quebec, Canada (see Figure 2.4). Other examples are given in Section 2.5. It is shown in that section that the scaling properties of rainfall are very useful for IDF modelling (Burlando & Rosso, 1996; Menabde *et al.* 1999a; Yu *et al.* 2004; Langousis & Veneziano, 2007). The scaling laws may hold over a wide range of scales, as was shown by Willems (2000). He found that Belgian rainfall is multi-scaling for the range between 10 minutes and 3.5 days, which makes the scaling laws useful for both urban drainage and river hydrological applications.

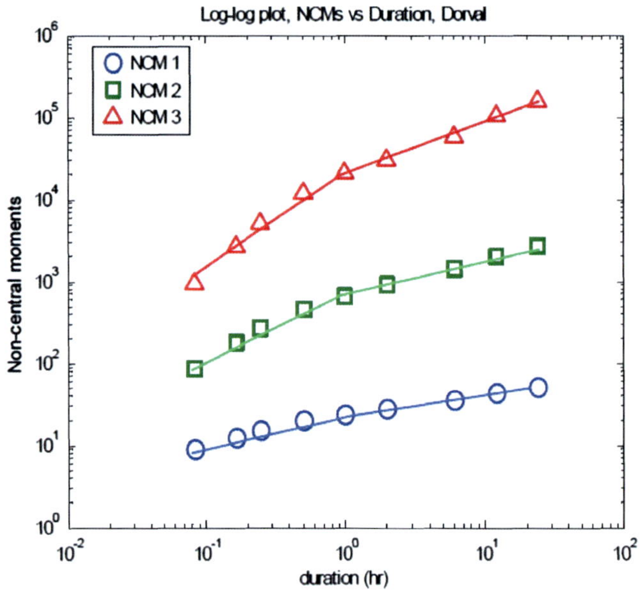

Figure 2.4 Log-log plot of first three non-central moments (NCM) of annual maximum point rainfall data versus time scale (Dorval station, Quebec) (after Nguyen *et al.* 2002).

2.3 RAINFALL DISAGGREGATION

When fine-scale rainfall data are required but historical rainfall series are unavailable at that scale, rainfall disaggregation can be an attractive option. One group of approaches is based on fitting theoretical probability distribution functions to variables such as number of events per day, starting times, and event volume and duration (e.g. Hershenhorn & Woolhiser, 1987; Econopouly *et al.* 1990; Connolly *et al.* 1998). Another group of methods has been developed from rectangular pulse stochastic rainfall models (described in Section 2.1) (e.g. Bo *et al.* 1994; Glasbey *et al.* 1995; Koutsoyiannis & Onof, 2001). Disaggregation also can be done using artificial neural network models based on theory of learning (Burian *et al.* 2000) or chaotic models (Sivakumar *et al.* 2001; Gaume *et al.* 2006).

The scaling properties of rainfall statistics (Section 2.2) provide a natural framework for rainfall disaggregation and are useful for making a scale shift from the range of scales spanned by the rainfall data to any scale needed in hydrological applications. In the simple scaling case, both scale magnification (upscaling) and scale contraction (downscaling) are possible; whereas only downscaling is possible in the multi-scaling case. For instance, if hourly rainfall data are available, rainfall statistics down to aggregation levels of 5–10 minutes can be derived in this way. It is clear that this has interesting practical applications in urban hydrology and can play an important role in statistical downscaling.

The scaling-based models described in Section 2.2 are generally applicable for disaggregation in time and/or space, but have not always been evaluated for this purpose. A number of random cascade models have, however, been developed with the specific objective of temporal rainfall disaggregation. Key aspects of the models are (1) the probability distribution of the cascade weights and (2) whether they are canonical (no exact conservation of mass in each cascade branching) or micro-canonical (exact conservation). According to Gaume *et al.* (2006), random cascade models "seem to be the only type of model able to simulate in a simple way the internal storm structures at small time steps (5–10 min)", hence useful for applications in urban hydrology. Less convincing were the results by Nguyen and Pandey (1994) who have assessed the feasibility of a time scale-independent probability model proposed by Schertzer and Lovejoy (1987) based on the multifractal multiplicative cascade model for the estimation of the distribution of hourly rainfalls from the distributions of daily and longer time-resolution rainfalls. This model was found to provide an underestimation of the hourly rainfalls as compared to the observed values.

Further developments were provided by Olsson (1998), who proposed a micro-canonical cascade-based temporal disaggregation model for daily rainfall using a uniformly distributed generator, dependent on rainfall intensity and position in the rainfall sequence. The method was further developed and tested in different climates by Güntner *et al.* (2001). Onof *et al.* (2005) used a canonical log-Poisson generator for disaggregation of hourly rainfall. Molnar and Burlando (2005) tried both a canonical and a micro-canonical model with a beta-distributed generator for disaggregation of daily values (see Box 2.2). Hingray and Ben Haha (2005) compared seven different disaggregation models for hourly disaggregation, including random cascades. Rupp *et al.* (2009) studied the dependence of cascade parameters on time scale and rainfall intensity to disaggregate daily data. Licznar *et al.* (2011) evaluated six cascade-based models for daily disaggregation.

Box 2.2 Temporal rainfall disaggregation by random cascade model: micro-canonical approach with beta-distributed generator

One useful application of random cascade models is for disaggregation of continuous rainfall time series. There are several variations on this theme but here we outline one of the main approaches currently used. Generally, a temporal cascade process involves successive branching which describes how some quantity (e.g. rainfall) is redistributed as the time resolution changes. The redistribution implies a successive fine-graining process that starts from an original, large-scale resolution r_L and continues until a target, small-scale resolution r_S is reached.

Commonly a branching number of 2 is used, which implies a redistribution of total rainfall in period i at resolution r, $R_{i,r}$, between the amount associated with the first and last half of the period, respectively. In a micro-canonical approach, the amount redistribution may be described by multiplicative weights $W_{i,r}$, $0 \leq W_{i,r} \leq 1$, that assigns $W_{i,r} * R_{i,r}$ to the first half of the period and $(1 - W_{i,r}) * R_{i,r}$ to the last half. In each branching, two principal possibilities exist: (1) $W_1 = 0$ or $W_1 = 1$ and (2) $0 < W_1 < 1$. The former option, where the total amount originates from one half of the period, is related to rainfall intermittency which is one key issue in cascade-based rainfall disaggregation. The occurrence of (1) and (2) may be expressed in terms of probabilities $Pr_{01} = Pr(W_1 = 0 \text{ or } W_1 = 1)$ and $Pr_{xx} = Pr(0 < W_1 < 1) = 1 - Pr_{01}$. In some applications a distinction has been made between $Pr(W_1 = 0)$ and $Pr(W_1 = 1)$, respectively. Depending on the range of resolutions considered, Pr_{01} may either be assumed resolution-independent or parameterised as a scaling law:

$$Pr_{01}(r) = c_1 r^{c_2} \tag{1}$$

where c_1 and c_2 are constants.

Another key issue concerns the characteristics of the weights $W_{i,r}$ related to the case $0 < W_1 < 1$, $W_{i,r}^{xx}$. Generally the probability distribution of $W_{i,r}^{xx}$ is assumed to follow some theoretical distribution ("cascade generator") and the one-parameter beta distribution:

$$p(W) = \frac{1}{B(a)} W^{a-1}(1-W)^{a-1}$$

where a is a shape parameter and B the beta function, is often found to be an acceptable approximation. There is empirical evidence that the shape parameter a of the distribution depends on the time resolution r and that generally this dependency may be well described by the scaling relationship:

$$a(r) = a_S r^{-H} \tag{2}$$

where a_S is the shape parameter related to the highest resolution r_S and H an exponent describing how rapidly a decreases with decreasing resolution.

In its basic version, the model is thus specified by four parameters: c_1, c_2, a_S and H. Parameter estimation may be performed by a coarse-graining procedure, a "reverse cascade" which starts from the highest resolution r_S. By gradually aggregating adjacent rainfall amounts two by two, breakdown coefficients (BDCs) may be calculated as:

$$\text{BDC}_{i,r} = \frac{R_{i,r}}{(R_{i,r} + R_{i+1,r})}$$

where $\text{BDC}_{i,r}$ thus corresponds to the weight $W_{i,r}$ used to redistribute the total rainfall amount in a fine-graining process. After the coarse-graining has proceeded to up to the lowest resolution r_L, the BDC-values may be used for estimation of all parameters.

Concerning the calibration and application of the disaggregation model to an existing time series at original resolution r_L, two principal situations may be considered. One is when some representative data at target resolution r_S are available, for example from a temporary measurement campaign at the same location or from a nearby high-resolution gauge. Then BDCs may be extracted and parameters estimated over the actual resolution interval $r_L \leq r \leq r_S$. The other situation is when no representative high-resolution data are available. Then parameter estimation may still be possible by coarse-graining from resolution r_L to gradually lower resolutions. Then Equations (1) and (2) may be evaluated and if the scaling laws are found to hold they may be extrapolated to estimate Pr_{01} and a for higher resolutions up to r_S. This is an attractive feature of scaling-based disaggregation but it must be emphasised that extrapolation is always associated with large uncertainties and that it should always be attempted to verify parameters using some surrogate data.

After calibration, Monte Carlo simulations are performed to gradually fine-grain the data and generate realisations at resolution r_S. In case r_S is not reached exactly by resolution doubling from r_L, a higher target resolution may be used followed by interpolation to r_S.

Alternative options include using a different generator, that is a theoretical distribution for $W_{i,r}^{xx}$, and for example normal and beta-normal distributions have been suggested. Further, a dependency of Pr_{01} on rainfall amount $R_{i,r}$ is empirically supported. Also a dependency on the position within the rainfall sequence may be considered. It should finally be mentioned that the intensity resolution of the measurement device may have a strong impact on the BDC characteristics which requires careful analysis.

For applications of random cascade micro-canonical disaggregation with particular focus on urban hydrological processes and time scales, see for example Molnar and Burlando (2005), Hingray and Ben Haha (2005) and Licznar *et al.* (2011).

Rainfall disaggregation based on random cascades has clear potential, but the most suitable model type remains to be identified as different designs have proved to perform differently for various data sets. The recent comparative evaluations cited above are important for clarifying this issue. It is clear that not only geographical and climatological factors influence model performance, but so do resolution and set-up of the measurement device (e.g. Licznar *et al.* 2011).

Although most applications focus on temporal disaggregation, spatial disaggregation examples exist as well (e.g. Kottegoda *et al.* 2003). van den Bergh *et al.* (2011) proposed a method for downscaling coarse scale radar images based on the simultaneous use of fractal-based scaling of the marginal probability distribution functions and a copula which describes the dependence between both scales. Through introducing this dependence, the proposed framework allows for a better estimation of the actual shape of sub-pixel probability functions compared to the scaled marginal distribution function. They demonstrated the method based on radar images for Belgium, for downscaling from 19.2 km coarse scale grids to 600 m fine scale grids.

2.4 STATISTICAL RAINFALL EXTREME VALUE ANALYSIS

PDS/POT based analysis

Regardless of the method employed to obtain rainfall series, through observations or through stochastic generation, and regardless of the temporal scale, statistical analysis is required to quantify the probabilities of the rainfall intensities. Probability distributions can be assessed for the mean rainfall intensities at the relevant time scales, but most important for urban drainage applications are the rainfall extremes. These extremes can be extracted from the full rainfall series using the classical approach of annual maxima (Coles, 2001) where the annual maximum within a (hydrological) year is included in the extreme value analysis. Traditionally, this approach has been used for analysing rainfall extremes (e.g. Schaeffer, 1990; Alila, 1999; Wallis *et al.* 2007). Another approach considers events above a threshold level in the extreme value analysis. This approach, referred to as the *Partial Duration Series* (PDS) or *Peak-Over-Threshold* (POT) method, has been used for analysing extreme rainfall at fine temporal scales in for example Madsen *et al.* (2002), Beguería and Vicente-Serrano (2006), Willems *et al.* (2007) and Willems (2009). The pros and cons of the *Annual Maxima Series* (AMS) approach versus the PDS/POT method have been discussed, amongst others, by Stedinger *et al.* (1993), Madsen *et al.* (1997) and WMO (2009c). The AMS method considers only the maximum event within a year although other events in the year may exceed annual maxima of other years. The POT approach provides a more consistent definition of the extreme values by considering all events above a threshold. However, as opposed to the AMS approach that generally assures independent events, independence criteria have to be defined to ensure independence between extreme events in the PDS/POT series. In addition, the PDS/POT method includes selection of a threshold level, which will introduce some sort of subjectivity in the extreme value analysis. Due to its simpler structure, the AMS-based method is more popular in practice. The PDS/POT analysis, however, appears to be preferable for short records, or where return periods shorter than two years are of interest (WMO, 2009c). Since the theory and application of the AMS approach have been well documented in hydrologic and engineering literature (Stedinger *et al.* 1993; WMO, 2009c), this section mainly focuses on the PDS/POT method.

Extreme value distributions

Several probability distributions have been applied to describe the distribution of extreme rainfall intensities at a single site (e.g. Chow, 1964; Benjamin & Cornell, 1970). Common distributions that have been applied to the analysis of AMS include the Gumbel (NRCC, 1989), Generalized Extreme Value (GEV) (NERC, 1975), Log-normal (Pilgrim, 1998), and Log-Pearson type 3 (Niemczynowicz, 1982; Pilgrim, 1998) distributions. Among these distributions the GEV and its special form, the Gumbel distribution, have received dominant applications in modelling the annual maximum rainfall series. The Gumbel distribution was found, however, to underestimate the extreme precipitation amounts in several cases (Wilks, 1993). Studies using rainfall data from tropical and non-tropical climatic regions (Wilks, 1993; Nguyen *et al.* 2002; Zalina *et al.* 2002) suggest also that a three-parameter distribution can provide sufficient flexibility to represent extreme precipitation data. In particular, the GEV distribution has been found to be the most convenient, since it requires a simpler method of parameter estimation and it is more suitable for regional estimation of extreme rainfalls at sites with limited or without data (Nguyen *et al.* 2002). When the return periods associated with frequency-based rainfall estimates greatly exceed the length of record available, discrepancies between commonly used distributions tend to increase.

For PDS/POT extremes, following the extreme value theory of Pickands (1975), the distribution's tail of PDS/POT extremes converges asymptotically to a Generalized Pareto Distribution (GPD). The cumulative distribution function $F(x)$ of the GPD is given by:

$$F(x) = 1 - \left(1 + \gamma \frac{x - x_t}{\beta}\right)^{-1/\gamma} \quad \text{for } \gamma \neq 0$$

$$F(x) = 1 - \exp\left(-\frac{x - x_t}{\beta}\right) \quad \text{for } \gamma = 0$$

where x_t is the threshold level above which the distribution is considered, β is the scale parameter and γ the shape parameter (also symbol κ is often used, where $\kappa = -\gamma$). For $\gamma = 0$ the exponential distribution is obtained as a special case. The parameter γ is also called the "extreme value index" and describes the shape of the tail of the distribution (heavy tail when $\gamma > 0$, normal (exponential) tail when $\gamma = 0$, light tail with an upper bound when $\gamma < 0$). According to the sign of the extreme value index, the following three classes are traditionally considered for extreme value distributions: class I (for $\gamma > 0$), class II (for $\gamma = 0$), and class III (for $\gamma < 0$) (Coles, 2001). The event with a return period or average recurrence interval of T years (referred to as the T-year event) is defined as the $(1 - 1/(\lambda T))$ – quantile in the GPD, for example:

$$x_T = x_t - \frac{\beta}{\gamma}\left[1 - \left(\frac{1}{\lambda T}\right)^{-\gamma}\right] \quad \text{for } \gamma \neq 0$$

$$x_T = x_t + \beta \ln(\lambda T) \quad \text{for } \gamma = 0$$

where λ is the mean annual number of exceedances of the threshold x_t.

The GPD for PDS/POT extremes is equivalent to the GEV distribution for annual maxima. It can be shown mathematically that, when PDS/POT extremes occur randomly distributed in time (as in a Poisson process), and the exceedances follow a GPD, the corresponding annual maxima are GEV distributed (e.g. Madsen *et al.* 1997).

Following the extreme value theory, the GPD only holds perfectly when considered asymptotically in the tail (towards values of $+\infty$) (Coles, 2001). This means that it may not hold exactly for the lower rainfall extremes. In most practical applications, however, it has been verified that the GPD fits well also for lower rainfall intensities (e.g. Madsen *et al.* 2002).

Two conditions have to be fulfilled for the asymptotic convergence of the GPD to apply: the PDS/POT extremes need to be independent and identically distributed (Coles, 2001). For the first condition to hold, independence criteria often have to be imposed on the threshold exceedances. For the second condition to be fulfilled a careful selection of the threshold level is required.

The criteria for independence between events was studied in detail in the literature review by Arnell *et al.* (1984). They found, that the criteria should be between 30 minutes and 12 hours depending on a number of variables. Willems (2000) defined, in the cases of temporal scales less than 12 hours, two successive PDS extremes to be independent if they are separated by at least a 12-hour time interval. For durations longer than 12 hours, independent events should be separated by a time interval larger than the considered duration. Madsen *et al.* (2002) also defined independent events according to their duration. Others split the time series into wet periods, by means of dry spell identification. Various approaches for this division exist in the literature. An early theoretical approach was proposed by Restrepo-Posada and Eagleson (1982), which essentially aims at identifying the minimum separation at which events become mathematically independent. Another example is Verworn *et al.* (2008), who defined dry spells as periods for which the mean rainfall intensity is less than 0.02 mm/min for at least 0.75 hours, while wet periods should have at least 0.1 mm of rain over the whole wet spell period.

Regarding the selection of the threshold level, different methods have been proposed. An overview of these methods is given in Lang *et al.* (1999) together with recommendations for operational use. One of the proposed techniques tests the stability of the distribution parameters for varying thresholds. Also Willems (2000) applied this method, where the optimal threshold was defined as the threshold above which the mean squared error of the estimated distribution is minimal. The mean squared error is calculated from the differences between the estimated quantiles and the corresponding empirical quantiles. It can be shown that if the threshold exceedances follow the GPD for a given threshold level, then for any higher threshold the exceedances will also follow the GPD with the same shape parameter (e.g. Madsen *et al.* 1997). This important property of the GPD can be used to select an appropriate threshold level. Instead of defining a threshold, the PDS/POT extremes can also be defined by including the n most extreme events corresponding to using a fixed average number of events per year (Mikkelsen *et al.* 1995).

Distribution parameter estimation

Having defined the PDS/POT extreme value series, the next step is to estimate the distribution parameters. The GPD includes different tail behaviours as described by the shape parameter. In general, a light tail distribution is not expected for rainfall given that a light tail implies that the distribution has an upper limit. Physically, an upper limit exists as defined by the Probable Maximum Precipitation (PMP) (Chow, 1951; Bruce & Clark, 1966; WMO, 2009b), but this limit is not relevant for urban drainage problems. Estimation of PMP is based on other methods than the extreme value methods presented here. Heavy and normal (exponential) tailed GPDs are found in the literature for rainfall extremes. Based on 10-minute rainfall series in Belgium, Willems (2000) concluded that the distribution's tail is normal, but that after mixing normal tail distributions for convective and stratiform rain events, a resulting heavier tailed distribution is obtained. They succeeded to model this tail behaviour with a two-component exponential distribution (thereby mixing two normal-tailed GPDs). In a regional analysis of Danish rainfall extremes, Madsen *et al.* (2002) found that intensities with durations between 1 minute and 48 hours can be described by a heavy-tailed GPD. In the analysis of daily rainfall extremes in the Ebro Valley, Spain, Beguería and Vicente-Serrano (2006) also found that the PDS/POT series could be described by a heavy-tailed GPD.

Extensive reviews of GPD parameter estimation methods that partly go beyond applications in hydrology are given by Madsen *et al.* (1997), Bermudez and Kotz (2010a) and Bermudez and Kotz (2010b). The traditional methods include: the maximum likelihood method (ML), methods of moments (MOM), and probability weighted moments (PWM) or L-moments (Stedinger *et al.* 1993 and Madsen *et al.* 1997). A relatively new method consists of a combination of likelihood and moment estimators (Zhang, 2007). For other advances, discussions and a comparison with the traditional methods, see Huesler *et al.* (2011), Mackay *et al.* (2011), Zhang and Stephens (2009) and Martins and Stedinger (2001). Another type of methods is based on regression in quantile-quantile (Q-Q) plots. These methods have smaller variance in the estimation of the extreme value index, but might be biased due to the asymptotic properties of the extreme value theory (Willems *et al.* 2007). L-moment and ML estimation of the GPD are described in Box 2.3 and Box 2.4.

Box 2.3 Parameter estimation in the GPD – the L-moments method

Assume that we have selected a threshold, x_t. From this a PDS of n extreme values are defined.

Estimates of β and γ are obtained by the following equations:

$$\widehat{\gamma} = -\frac{\widehat{\lambda}_1 - x_t}{\widehat{\lambda}_2} + 2$$

$$\widehat{\beta} = \left(\widehat{\lambda}_1 - x_t\right)(1 - \widehat{\gamma})$$

where $\widehat{\lambda}_1$ and $\widehat{\lambda}_2$ are the first and second L-moment estimates, respectively. These are, in turn, related to the first and second PWM estimates:

$$\widehat{\lambda}_1 = \widehat{\beta}_1, \qquad \widehat{\lambda}_2 = 2\widehat{\beta}_1 - \widehat{\beta}_2$$

$$\widehat{\beta}_1 = \frac{1}{n}\sum_{i=1}^{n} x_i, \qquad \widehat{\beta}_2 = \frac{1}{n}\sum_{i=1}^{n-1}\frac{n-i}{n-1}x_{(i)}$$

The notation $x_{(i)}$ indicates that the extreme values have been sorted in a descending order. L-moments are linear combinations of the PWM and thereby linear combinations of the ranked observations. For more details on the estimation procedure and a discussion of the method the reader is referred to Stedinger *et al.* (1993) and Hosking (1990).

Computations of the uncertainty in the parameter estimates are more complicated. They can be derived from asymptotic theory, where approximate equations for the variances are obtained from Taylor series expansion (Rosbjerg *et al.* 1992). The resulting equations can be found in Hosking and Wallis (1987). They have been implemented in many common software tools. Appendix A contains a practical guide on L-moment based GPD parameter estimation, using the open source software R.

Box 2.4 Parameter estimation in the GPD – the ML method

Again we have selected an adequate threshold, x_t and from this a PDS of n extreme values are defined.

In the ML theory we explicitly utilize that x follows a GPD. As the parameters β and γ are unknown, the density function $f(x)$ is used to predict the likelihood of different parameter values for all x. The parameter set with the highest likelihood will give the best description of the data. The likelihood function reads:

$$L(\beta, \gamma) = \prod_{i=1}^{n}\frac{1}{\beta}\left(1 + \gamma\frac{x_i - x_t}{\beta}\right)^{-1/\gamma - 1}$$

For practical reasons it is often the log-likelihood function that is maximized:

$$l(\beta, \gamma) = -n\log(\beta) - (1/\gamma + 1)\sum_{i=1}^{n}\log\left(1 + \gamma\frac{x_i - x_t}{\beta}\right)$$

given that $\left(1 + \gamma\frac{x_i - x_t}{\beta}\right) > 0$ for $i = 1 \ldots n$.

When $\gamma = 0$ (exponential distribution), the ML estimate of β equals the mean value of the exceedances (similar to the MOM and L-moment estimates). When $\gamma \neq 0$, numerical techniques are required for estimation of the parameters. Methods for numerical maximization are implemented in most statistical softwares (see Appendix A using the open source software R). Convergence problems can occur leading to insensible parameter values, especially when $\gamma \approx 0$ (Coles, 2001). The ML estimator is asymptotically the most efficient estimator (has the lowest mean squared error). However, for small sample sizes the ML method may results in unreasonable estimates of the shape parameter (e.g. Madsen *et al.* 2007; Martins & Stedinger, 2001).

Several methods exist for estimation of the uncertainty in the parameters obtained by the ML method. Basically they all evaluate the second order partial derivatives of the function $l(\beta,\gamma)$ as they express how well the maximum of $l(\beta,\gamma)$ is defined in term of the curvature. The matrix of second order partial derivatives is known as the Information matrix (J), the Hessian matrix or the Fisher matrix. A central assumption in standard likelihood theory is that the ML estimator asymptotically follows a multivariate normal distribution with J^{-1} as variance-covariance matrix. For details on estimation procedures see Coles (2001).

Estimated T-year rainfall intensities may have large sampling uncertainties, especially when extrapolated far beyond the observation period. To obtain more accurate estimates, additional information should be included in the estimation process. A common approach is the use of regional information where information from several rainfall records that can be assumed to have similar extreme rainfall characteristics is combined. Importantly, regionalization also allows estimation at ungauged locations (WMO, 2009c).

Regional analysis

Regional frequency analysis involves the following basic steps: (i) identification of a homogenous set of stations with similar extreme value characteristics, (ii) determination of a regional extreme value distribution, and (iii) combination of extreme rainfall records from the different sites in the region for estimation of the regional distribution. The regional frequency analysis approach based on L-moments has been widely applied in hydrology since it was first introduced by Hosking and Wallis (1993). The method is based on an "index flood" approach. The key assumption of this approach is that data from different sites follow the same distribution except for a site-specific scaling factor, the index parameter. Usually the mean of the distribution is taken as the index parameter. Homogeneity is then defined in terms of constant second and higher order moments.

The regional frequency analysis approach proposed by Hosking and Wallis (1993) includes a homogeneity test based on L-moments where the dispersion of L-moment ratio estimates from the group of sites is compared to the expected dispersion in a homogeneous region (i.e. due to sampling uncertainty). Similarly, an L-moment based goodness-of-fit test can be used for choosing an appropriate regional distribution. Regional distribution parameters corresponding to second and higher order moments are estimated by weighting the site-specific L-moment estimates from the group of sites.

The L-moment based regional estimation procedure has been widely used in extreme rainfall analysis. In a regional analysis in Canada, Alila (1999) found that the L-skewness of annual maxima precipitation could be assumed constant in the entire country, and homogenous regions with constant L-coefficient of variation (L-CV) could be defined according to the mean annual precipitation (MAP). Di Baldassarre *et al.* (2006) detected for a dense network of rain gauges in northern central Italy significant relationships between the L-moments of annual maximum rainfall intensities and the MAP. These relationships were valid for durations ranging from 15 min to 1 day. Wallis *et al.* (2007) divided Washington State into 12 regions, and within each region L-CV and L-skewness were found to vary systematically with MAP. A similar approach was applied by Haddad *et al.* (2011) for rainfall extremes in Australia where regression equations were developed for L-skewness, L-CV and mean of annual precipitation maxima. However, the use of MAP as an "index variable" may not be appropriate for other regions with different climatic or topographic conditions. For instance, the median of annual maximum rainfalls at a site was recommended as the index variable for regional estimation of extreme rainfalls by the UK Institute of Hydrology (1999). In general, one of the main difficulties in the application of this technique is related to the definition of "homogeneous" regions. Various methods have been proposed for determining regional homogeneity, but there is no generally accepted procedure in practice (Fitzgerald, 1989; Schaefer, 1990; Hosking & Wallis, 1993; Fernandez Mills, 1995; Nguyen *et al.* 2002).

Madsen *et al.* (2002) developed a regional PDS/POT model for rainfall extremes in Denmark and found that the GPD shape parameter could be considered constant for the entire country. Sub-regions were defined with constant PDS/POT mean value, and a regression model was developed for describing the regional variability of the Poisson parameter from MAP. For rainfall extremes in the Ebro Valley, Spain, Beguería and Vicente-Serrano (2006) developed a regional PDS/POT model using regression relations of the threshold level and GPD parameters with location and relief variables.

2.5 IDF RELATIONSHIPS

The previous section presents the basic considerations of frequency analysis of extreme rainfalls in order to estimate the amount of precipitation falling at a given point or over a given area for a specified duration and return period. Results of this analysis are often summarized in the form of rainfall *Intensity-Duration-Frequency* (IDF) or *Depth – Duration – Frequency* (DDF) relationships for a given site, or are commonly presented in the form of a "rainfall frequency atlas", which provides information on rainfall depths accumulated or averaged over various time durations and for different return periods (or average recurrence intervals) (e.g. Chow *et al.* 1988; WMO, 1994, 1983, 1994, 2009c). They thus combine information on both the frequency and intensity of the rainfall events.

Typical empirical IDF relationships found in the literature are of the form (Bilham, 1962; WMO, 2009c):

$$i(D) = \frac{\alpha}{(D^{\varsigma} + \theta)^{\eta}}$$

$$i(D, T) = \frac{\alpha T}{(D^{\varsigma} + \theta)^{\eta}}$$

$$i(D, T) = \frac{\alpha + \beta \log T}{(D^{\varsigma} + \theta)^{\eta}}$$

where $i(D)$ or $i(D,T)$ is the intensity for a given aggregation level or duration D and/or return period T, $\alpha > 0$ and $\beta > 0$ are scale parameters, and $\theta \geq 0$ and $0 < \eta < 1$ and ς are shape parameters (ς is most often taken equal to 1 and also η has for many cases the value 1). Other equations, however, exist as well. When the return period is excluded from the equation, the parameters depend on the return period. An example of fitted IDF curves to estimated rainfall intensities for different durations and return periods based on the regional estimation procedure in Denmark described in Madsen *et al.* (2002) is shown in Figure 2.5.

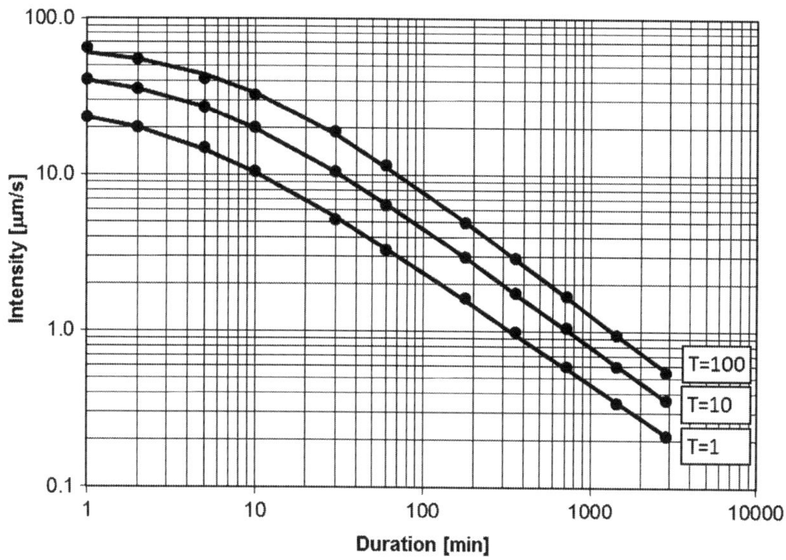

Figure 2.5 Example of fitted IDF curves for three different return periods (*T* in years) (after Madsen *et al.* 2002).

For urban drainage applications, IDF relationships down to aggregation levels of 5–15 minutes are required, while the number of rain gauges with records at such short duration is sparse. Overeem *et al.* (2009) solved this problem by using radar data but reported two additional problems. Radar data need to be adjusted to rain gauge data because of their lower accuracy in the quantitative rainfall estimate. Moreover, radar data are for most areas only available since recent years. For The Netherlands, Overeem *et al.* (2009) could make use of 11 years of good quality radar data for durations of 15 minutes or longer.

Several authors have made use of the scaling properties of rainfall, as discussed in Section 2.2, to obtain smooth IDF relationships. Scaling functions can be used to fit/smooth the IDF relationships for the range of time scales considered. Scaling properties can also be used to downscale IDF statistics. For instance, if an hourly rainfall time series is available, IDF relationships for hourly or coarser time steps can be derived directly from the data, and indirectly down to aggregation levels of, for instance, 10 minutes based on the scaling properties.

Burlando and Rosso (1996) developed scaling-based models of DDF curves and generalized IDF relationships were formulated and tested by Menabde *et al.* (1999a) for two different sets of data from Australia and South Africa. Nguyen *et al.* (2002) have shown for annual maximum point rainfall data in Quebec, Canada, that the parameters and moments of the rainfall distribution (GEV in their case) are a function of the time scale. They found two different scaling functions: one for durations between 5 min and 1 hour, and one between 1 hour and 1 day (Figure 2.4). Yu *et al.* (2004), Pao-shan *et al.* (2004), Kuzuha *et al.* (2005) and Nhat *et al.* (2007) applied scaling properties for IDF-curve regionalization. Another approach to scaling-based IDF modelling was undertaken by Langousis and Veneziano (2007), who separated the exterior (storm) and interior (substorm) rainfall processes. Bendjoudi *et al.* (1997)

have shown how formal equations can be derived for the IDF relationships on the basis of the scale invariant properties of multifractals.

2.6 DESIGN STORMS

IDF relationships are typically used for the estimation of the "design storm" for the design of urban drainage networks (Yen & Chow, 1980; Wenzel, 1982; Willems, 2000; Madsen *et al.* 2002). The design storm represents the temporal pattern of the design rainfall hyetograph, which is commonly used to overcome the long computational times of continuous urban drainage and hydrodynamic sewer model simulations. Instead of simulating the full rainfall series (observed or stochastically generated) in the urban drainage model and statistically post-processing the simulation results, the rainfall series is effectively pre-processed. The statistical analysis is thus carried out prior to the model simulations. The method assumes that the frequency of the urban drainage flow equals the frequency of the rainfall event. This is the case for many urban drainage systems. The rainfall intensity averaged over the concentration time and the effective urban catchment area mainly control the peak flows at a given location in the urban drainage system. This is the case when the urban runoff mainly originates from the paved areas, when there is no strong seasonal variation, for example in soil saturation level, as in watershed hydrology, and when the concentration time moreover is nearly constant for a given location in the urban drainage system. Under these conditions, the sewer flow at a given location is typically higher for higher rainfall intensities averaged over the concentration time.

The main difficulty related to the determination of a suitable storm pattern for design purposes has been confirmed by the availability of various synthetic design storm models developed and used around the world (NERC, 1975; Arnell *et al.* 1983; ASCE, 1983). The Chicago model was the first developed in the USA by Keifer and Chu (1957), and it was followed by other alternatives such as the pattern proposed by Sifalda (1973), Pilgrim and Cordery (1975), Desbordes (1978), Yen and Chow (1980) and the balanced model suggested by the US Army Corps of Engineers (1982). Nguyen *et al.* (2002) have performed a systematic evaluation of various existing design storm models and have proposed guidelines regarding how to select an appropriate design storm for a particular site (see also Peyron *et al.* 2005).

Figure 2.6 shows an example of the Chicago design storms, also called composite storms, estimated from the IDF curves given in Figure 2.5 for three different return periods. In these storms, the rainfall intensities averaged over a range of durations (aggregation times) equal the values obtained from the IDF relationships for a given return period (the return period of the intensity). The range of aggregation times to be considered should be equal to the concentration time along the system, that is, the time of travel from the most remote point of the catchment to any design location in the system.

Recent work by Vandenberghe *et al.* (2010) proposed a methodology for stochastic design storm generation based on copulas and mass curves. Copulas can be used to describe the dependence structure between rainfall intensities at different scales or, as the authors did, between rain storm duration and total storm depth. Mass curves were used to construct design storms with a realistic storm structure, hence to construct shapes based on rain storm duration and total storm depth.

From the above discussion, it becomes clear that in urban drainage applications, IDF relationships and design storms (or rainfall statistics in general) do not consider the seasonal variations in the occurrence of events. This is because rain storms occurring in winter and summer lead to more or less the same stormwater runoff from paved areas. This is clearly different for unpaved areas or river catchments. When the urban drainage system receives significant inflows from unpaved areas or groundwater infiltration, the time of year when the extreme events occur would also have to be taken into account.

This means that separate statistics have to be calculated for different months or seasons (as shown by Willems, 2000, for winter and summer), or complete time series have to be employed.

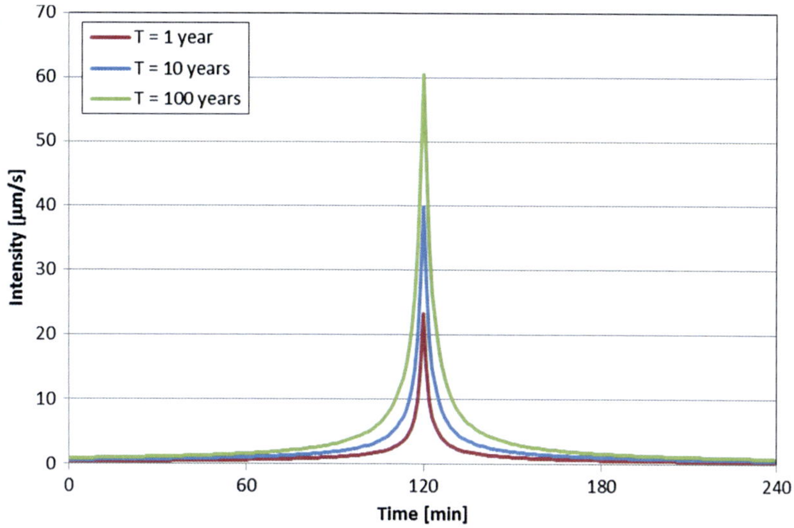

Figure 2.6 Example of estimated Chicago design storms for three different return periods based on the IDF curves shown in Figure 2.5 (after Madsen *et al.* 2002).

2.7 POINT VERSUS AREAL RAINFALL

Referring to the difference between point rainfall and areal rainfall, Section 2.3 described disaggregation techniques to support the downscaling. Alternatively the point rain gauge observations can be upscaled. Upscaling of point observations may be useful if statistical information on areal rainfall intensities needs to be obtained, but no long records of spatial rainfall are available. Point scale rain gauge observations are often the most accurate source of rainfall data and may be available for sufficiently long periods. The upscaling is also needed when rain gauge data are transferred to rainfall inputs in urban drainage models. This input can be provided in a spatially variable way or it can be lumped over given subcatchments. In both cases, spatial interpolation needs to be undertaken.

Areal rainfall estimates can be obtained from rain gauge observations by means of several types of interpolation methods. These include the Thiessen polygon method, the inverse (square) distance method, the isohyetal method, kriging or other geostatistical techniques based on spatial correction functions, spline functions, etc (McGuinness, 1963; Rodriquez-Iturbe & Mejia, 1974; Bastin *et al.* 1984; Lebel & Laborde, 1988). These often have a significant level of uncertainty, depending on the density of the rain gauge network, the spatial rainfall variability and the quality of the data (Wilson *et al.* 1979; Bastin *et al.* 1984; Lebel *et al.* 1987; Fontaine, 1991; Peters-Lidard & Wood, 1994; Roux *et al.* 1995; Schilling, 1984; Willems, 2001; Yoo & Ha, 2002; Lau & Sharpe, 2003). When areal and point rainfall statistics are studied, systematic differences exist, which for a given region depend on the rainfall intensity or return period, the aggregation level, the area and the distance between the point rainfall location and the centre of the area considered (Willems & Berlamont, 2002a). The ratio between the mean rainfall over an area and the point rainfall intensity is called the *Areal Reduction Factor* (ARF) (Nguyen *et al.* 1981). Vaes *et al.* (2005) considered two types of areal reduction factors, namely those based on individual historical events and those based on rainfall intensities for specific return periods. They studied the relationship between these factors and the rainfall intensity, the aggregation level, the area and the distance to the rain gauge. According to the results by Einfalt *et al.* (2004) and Vaes *et al.* (2005), for the typically small size of urban drainage catchments, areal correction factors may also have an increasing effect on the point rainfall statistics rather than a reduction effect.

More details on the derivation of ARFs and upscaling of point rainfall can be found in NSSP (1961), NERC (1975), Rodriquez-Iturbe and Mejia (1974), Sivapalan and Blöschl (1998), Nguyen *et al.* (1981), WMO (2009c) and other references. Literature on the inverse problem of assessing point precipitation (extremes) based on rainfall intensities defined at large spatial scales can be found in for example Booij (2002).

Another aspect related to the spatial scale issue is the difference between the return period or frequency of exceedance at a given point in space and the frequency of exceedance elsewhere in the system. In urban drainage design, frequency of exceedance usually refers to either rainfall intensity or sewer overflow frequency. For the City of Seattle in the USA, Schimek *et al.* (2008) estimated that there is a 31% probability that a 100-year point rainfall intensity is exceeded somewhere within the city in a given year. This makes a 100-year event (at a given location) a 3.2-year event for the entire area of the city. This difference between point scale based exceedance statistics and areal statistics has to be taken into account when evaluating urban flood frequencies based on historical records (i.e. counting flood events that have occurred throughout the entire city).

2.8 DISCUSSION

This chapter has shown how rainfall extremes can be statistically analysed and stochastically modelled in support of urban drainage studies. The short response times and small spatial scales of these studies require emphasis to be put on the rainfall properties at the small temporal, that is sub-daily, and spatial, that is city area, scales. As mentioned in Chapter 1, and further elaborated in the next chapters, the atmospheric processes that underlie these smaller scale rainfall properties differ from the ones that explain large scale rainfall. This may lead to a potentially large impact of climate change.

As was shown in this chapter, the stochastic rainfall properties, the rainfall extreme value distributions, IDF curves, design storms and areal correction factors, are typically derived from historical series. While presenting the methods discussed in this chapter, the intrinsic assumption was that these series are stationary. This assumption, however, is likely to be violated in a changing climate. Question is whether climate change so far has led to significant trends in the statistical properties of extreme rainfall. Recent extreme events may be an indicator of increasing intense rainfalls and urban floods, but is this supported by the observations and statistics? Has climate already started to change, and, if so, by how much did extreme rainfall statistics change over the past few decades? Methods for analysing such trends or other non-stationarities are presented in the next chapter, illustrated with some recent results.

Chapter 3

Variability, trends and non-stationarity in extreme rainfall and runoff

The techniques described in the previous chapter assume that the rainfall series have stationary properties. However, climate change introduces non-stationarities. This chapter discusses the study of non-stationary properties, both for historical periods and future conditions.

Changes over time in historical periods can be assessed using statistical trend analysis, which allows investigation of whether recent historical changes in the frequency and amplitude of rainfall extremes can be detected (Section 3.1), and whether these can be considered statistically significant in comparison with the natural temporal variability of rainfall intensities (as observed in long series) (Section 3.2). The analysis of trends and natural variability can be carried out for various time scales corresponding to the range of aggregation levels covered by IDF relationships. Results on how climate trends affect urban drainage systems are discussed in Section 3.3, which also discusses the importance of climate trends versus other anthropogenic trends (such as urbanization) in the urban catchment itself.

3.1 TRENDS IN RAINFALL PROCESSES AND EXTREMES

Methods

Several authors have investigated trends in rainfall processes and rainfall extremes. They have applied a broad range of methods, analysing a wide range of variables. Some apply standard frequency analysis methods (see Section 2.4) assuming stationarity within different sub-periods (e.g. decades by Verworn *et al.* 2008; moving periods of 10 to 15 years by Ntegeka & Willems, 2008). Trends or temporal changes are checked visually or are based on statistical hypothesis testing. Classical non-parametric trend tests are often applied. The most common is the Mann-Kendall or Kendall's tau test (see Box 3.1; Mann, 1945; Kendall, 1975; Hirsch *et al.* 1992). Other commonly applied tests are the Cox-Stuart sign test (e.g. Verworn *et al.* 2008) and non-parametric tests based on rank statistics (e.g. Arnbjerg-Nielsen, 2006) including the Spearman's rank correlation coefficient tests for trends in the mean value or the variance. Other non-parametric tests include the Kruskal-Wallis test, the non-parametric linear Theil-Sen slope estimator and normal scores regression (see Kundzewicz & Robson, 2000, for an overview). Khaled (2008) proposed a modification to the Mann-Kendall trend test to account for the effect of the scaling properties, as discussed in Section 2.2.

Non-parametric tests have the advantage that there are no or only very few inherent assumptions about the shapes of the underlying population distributions. They can be applied directly to the data, providing the data meet assumptions of independence and constancy of distribution. However, for the same reason the power of the tests is often relatively weak, meaning that the tests often fail to reject the H_0 hypothesis even if the alternative hypothesis is true. The most well-known parametric method for trend testing is linear regression (e.g. Schmidli & Frei, 2005; Fujibe, 2008; Quirmbach *et al.* 2009; Einfalt *et al.* 2011), where the modelled response variable is assumed to follow a normal distribution. One extension is the GLM where the possible stochastic behaviour of the response variable is extended to all distributions that belong to the exponential family, for example the Poisson distribution as applied in Villarini *et al.* (2011) and Gregersen *et al.* (2012). Another extension is quantile regression (e.g. Villarini *et al.* 2011).

Box 3.1 Mann-Kendall trend test

The Mann-Kendall trend test (Mann, 1945; Kendall, 1975) is a rank-based non-parametric test for assessing the significance of a trend. It has been widely used in hydrological trend detection applications. It is valid for a sequential dataset of independent and identically distributed values x_i, $i = 1$, n. It is based on the following test statistic:

$$S = \sum_{i=1}^{n} \sum_{j=i+1}^{n} \mathrm{sign}(x_j - x_i)$$

where:

$$\mathrm{sign}(x_j - x_i) = \begin{cases} 1 & \text{if} \quad x_j > x_i \\ 0 & \text{if} \quad x_j = x_i \\ -1 & \text{if} \quad x_j < x_i \end{cases}$$

When $n \geq 8$, the statistic S is approximately normally distributed with the mean and variance as follows:

$$E[S] = 0$$

$$\mathrm{Var}[S] = \frac{n(n-1)(2n+5) - \sum_{m=1}^{n} t_m m(m-1)(2m+5)}{18}$$

where t_m is the number of ties of extent m.

For independent sample data without trend the P value should be equal to 0.5. For sample data with a large positive trend the P value should be closer to 1, whereas a large negative trend should yield P value closer to 0.

In case the data series is subject to lag-1 autocorrelation, the series can be pre-whitened following the method by von Storch (1995):

$$y_j = x_j - \rho x_{j-1}$$

where y_i, $i = 1,n$ is the pre-whitended data series and ρ is the lag-1 autocorrelation coefficient.

In the alternative method by Hamed and Rao (1998), the influence of serial dependence is accounted for replacing Var[S] by:

$$\frac{\mathrm{Var}[S]n}{n^*}$$

where n^* is the effective sample size. The correction factor $\frac{n}{n^*}$ is calculated based on the lag-k autocorrelation coefficient of the ranks ρ_k^R of the sample data:

$$\frac{n}{n^*} = 1 + \frac{2}{n(n-1)(n-2)} \sum_{j=1}^{n-1} (n-k)(n-k-1)(n-k-2)\rho_k^R$$

Other tests are available that do not estimate the significance of trends but detect other types of changes or non-stationarities, for example sudden temporal changes or steps. Examples of methods for detecting such non-stationarities are Pettitt's test, Wilcoxon-Mann-Whitney or rank-sum test, Kruskal-Wallis test, Lombard change-point test, distribution-free CUSUM test, cumulative deviations test, etc (Pettit, 1979; Lombard, 1988; Siegel & Castellan, 1988; Helsel & Hirsch, 1992). Most of these are non-parametric tests based on rank statistics. They detect changes in the median of a series with the exact time of change unknown, or differences between two independent sample groups. Parametric change tests include the t-test for testing changes in the mean, the F-test for testing changes in the variance, and likelihood-ratio tests (e.g. Worsley, 1979). See Kundzewicz and Robson (2000) for an overview.

Another useful approach is exponential smoothing using techniques such as the Holt-Winters seasonal forecasting method, as described by Kamruzzaman *et al.* (2011). This forecasting procedure is applied iteratively and is well suited to identifying any changes in seasonal pattern as well as any change in the underlying level. The Holt-Winters method relies on exponential smoothing and assumes an underlying level with the addition of a possible trend and seasonal effects. Its versatility is that the level, trend and seasonal effects are allowed to change over time. Thus it is a non-stationary model, and is useful for tracking changes in the underlying parameters of a time series.

The tests described above are for data series at single locations. The tests can, however, be applied to several locations in a given region to analyse *field significance*. This means that it is tested whether the rainfall extremes from all sites in a given region are stationary or subject to trends. Such regional trend analysis is an area which has gained increasing focus over the last years. The estimation of a regional trend can decrease the uncertainty on the estimation because of the increase in the amount of data, but spatial dependence must be accounted for. This can be done by combining the data from different stations under the assumption that they come from a reduced (effective) number of independent stations (Matalas & Langbein, 1962). Other methods for testing field significance under spatial correlation are based on resampling such as bootstrapping (Livezey & Chen, 1983; Renard *et al.* 2008). A regional version of the Mann-Kendall test was developed by Yue and Wang (2002) and applied by for instance Sadri *et al.* (2009). Another option to deal with the spatial correlation is to apply the tests based on single data series using regional data (e.g. mean regional values). Renard and Lang (2007) implemented spatial dependence in their Bayesian framework using copula functions. Gregersen *et al.* (2012) applied generalized estimation equations in the GLM to describe spatial dependence for a Poisson process. In most methods for statistical significance testing, the probability of false rejection of the null hypothesis of non-significance when it is really true is controlled (by the choice of the significance level, e.g. 5%). This is done per test. When several tests are applied or the test applied to several stations to detect field significance, the probability of falsely rejected tests out of all rejected tests (i.e. stations or methods) is no longer controlled. Ventura *et al.* (2004) therefore developed an approach that controls this probability, accounting for the spatial correlation between the data at different stations.

Looking specifically at trends in extremes, a non-stationary extreme value distribution with time-varying shape and scale parameters discussed in Section 2.4 can be applied. The time dependency can be modelled by a high variety of linear or non-linear relations and some authors prefer to model transformed values of the scale parameter as a linear function of time (e.g. Coles, 2001; Kysely *et al.* 2010). Galiatsatou and Prinos (2007) applied simple parametric (polynomial, sinusoidal) models for the temporal trend in the location and scale parameters of the GEV and GPD. Parameter estimation is obtained through ML (Coles, 2001; Kysely *et al.* 2010; Parey *et al.* 2007; Strupczewski *et al.* 2001), L-moments (Fowler & Kilsby, 2003), weighted least squares estimation (Strupczewski & Kaczmarek, 2001) or using Bayesian methods (Renard *et al.* 2006). All the mentioned methods apply both to annual maxima and PDS/POT extremes. A short introduction to applied non-stationary frequency analysis in R is given in Appendix A. Recent advances include application of a non-stationary POT threshold (Kysely *et al.* 2010) and a vector GLM (Maraun *et al.* 2010).

Whatever trend testing method is selected, it has to be applied in a careful way. Each method has a number of underlying assumptions, which have to be checked, and which limit their applicability. Biased results will be obtained when the assumptions are violated. One of the assumptions, which most trend tests consider, is independence in the data. Section 3.2 will show that this assumption is often not valid, unless the serial dependence is removed from the data (examples were given in Box 3.1). The results of trend testing methods can also be biased due to the effect of scaling (see Section 2.2), as was shown by Khaled (2008). Based on his modified Mann-Kendall test, he showed that the number of sites with significant trends is considerably reduced when the effect of scaling is taken into account. Scaling thus appears to be another factor that weakens the evidence of real trends in hydrologic data as compared to classical trend testing methods. He also showed that taking into account scaling in the modified test helps to avoid discrepancies found in some previous studies, such as the existence of significant opposite trends in neighbouring sites, or in different segments of the same time series.

Due to the natural climatic and sampling variability discussed, it becomes clear that regardless of the method employed for trend testing, it is difficult to distinguish trends from variability (e.g. Frei & Schär, 2001; Rauch & de Toffol, 2006; Verworn *et al.* 2008). The smaller the scale or the more extreme the values, the more difficult this becomes. Averaging over larger regions reduces the natural variability more than averaging over smaller areas. Due to the climate oscillations, trend testing results may be biased (as discussed previously) but may also strongly depend on the period selected.

Results

In Italy, Pagliara *et al.* (1998) found based on the Mann-Kendall test that for annual maxima at Tuscany, short-duration rainfall extremes at 1, 3 and 6 hours have increased for Firenze since the mid-20th century, whereas the increase was less pronounced for the longer duration extremes of 12 hours. They could model the increase with a linear trend and a lag-one autocorrelation. Increase in extreme rainfall for Italy was confirmed by Brunetti *et al.* (2001) based on the same type of test and seven stations in northern Italy for the period 1920–1998. They found a negative trend in the number of wet days associated with an increase in the contribution of heavy rainfall events to total rainfall volumes. They also observed a reduction in return period for extreme events since 1920.

In Germany, Einfalt *et al.* (2011) analysed trends based on linear regression and the Mann-Kendall test for 750 rain gauges in North Rhine, Westphalia for the period 1950–2008. The results showed changes in the past climate that vary regionally. For annual values, half of the stations showed a significant increase. For half-yearly and seasonal time intervals, significant trends were found for winter rainfall, but no significant trend was observed for summer rainfall. Monthly values showed significant increases for March and significant decreases for August at the majority of the stations. No significant trend could, however, be observed for the IDF curves down to 5 minutes over the time interval 1950–2008.

Also in Germany, Verworn *et al.* (2008) analysed 15-minute rainfall series at 15 stations within the Emsher-Lippe region based on linear regression. Only the changes in wet spell amounts within the most recent decade studied (1997–2006) were found to be noticeable. Events with 5 mm or higher, which were detected 20 times a year in all decades up to 1996, occurred more than 30 times a year in the last decade. For the same frequency (20 times a year) and decades, the event threshold increased from 5 mm to 6 mm. The results for changes in wet and dry spell durations were less clear, although some stations showed clear increases in wet spell durations, especially for durations less than 5 hours.

In Denmark, changes in regional IDF relationships have recently been investigated. A set of regional IDF curves down to time scales of 1-minute were constructed by Madsen *et al.* (2002) for extreme rainfall based on 650 station-years of data from 41 stations. The analysis was updated by Madsen *et al.* (2009) to include recent station data in the analysis, based on 66 stations and 1250 station-years. The datasets covered the periods 1979–1997 and 1979–2005, respectively. They concluded that for the durations and return periods typical for most urban drainage systems (i.e. durations between 30 and 180 minutes and return periods of around 10 years) the increase in rainfall intensity is more than 15%. No changes or decreasing tendencies were seen for extreme rainfalls with durations of 24 hours or more. Given that their analysis was based on relatively short historical periods and that the trends were not statistically significant, the authors implicitly assumed that the increases could be seen as an adjustment rather than a clear sign of change. This allowed rapid and easy implementation of the new design intensities into national design practice. The observed changes in rainfall intensities are shown in Figure 3.1. An analysis of the same dataset including data up to year 2009 indicated that the change in the number of occurrences of rainfall extremes were significant, while changes in the rate of rainfall were much less apparent in the dataset (Gregersen *et al.* 2010). Sadrie *et al.* (2009) analysed the trends in the Danish rainfall records using a regional version of the Mann-Kendall test (Douglas *et al.* 2000). They found trends consistent with the results reported in Madsen *et al.* (2009).

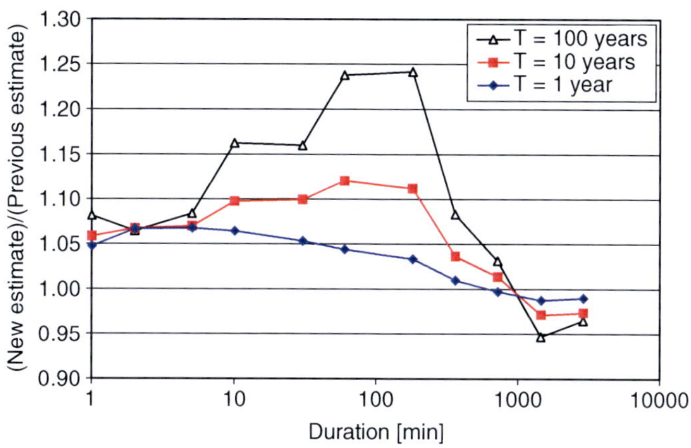

Figure 3.1 Change in regional *T*-year event rainfall intensity estimates in the Western part of Denmark when comparing estimates from the period 1979–1997 to 1979–2005 for durations between 1 minute and 48 hours (after Madsen *et al.* 2009).

In Sweden, Bengtsson and Milotti (2008) found an increasing trend of annual 10-minute maxima in Malmö but no trend in Stockholm and overall concluded that the 25-year time series they considered (1980–2007) were too short for meaningful trend detection for single rainfall series (see also the following Section 3.2). The findings are similar to the Danish study by Madsen *et al.* (2009) cited above.

In Canada, based on annual maxima of short-duration precipitation (from 5 minutes to 12 hours) intensities from 44 stations in the province of Ontario, Adamowski and Bougadis (2003) found for a limited time frame of 20 years significant positive as well as negative trends. The trends, which were computed with the help of the Mann-Kendall test, were more pronounced for short durations. In another study by Cobbina *et al.* (2008) based on an hourly rainfall series at Toronto for the period 1954–1998, it was found that the annual number of rainfall events is increasing while the mean event duration and inter-event time are decreasing since the 1990s. In Vancouver, Canada, Denault *et al.* (2006) found pronounced and statistically significant increases of annual short-duration maxima (from 5 minutes to 2 hours) since the mid-1960s. Daily maxima, however, slightly decreased during this period.

Rosenberg *et al.* (2010) analysed hourly precipitation records for the time period 1949–2007 from weather stations surrounding three major metropolitan areas of the state of Washington. They found significant differences between the 25-year periods 1956–1980 and 1981–2005 using the Komolgorov-Smirnov test for differences in distribution, the Wilcoxon-Mann-Whitney rank-sum test for differences in the mean, and for trends in the entire time series using the Mann-Kendall test. The largest change was seen for the 24-hour duration at SeaTac where the 50-year return period rainfall intensity increased by 37% from 1956–1980 to 1981–2005. What was a 50-year intensity based on 1956–1980 became a 8.4-year intensity for 1981–2005, thus about six times more frequent. The trends in the event frequency or rainfall threshold exceedances were negative. Based on linear regression, statistically significant trends were found for events exceeding several of the thresholds, such as 0.2 inch at SeaTac displaying a 15% decrease over the 59-year period.

In eastern Japan, recent trends in heavy rainfall were investigated by Iwasaki (2012), using 31-year series of hourly rainfall for the period 1976–2006 for a network of over 600 stations having fine spatial resolution. They applied the Wilcoxon rank-sum test. Two regions were found to be representative of a significant positive trend in high-intensity rainfall. One is the region around the southeastern foot of the mountains, where the evening convective maximum was naturally dominant, and the other is the region around the Pacific coastline, where the early morning convective maximum was naturally dominant. For the same data, but for the very intense 1- and 6-hour rainfall extremes ($\geq$100 mm/h and $\geq$300 mm/6 h), Fujibe (2008) concluded based on linear regression an increase in these extremes. They also showed that the trend in the spatial concentration of rainfall during the last century has been increasing, as have changes in the pattern of diurnal variation of rainfall, characterized by a relative increase in rainfall amounts in early morning and a decrease in early afternoon.

Many other authors studied trends in rainfall extremes, but without an explicit focus on urban hydrology time scales, hence on daily or supra-daily data. Some studies – mainly the ones with focus on daily rainfall extremes – are reported hereafter, because they give an idea of the changes in rainfall extremes in other regions of the world.

Using a global dataset they developed including about 3000 daily precipitation series of 40 years or more covering the period 1946–1999 (Global Climate Observing System (GCOS) surface network; http://gosic.org/gcos/GSN-data-access.htm), Frich *et al.* (2002) did not find any clear patterns of trends, but significant increases in the frequency and intensity of daily precipitation extremes were observed. It was also found that the trend results vary highly from region to region. Weighted linear regression was applied by weighting the values according to the number of stations with data available each year. For the same extremes data, but after interpolation on a regular grid, Kiktev *et al.* (2003) found significant increase in the frequency of wet days with 10 mm rainfall or more, with field significance over east Europe, the Sahel, and Japan, and over central and northeast North America and Alaska.

Groisman *et al.* (1999) found increasing trends in heavy precipitation during (parts of) the 20th century at middle to high northern latitudes during the winter half year. Observations show that in each of the countries they considered, Canada, the United States, Mexico, the former Soviet Union, China, Australia, Norway, Poland, except China, mean summer rainfall has increased by at least 5% in the past century. In the USA, Norway, and Australia the frequency of summer rainfall events has also increased, but there is little evidence of such increases in any of the countries considered during the past 50 years. Based on a statistical model, assuming no change in the number of rainy days, they have shown that 5% increase in mean summer rainfall leads to an increase in the probability of daily rainfall exceeding 25.4 mm (1 inch) in northern countries (Canada, Norway, Russia, and Poland) or 50.8 mm (2 inches) in mid-latitude countries (the USA, Mexico, China, and Australia) by about 20% (nearly four times the increase in mean rainfall). Groisman *et al.* (1999) pointed out that this relationship between rainfall extremes and total rainfall is, however, not universal: in some places, like Siberia, an increase in summer rainfall extremes was observed together with a tendency toward a decrease in total rainfall.

Using the European Climate Assessment and Dataset (http://eca.knmi.nl/), containing daily precipitation records from more than 200 stations in Europe and the Middle East, both negative and positive trends were found by Klein Tank *et al.* (2002) and Klein Tank and Können (2003) in the number of days with extreme precipitation (daily intensity higher than 20 mm) and in the annual maximum 5-day precipitation depth. They applied a linear regression based trend testing method. For only approximately 20% of the stations the trend between 1946 and 1999 was found to be significant at the 5% level. Large differences in trend were found for many nearby stations. Using another European dataset (STARDEX) of 470 stations for the period 1958–2000, Haylock and Goodess (2004) found a coherent region over northern and central Europe with an increase of the 90% quantile of daily precipitation and the maximum 5-day precipitation in winter and autumn. The trend magnitude is particularly large north of the Alpine region. Also Hundecha and Bárdossy (2005) found for that period and based on the same STARDEX database an increase in heavy daily precipitation both in magnitude and frequency of occurrence in all seasons except summer. Their results were for southwestern Germany, based on 611 stations. They also interpolated the daily precipitation on a regular 5 km grid so that the changes could be investigated on areal precipitation. Although the spatial pattern remained more or less similar with that of the point-scale trends, the average trend magnitude showed an increase with the scale of the area on which the precipitation was aggregated.

Almost opposite conclusions were formulated for the UK by Fowler and Kilsby (2003). They found little change in the 1- or 2-day rainfall depths but significant decadal changes in 5- and 10-day rainfall for many regions in the UK. The temporal changes were examined by calibration of the GEV distribution to annual maxima using L-moments. This was done for 10-year moving windows and fixed decades in the period 1961–2000 using observations from 204 sites across the UK. Another study for the UK, by Osborn *et al.* (2000), indicated increases in the number of heavy rainfall days along with an increase of the daily rainfall intensity for the period 1961–1995, particularly in winter months. Their analysis was based on empirical analysis of daily rainfall intensity classes, which correspond to percentages contribution in the total monthly rainfall total. The heavy rainfall class was based on the most intense days per month, contributing to 10% of the monthly total.

In Sweden, Bengtsson (2008) investigated long daily rainfall series (30–89 years) from southern Sweden, among which 230 stations with data for the period 1961–1990; and 89-year series from Malmö, Halmstad and Göteborg. They did not find significant temporal trends in the annual daily maxima based on linear regression.

In Switzerland, Schmidli and Frei (2005) applied linear and logistic regression on daily rainfall series of 104 stations covering the entire 20th century (1901–2000). The linear regression was used for statistics with a continuous value range (e.g. rainfall quantiles), and logistic regression using a GLM (see Section 2.1) for statistics with a discrete value range (e.g. counts of threshold exceedances and frequencies). They found a centennial increase of 10 to 30% for the high quantiles of 1-day to 10-day extremes in winter. In autumn, statistically significant increases were found as well, but for spring and summer the heavy rainfall statistics did not show statistically significant trends. Schmidli and Frei (2005) also conducted sensitivity tests to indicate that the winter and autumn trends are robust with respect to inhomogeneities in the rain-gauge time series. Some of the series they used were prior homogenized using monthly correction factors based on neighbouring stations. Thus was done after homogeneity tests detected shift inhomogeneities or gradual shifts in the mean or changes in the variance of the time series. Comparison of the trend testing results using the homogenized versus the raw series did not lead to different conclusions.

Kyselý (2009) analysed trends in multiple characteristics of heavy rainfall over the Czech Republic using a dataset from 175 rain gauges. He identified spatially coherent increasing trends in winter in the western region, the relative trend magnitudes being mostly between 20 and 30% over the the 45-year period 1961–2005. Increasing but insignificant and spatially less coherent trends prevailed for the eastern region in winter and for summer. The author indicated that his findings suggest that the pattern of changes is more complex and less coherent in eastern than in western Europe.

For southern Poland and central-eastern Germany, Łupikasza *et al.* (2011) concluded based on the Mann-Kendall test that trends in 90th and 95th daily rainfall extremes in 1951–2006 differ between the eastern and the western part. In all seasons, increasing trends in extreme rainfall dominated in central-eastern Germany, whereas opposite trends prevailed in southern Poland. This pattern was particularly prominent in winter. Summer showed the highest trend magnitude of all seasons.

For Canada, Vincent and Mekis (2006) examined trends and variations in daily extreme precipitation for the period 1900–2003. The Man-Kendall trend analysis revealed more days with precipitation and a decrease in the daily intensity, but no consistent changes were found in the extreme intensities. However, a small significant increase was found in the number of days with heavy rainfall ($\geq$10 mm).

Across the contiguous USA, Karl and Knight (1998) and Groisman *et al.* (2001) concluded based on linear regression and the Mann-Kendall test applied to daily rainfall series for 182 stations since 1910 that 90% and 95% quantile rainfall extremes show a significant increasing trend in both the intensity and the frequency. The annual rainfall volume increased by about 10%, of which over half (53%) is due to the positive trends in the upper 10% quantiles of the daily rainfall values. Similar results were obtained by Kunkel *et al.* (1999) and Kunkel (2003) for a larger set of stations covering the USA and Canada. They analysed 1- to 7-day precipitation extremes for return periods of 1 year or longer. They show local statistical significance of linear increasing trends in 6.9% of the area of the USA: in the southwest and across the central Great Plains. They reported negative trends for the northwest and southeast of the USA only. For the stations in Canada, they did not find a discernible trend in the frequency of rainfall extremes. A more recent study on century-long rainfall records throughout the contiguous USA by Pryor *et al.* (2009) concluded that there is substantial variability in terms of the magnitude, significance and sign of the linear trends, but the majority of stations that exhibit significant linear trends show evidence of increases in the intensity of events above the 95% quantile. They moreover observed that the resolved trends tend to have a larger magnitude at the end of the last century. Another study by Madsen and Figdor (2007) confirmed that an unusual increase in precipitation intensity has occurred over the contiguous USA since 1970 when compared to the 1948–1969 period. They found a statistically significant increase of 30% in the frequency of extreme rainfall in northern Washington, and of 45% in the Seattle-Tacoma-Bremerton area. Interestingly, however, trends in neighboring states were widely incongruent, with a statistically significant decrease of 14% in Oregon and a non-significant increase of 1% in Idaho. Also Gleason *et al.* (2008) noted for the contiguous US a steady increase in much above-normal proportion of 1-day rainfall events to total rainfall from the early 1970s to about 1998. They observed that 1-day rainfall extremes have been above the expected value of 10% in 10 of the 12 years over the period 1995–2006. These extremes are risen since the early 1970s yet has large multidecadal variability.

DeGaetano (2009) tested the trends for continental USA by calibration of GEV distribution to POT rainfall extremes for overlapping 30-year or lengthening intervals. They used daily rainfall data obtained from the 1061 stations that compose the daily Historical Climatology Network (HCN) of Easterling *et al.* (1999). The POT values were selected after screening for spatial consistency. Their method involves selection of POT values in excess of a given threshold under the condition that a second rainfall excess value be observed at a station located within a given radius. Resampling procedures are used to assess both trend and field significance. Nearly two-thirds of the trends in the 2-, 5-, and 10-year return-period daily rainfall intensities, as well as the GEV distribution location parameter, were positive. Significant positive trends in these values were observed to cluster in the Northeast, western Great Lakes, and Pacific Northwest. Slopes were more pronounced in the 1960–2007 period when compared with the 1950–2007 interval. In the Northeast and western Great Lakes, the 2-year rainfall value increased at a rate of approximately 2% per decade, whereas the change in the 100-year intensity was between 4% and 9% per decade. These changes result primarily from an increase in the location parameter of the fitted GEV distribution. Collectively, these increases result in a median 20% decrease in the expected recurrence interval, regardless of interval length. DeGaetano (2009) concluded that, at stations across a large part of the eastern USA and Pacific Northwest, the 50-year rainfall intensity based on 1950–79 data can be expected to occur on average once every 40 years, when data from the 1950–2007 period are considered.

Using linear regression, Schimek *et al.* (2008) found for Seattle a weak increasing trend for extreme daily rainfall intensities (of 25-year return period or higher). They considered point rainfall at 17 rain gauges since the 1970s. The number of rain gauges recording these extreme events appears to have declined, however, suggesting a tendency towards increased frequency of more localized extreme rain events. The increasing frequency of extreme rainfall in the southeastern USA over the period 1994–2008 was by Kunkel *et al.* (2010) linked to changes in the frequency of tropical cyclones and more landfalling of hurricanes.

For the North American monsoon region in Northwest Mexico, Cavazos *et al.* (2008) found linear upward trend in extreme daily rainfall, defined as 95% quantile values, in the mountain sites during the period June-Sept. for 1961–1998. This increase was explained by an increased frequency of tropical cyclones. The frequency of extreme rainfall, the total monsoon precipitation and the extreme rainfall intensity in the coastal stations did not change significantly.

Peterson *et al.* (2008) extended the trend analysis for daily rainfall extremes towards the entire North American region from Canada to the United States and Mexico. This was done for 90th, 95th and 97.5th quantiles since the late 1960 s. They used data from 210 stations in Canada, the GHCN dataset (see Section 4.1) for the USA and Mexico, extended with 163 stations for Mexico. Two trend testing methods were applied: the Mann-Kendall method and the rank-statistics based multiresponse permutation procedure by Mielke and Berry (2001). The tests learned that the extreme quantiles and the average amount of rainfall falling on wet days has also been increasing since the late 1960s.

In South America, daily gridded rainfall data for the period 1940–2004 analyzed by Khan *et al.* (2007) did show increasing trends in the Amazon basin except eastern parts, few parts of the Brazilian highlands, north-west Venezuela including Caracas, north Argentina, Uruguay, Rio De Janeiro, São Paulo, Asuncion and Cayenne. This was based on trends in the ratio of POT rainfall quantiles, estimated from a GPD, for 25-year moving periods. This is consistent with the specific results obtained for São Paulo, Brazil, by Sugahara *et al.* (2009). Also these authors calibrated GPDs to POT daily rainfall extremes. They show an increase of 99% quantiles by about 40 mm between 1933 and 2005. This trend is significant following the Mann-Kendall trend test. In the analysis by Haylock *et al.* (2006), trends were identified in 95% and 99% daily rainfall quantiles in Ecuador and northern Peru and the region of southern Brazil, Paraguay, Uruguay, and northern and central Argentina, whereas a decrease was observed in southern Peru and southern Chile. Several stations show a significant trend based on the Mann-Kendall test.

Data in South Africa, analysed by Mason *et al.* (1999), show significant increases in the intensity of extreme rainfall events between 1931–1960 and 1961–1990 over about 70% of the country. The significance of differences in the rainfall distributions for these periods was tested by resampling. The intensity of the 10-year high rainfall events has increased by over 10% over large areas, except in parts of the country. Percentage increases in the intensity of high rainfall events were found largest for the most extreme events. These results are consistent with the findings by New *et al.* (2006) for 14 countries in south and western Africa. Increases in regionally-averaged daily extreme rainfall intensities as well as the dry-spell duration over the period 1961–2000 were statistically significant. For the northwestern highlands of Ethiopia, Shang *et al.* (2011) did not find significant trends in daily rainfall extremes over the period 1953–2006. Also Seleshi and Camberlin (2006) did not find clear trends in extreme Ethiopian rainfall for 11 key stations over the period 1965–2002. The Mann-Kendall and linear regression show decreasing trends in 6-daily annual maximum rainfall over eastern, southwestern and southern Ethiopia, whereas no trends are found for the remaining part of Ethiopia. Aguilar *et al.* (2009) report that decreasing trends are found for several areas in central Africa. Daily rainfall extremes over 1955–2006 did show decreasing trends for western central Africa and non-significant increasing trends for Guinea Conakry and Zimbabwe.

For Southeast Asia and the South Pacific, Manton *et al.* (2001) analysed trends in extreme daily rainfall for the 38-year period 1961–1998 from 91 stations in 15 countries. They found that the number of days with 2 mm of rain or more decreased significantly for the tropics, throughout southeast Asia and the western and central-south Pacific, but increased in the north of French Polynesia, in Fiji, and at some stations in Australia. The frequency of extreme rainfall events has declined at most stations (but not significantly), although significant increases were detected in French Polynesia. Trends in the average intensity of the wettest rainfall events each year were generally weak and not significant.

Endo *et al.* (2009) focused specifically on southeastern Asian countries and found for daily rainfall data from the 1950s to the 2000s a decrease in the number of wet days, while the average rainfall intensity of wet days shows an increasing trend for many regions. They concluded that heavy rainfall increases in southern Vietnam, northern part of Myanmar, and the Visayas and Luzon Islands in the Philippines, while the same type of rainfall decreases in northern Vietnam.

For Japan, Fujibe (2008) studied daily rainfall series from 51 stations for the period 1901–2006 using linear regression. They found that heavy daily precipitation ($\geq$200 mm and $\geq$100 mm) has significantly increased during the last century whereas weak to moderate precipitation ($\geq$1 mm to $\geq$10 mm) has decreased.

In Australia, Chowdhury and Beecham (2010) used the Mann-Kendall test to identify statistically significant trends in monthly rainfall data starting from the 1960s for ten cities. The trend free pre-whitening approach was used to remove the effects of serial correlation in the dataset. The trend beginning year was identified using the CUSUM test and the influence of the Southern Oscillation Index on the identified trends was determined using a graphical representation of the wavelet power spectrum. They found that the trend starting year depends on the rainfall time scales. This was consistent with the findings of other Australian studies (Simmonds & Hope, 1997; Smith *et al.* 2000; Smith, 2004; Murphy & Timbal, 2008). For 10 cities in Australia, Beecham and Chowdhury (2012) extended this work to examine severe rainfall intensities from events corresponding to return periods of 1 month to 10 years for both pre-trend and post-trend conditions (using the trend starting years identified by Chowdhury & Beecham, 2010). Between 20 and 40% increases in severe rainfall (event peak, daily peak and daily rainfall) were observed in Perth and Darwin (Figure 3.2). These increases were found to be more significant for the 6-month to 5-year return periods.

While the total rainfall decreased in Perth, severe rainfall intensities were found to increase from the pre-trend period (1961–1974) to the post-trend period (1975–2000). No significant change was found for severe rainfalls in Cairns and Brisbane, although their total rainfall had an increasing trend. For the Hume Dam station in the Murray Darling Basin, Kamruzzaman *et al.* (2011), using a regression analysis followed by CUSUM plots on 100 years of data, found evidence of shifts in the underlying mean rainfall.

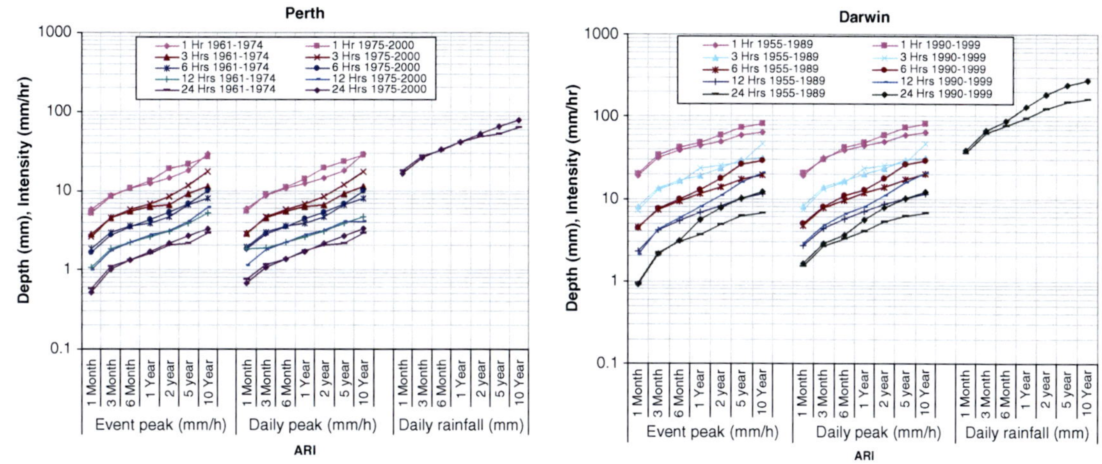

Figure 3.2 Changes in severe rainfall intensities from the pre- to post-trend periods. The IDF curves were estimated from 6-minute continuous observed rainfall data (after Beecham & Chowdhury, 2012).

In Maleysia, Mann-Kendall and linear regression trend tests show increasing trends in the extreme wet day intensities for 95% and 99% quantiles for 10 rain gauges over the period 1971–2005 (Wan Zin *et al.* 2010). A significant decrease in the frequency of 99% quantiles, was observed at 60% of the stations. Change points were found in the 1980s.

For China, Zhai *et al.* (2005) analysed trends in daily rainfall extremes, defined as 95% quantiles, for a dataset they developed for 740 stations in China, covering the period 1951–2000. They applied the Mann-Kendall method for the rainfall values and logistic regression using a GLM for the frequency of rainfall events. Significant increase in rainfall extremes have been found in western China and in parts of southwest and southern China. A significant decrease in extremes is observed in northern China. Similar results were found by Yao *et al.* (2008) using a gridded dataset of daily rainfall constructed by interpolating more than 2,200 gauges over the Asian continent and its surrounding islands. They tested trends for different classes of rainfall intensities based on linear regression and the Mann-Kendall test. During the period 1978–2002, rainfall exhibits positive trends over southeastern and northwestern China, separated by negative trends over central China and southwestern and northeastern Asia. The changes are largest over the entire tropical-subtropical Asia region. The changes over India are much smaller. Another study for China was carried out by Su *et al.* (2006). They applied both linear regression and the Mann-Kendall test to daily rainfall extremes for the period 1960–2002 at 108 meteorological stations. They concluded that the monsoon summer rainfall revealed a positive trend, but this was statistically not significant. However, a significant positive trend was found for the number of rain storm days (daily rainfall $\geq$50 mm). You *et al.* (2010) analysed daily maximum rainfall intensities at 203 meteorological stations in China for 1961–2003 and did find regional but non-significant trends. The positive trends were strongest in southeastern and northwestern China and, while the largest negative trends were for northern China. In the eastern and central Tibetan Plateau, non-sigificant changes were found in the daily maximum rainfall at 71 meteorological stations with elevation above 2000 m above sea level during 1961–2005 (You *et al.* 2008). The trends were increasing in the southern and northern Tibetan Plateau and decreasing in the central Tibetan Plateau.

For central India (the so-called monsoon belt), Goswami *et al.* (2006) considered daily gridded data for 1951–2003 and reported an increasing trend in the frequency and intensity of 99.5% and 99.75% quantiles.

Using the same dataset, Krishnamurthy *et al.* (2009) concluded that 90% and 99% rainfall quantiles exhibit field significance of increasing trends (Mann-Kendall method). Another study for central India was conducted by Rajeevan *et al.* (2008) using 104 years (1901–2004) of high resolution daily gridded rainfall data. They show that the frequency of extreme rainfall events has a statistically significant long-term trend of 6% per decade. Sen Roy and Balling (2004) studied daily rainfall records throughout the entire India territory. This was done for 129 stations over the period 1910–2000. 90%, 95% and 97.5% quantiles of daily rainfall were found to have significant upward trends in a contiguous region extending from the northwestern Himalayas in Kashmir through most of the Deccan Plateau in the south and decreasing values in the eastern part of the Gangetic Plain and parts of Uttaranchal. More recently, Guhathakurta *et al.* (2011) considered daily rainfall data at 2,599 stations all over India, having 30 years or more data in the period 1901–2005. They confirmed that the frequency of heavy rainfall events are decreasing in major parts of central and north India while they are increasing in peninsular, east and north east India. The trends were found significant for several regions after application of the Mann-Kendall test.

For Russia, Bulygina *et al.* (2007) applied linear trend testing on precipitation data from 857 stations across Russia during the past six decades. They found an increasing number of days with precipitation exceeding the 95% quantile in winter at stations in the north of European Russia and in Siberia. Negative linear trend coefficients in the time series of the number of days with heavy precipitation were only observed at individual stations in eastern and European Russia (for 1977–2006). In summer, no significant linear trend coefficients were found for the period 1951–2006. According to the data for the period 1977–2006, a reduced number of days with heavy precipitation was found in the south and southwest of European Russia, the north of western Siberia, in Chukotka, and the southern Far East. Individual stations in western Siberia show a tendency for an increase.

The above-cited studies on both daily or supra-daily and sub-daily rainfall data show that patterns of trends exist. These results suggest that urban drainage systems probably have already experienced changes in stormwater flows over the past decades. The trends are, however, not always clear and are subject to strong regional differences.

3.2 TRENDS VERSUS CLIMATE OSCILLATIONS

In the trend analysis of rainfall processes and extremes, the effect of clustering in time of the frequency and amplitude of the rainfall extremes has to be taken into account. This requires an analysis to be carried out to describe the short- and long-term time dependence induced by annual, decadal and multidecadal climate oscillations.

Large-scale, low-frequency oscillations have been observed in for example the Pacific and North Atlantic regions. The most well-known are the El Niño southern oscillation (ENSO, e.g. Philander, 1990) in the Pacific and the North Atlantic oscillation (NAO, e.g. Hurrell, 1995). They were represented by various types of indices. Some common indices considered by climatologists for the Pacific are the Pacific Decadal Oscillation (PDO) index (Mantua *et al.* 1997), the Southern Oscillation Index (SOI) (Trenberth, 1984), the Indian Ocean Dipole (IOD) index (Saji *et al.* 1999) and the Southern Annular Mode (SAM) (Thompson & Solomon, 2002). For the North Atlantic region, common indices next to the NAO are the AMO (Atlantic Multidecadal Oscillation) index (Kerr, 2000; Enfield *et al.* 2001; Knight, 2005), the Arctic Oscillation (AO) index (Noren *et al.* 2002) and the Antarctic Oscillation index (AOI) (Gong & Wang 1999).

Several authors have detected such oscillations also in series of (extreme) rainfall observations. May and Hitch (1989) and Blanckaert and Willems (2006) detected cycles in short-duration rainfall extremes after use of spectral (e.g. Fast and Windowed Fourier Transform (WFT) methods) and wavelet analysis. They considered time series of annual maxima of 1-hour extreme rainfall for 20 sites in lowland England for the 50-year period 1948–1997 (May & Hitch, 1989) and 1-hour POT rainfall extremes for the Uccle station in Belgium since 1898 (Blanckaert & Willems, 2006). May and Hitch (1989) described periodic variations with periods of 7, 11, 20 and 50 years. Blanckaert and Willems (2006) identified cycles with periods between 2 and 4 years for the summer season and between 4 and 8 years for the winter season, but only during limited time periods. Similar conclusions were obtained by De Jongh *et al.* (2006) based on the same data series but after aggregation to monthly, seasonal and annual time steps. In the WFT and wavelet spectra by Blanckaert and Willems (2006), the 1930s, the 1960s and the 1990 s were shown to have a higher energy density. The hypothesis of a random signal could be rejected for these periods at the 5% significance level. Based on these results, there is strong evidence that clustering of rainfall extremes occurs with both short and longer time intervals; thus showing inter-annual dependence in rainfall series. For an overview of the spectral analysis and wavelet based methods, the reader is referred to Ghil *et al.* (2002).

Gershunov and Cayan (2003) have shown that the frequency of heavy daily rainfall is linked to ENSO as well as to non-ENSO interannual and interdecadal variability in the North Pacific. Schreck and Semazzi (2004) came to similar conclusions for eastern Africa; as well as Haylock *et al.* (2006) and Grimm and Tedeschi (2009) based on extreme rainfall data in South America; and Gershunov (1998) and Cayan *et al.* (1999) for extreme rainfall frequencies in contiguous and western USA. Grimm and Tedeschi (2009) had strong indications that there is more sensitivity to ENSO in the extreme ranges of daily rainfall, which undergo relatively stronger changes in frequency, and that this sensitivity does not necessarily translate into changes of monthly total rainfall. Same conclusions were made by Gershunov (1998) and Cayan *et al.* (1999) for the USA. In Australia, Aryal *et al.* (2009) found linkage between temporal trends in daily rainfall extremes and those of the SOI and AOI. Also for Australia, Kamruzzaman *et al.* (2011) presented two analyses on rainfall series from six stations in the Murray Darling Basin and four cities in eastern Australia. They did not consider daily or sub-daily rainfall extremes but monthly rainfall series. The record lengths were between 38 and 108 years. The first analysis was based on multiple regression and the other used the Holt-Winters algorithm, for investigating non-stationarity in environmental time series. The first analysis focused on the residuals after fitting regression models which allowed for seasonal variation and for longer term oscillations reflected by the indices PDO and SOI. The regression models provided evidence that rainfall is reduced during periods of negative SOI, and that the interaction between PDO and SOI makes this effect more pronounced during periods of negative PDO. The residuals were also analysed for volatility, autocorrelation, long range dependence and spatial correlation. For all ten rainfall series, cumulative sum plots of the residuals provided evidence of non-stationarity, after removing seasonal effects and the effects of PDO and SOI. Rainfall was generally lower in the first half of the twentieth century and higher during the second half of the century. However, it decreased again over the last ten years. This pattern was highlighted with plots of five year moving averages. The second analysis decomposed the rainfall series into random variation about an underlying level, trend, and additive seasonal effects using the Holt-Winters algorithm. The results of the Holt-Winters analysis were qualitatively similar to those of the regression based analysis.

Other authors have made use of quantile-based methods to study rainfall dependence at longer (decadal and multidecadal) time scales. Ntegeka and Willems (2008) compared quantiles in moving block periods of 5, 10 and 15 years length, with quantiles derived from the entire series (in their case a more than 100-year long series of 10-minute intensities). They found that rainfall extremes over Belgium show multidecadal oscillations, with oscillation peaks in the 1910s–1920s, the 1960s and recently during the past 15 years (Figures 3.3 and 3.4). For the winter season, the increase in heavy rainfall extremes during the past

15 years is explained in part by a climate oscillation peak (Figure 3.3 around half of the increase is accounted for by the climate oscillation, and around half by climate change). For the summer season, the increase could entirely be explained by the climate oscillation peak (Figure 3.4). Some oscillation peaks were found significant at the 5% significance level (Figure 3.5). Box 3.2 explains how anomalies in extreme quantiles and their significance can be identified.

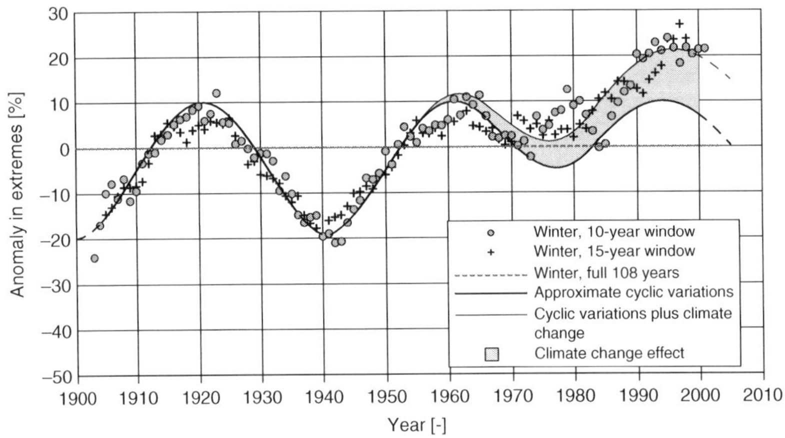

Figure 3.3 Anomaly in quantiles of 10-minute rainfall extremes (higher than 1 mm/10 min) at Uccle, Brussels, in the winter season, for moving windows of 10 and 15 years in comparison with the full period 1898–2005 (adapted from Ntegeka & Willems, 2008).

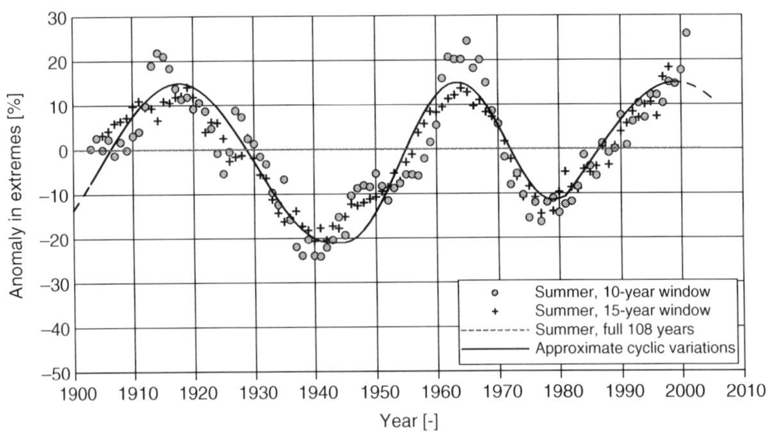

Figure 3.4 Anomaly in quantiles of 10-minute rainfall extremes (higher than 0.8 mm/10 min) at Uccle, Brussels, in the summer season, for moving windows of 10 and 15 years in comparison with the full period 1898–2005 (adapted from Ntegeka & Willems, 2008).

Box 3.2 Identification of anomalies in quantiles to detect multidecadal climate oscillations

The anomalies in extreme rainfall quantiles shown in Figure 3.3 and Figure 3.4 were computed based on the quantile perturbation method by Ntegeka and Willems (2008). The method can be used to analyse trends and anomalies in hydro-climatic extremes based on quantile perturbations on (multi-) decadal time scales. The perturbations refer to relative changes between two series. One of the series is taken as reference or baseline series while the other is a subseries of interest. For historical analysis, the reference series are taken as the full available historical series, while the subseries are block periods of interest (i.e. 10 years). The quantile perturbation approach can be applied to series at any time scale.

The extreme values within a subseries (particular block of L years) are ranked in descending order, where i is the rank of a given value in the series. Based on this rank, the empirical return period $T_{L(i)}$ can be estimated according to Equation (3) for each time step in the subseries. The same calculations are applied to the full long-term series (N years), Equation (4).

$$T_{L(i)} = \frac{L}{i} \tag{3}$$

$$T_{N(i)} = \frac{N}{i} \tag{4}$$

The values that correspond to these return periods are denoted as the quantiles $x_L, x_{L/2}, \ldots, x_{L/i}$ for the subseries and $x_N, x_{N/2}, \ldots, x_{N/i}$ for the full series. The perturbation factor $P(i)$ for a given return period then corresponds to the ratio indicated in Equation (5).

$$P_{(i)} = \frac{x_L(T_{L(i)})}{x_N(T_{N(i)})} \tag{5}$$

It is clear that the empirical return periods $T_L(i)$ of the subseries do not necessarily coincide with the empirical return periods of the full series $T_N(i)$. In that case, the $x_N(T_{L(i)})$ values are derived by linear interpolation from the values with closest empirical return periods. From the perturbation factors of all empirical return periods, an average perturbation factor can be calculated for all quantiles above a given threshold (or a given threshold for the return period).

These average perturbation factors represent anomalies in quantiles, and are calculated for moving block periods (subseries of L years length, moved with 1-year step from the beginning, towards the end of the full available series).

For a full series of N years [0;N], the following moving N-year subseries thus are considered: [0;N], [1;$N+1$], ..., [$N-9$;N]. The moving procedure causes dependency among the calculated anomalies (over periods of N years length), but allows easier visual interpretation of the temporal variations in these anomalies. The method allows to detect decadal and multi-decadal oscillations in extreme quantile anomalies (Ntegeka & Willems, 2008).

The same procedure can be applied but limiting the return period and perturbation factor calculations to a given season. In this case the particular block period (subseries) of interest will be limited to the months in the given season.

To test whether the trends and anomalies identified in the historical series are significant, Ntegeka and Willems (2008) calculated confidence intervals on the perturbation factors. This was done under the null hypothesis of no trend or serial dependence in the occurrence of rainfall extremes in time. They calculated the confidence intervals by a non-parametric bootstrapping technique. The values in the full series were randomly resampled a large number of times (N_{BR} bootstrap runs; each run containing the same number of values as in the full series) without replacement. For each run, perturbation factors are recomputed, where after confidence intervals are calculated for each time step based on the ranked N_{BR} perturbation factors per time step.

Since these confidence intervals define regions of expected variability under the null hypothesis, any anomaly value outside the confidence bounds is considered to be statistically significant (hypothesis of no trend or serial dependence is rejected), while the region inside the confidence bounds is considered insignificant (hypothesis accepted). See Figure 3.5 for an example of these confidence bounds. The figure shows significant high anomalies for the 1910s, 1960s and 1990–2000s, and significant low anomalies for the 1930–1940s and 1970–1980s.

In Willems and Yiou (2010), these anomalies were calculated in a similar way with data corresponding to regional climate influences as NAO. The aim of that comparison is to identify the external influences of the anomalies in rainfall and temperature. The method was also applied by Taye and Willems (2011, 2012) for the Blue Nile basin, where climate influences were found with the SOI and sea surface temperature differences over the Equatorial Pacific. Mora and Willems (2012) have shown for the Paute basin in Ecuador that daily rainfall extremes are influenced by the ENSO. Sites with similar correlations can be grouped to obtain regionalized estimates of the external influences. Taye and Willems (2012) have shown that extreme value statistics (e.g. IDF) computed from historical series with limited lengths (e.g. less than 30 years) might be biased from the same statistics but computed based on long periods of several decades. They suggested that the large-scale atmospheric indicators can be used for correcting that bias in the statistics estimated from short periods.

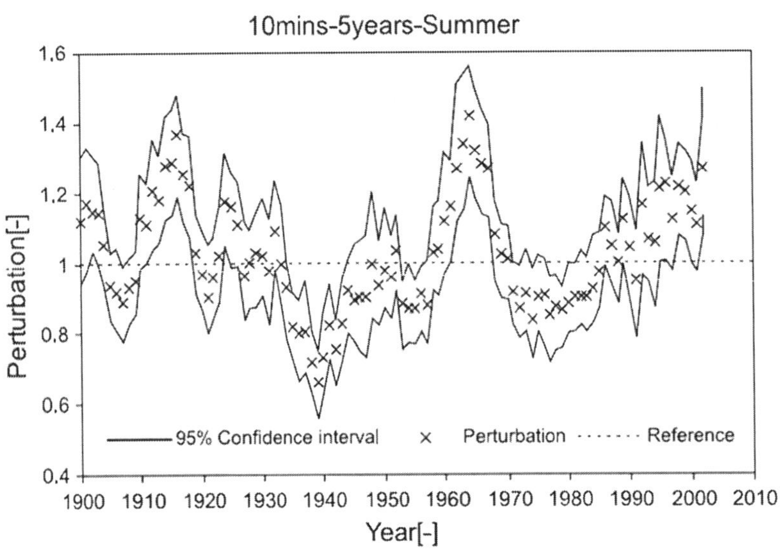

Figure 3.5 95% confidence bounds for the anomaly in quantiles of 10-minute rainfall extremes (higher than 1 mm/10 min) at Uccle, Brussels, in the summer season, for moving windows of 5 years in comparison with the full period 1898–2005 (adapted from Ntegeka & Willems, 2008).

The multidecadal oscillations in rainfall extremes, as shown in Figures 3.3 and 3.4, explain why trend analysis of periods of less than 30 years might wrongly identify increasing or decreasing parts of longer-term climate oscillations as trends, even when merging information from several series in a region. Moreover, when short- and long-term persistence is present in the series, classical trend testing techniques such as the linear regression, Mann-Kendall, Spearman's rank and other tests cannot be directly applied. They assume independence in the data, which may not be valid for climatologic series when short- and/or long-term persistence is present in the series.

One solution is to remove autocorrelation in the series by pre-whitening before application of a trend test. Details of pre-whitening techniques can be found in von Storch (1995), Kulkarni and von Storch (1995), Zhang *et al.* (2000), Wang and Swail (2001), Zhai *et al.* (2005), Kamruzzaman *et al.* (2012) and Chowdhury and Beecham (2010). These techniques intend to remove a serial correlation component such as a lag-one autocorrelation from the time series (see also Box 3.1). Another method would be to replace the real sample size by an effective sample size to account for the effect of serial dependence on the variance (see e.g. Hamed & Rao, 1998). However, Yue *et al.* (2002) have shown that such correction methods may lead to incorrectly identified significant trends, because they do not address the potential interaction between a trend and the autocorrelation when both exist within a time series. Another option is the use of the modified Mann-Kendall test by Yue and Wang (2002) for auto-correlated data. A comparison of methods has been made by Renard (2006), who developed a general framework for the selection of trend tests based on the degree of autocorrelation, the type of probability distribution, the change type and the length of the available historical series.

Several authors, although they did not explicitly study climate oscillations, reported on differences in trend testing results when the tests are applied to different periods, or to subperiods of the full available series. Some examples are shown hereafter.

In the study by Einfalt *et al.* (2011) with focus on in North Rhine, Westphalia, Germany, trend testing was done for 750 rain gauges based on linear regression and the Mann-Kendall test. It was found that the number of heavy precipitation events had increased in the past ten years as a function of the sampling interval. For short intervals ($\leq$1 hour), the increase was higher than for longer sampling intervals. They also confirmed that trends strongly depend on the decadal period selected. They found that in the period from 1971 to 1980 less rainfall was recorded than in any other ten year period between 1950 and 2008, and that this time period played a major role in the trend analysis when based on time periods shorter than 50 years. Trend analysis over only 30 years, although not for extremes but for annual rainfall sums, showed very different results depending on the time interval chosen, especially if the decade 1971–1980 was included. Figure 3.6 shows a decrease of the yearly sums for the period 1950–1979, while the period 1971–2000 shows a significant increase. When the analysis is undertaken for the full period 1950–2008, most stations show an increase in the annual rainfall, this being significant in more than 50% of the stations (Figure 3.6).

Bengtsson (2008) did not find trends for 89-year time series of annual maxima of daily rainfall for stations in southern Sweden, but noticed that there have been trends over shorter periods.

For Emilia-Romagna, a region in northern Italy, Pavan *et al.* (2008) did not find clear trends in the frequency daily rainfall events for a dense network of stations over the period 1951–2004. They rather identified the presence of a clear decadal variability (even periodicity) in the frequency of such intense daily rainfall.

Also Manton *et al.* (2001), who conducted trend analysis on extreme daily rainfall series for Southeast Asia and the South Pacific for the period 1961–1998, warned that such time frame is relatively short and may produce trends that are sensitive to the sampling period. They observed that trends in total rainfall, dry days and extreme events in Australia from 1910–1995 tend to be in the opposite direction to those for 1961–1998. However, for some countries in the region, appropriate data do not exist from before the late 1950s. So extending the analysis (across the entire region) backwards was not possible.

For south China, based on daily rainfall data for 1961–2003 at 17 stations, Ning and Qian (2009) found strong interdecadal changes. They were found consistent with the changes in latent heat flux over the South China Sea and the sensible heat flux over the Indochina peninsula through the South China Sea summer Monsoon. In Central India, Rajeevan *et al.* (2008) shows that the frequency of extreme rainfall events has inter-decadal variations, modulated by the SST variations over the tropical Indian Ocean.

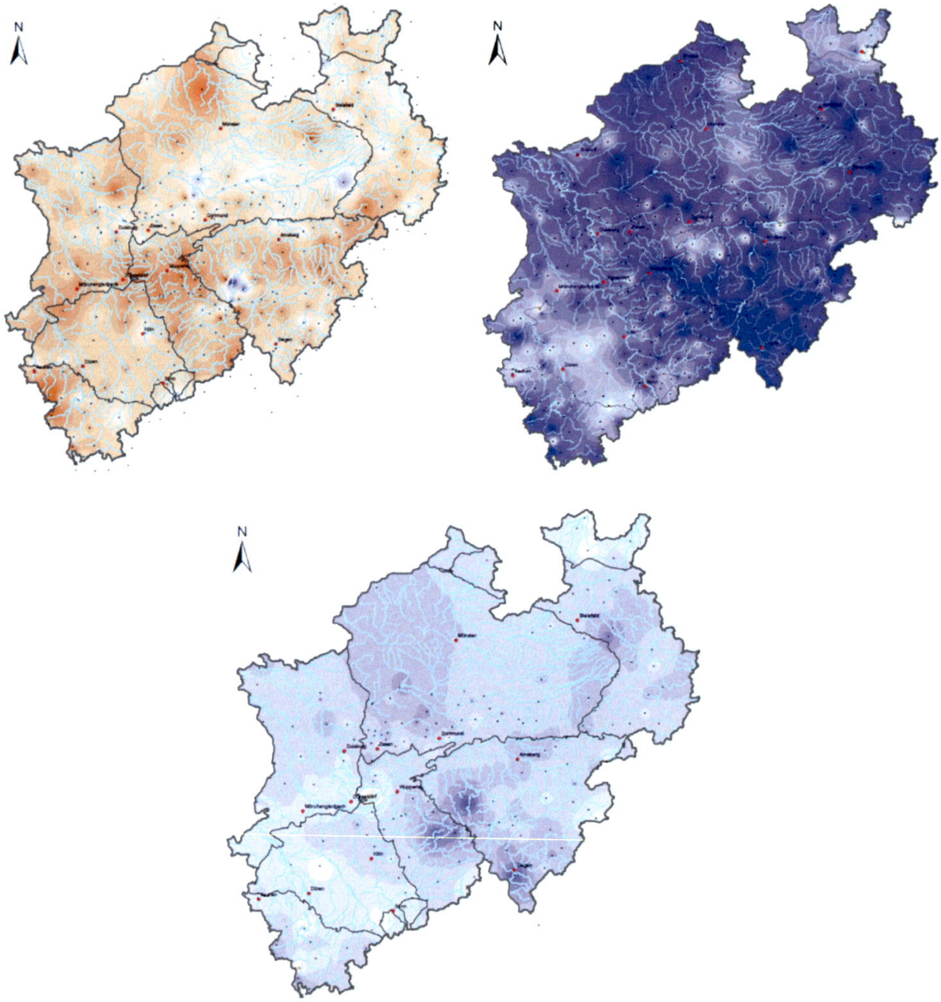

Figure 3.6 Trend results by linear regression on annual rainfall sums over North Rhine, Westphalia, Germany, for 1950–1979 (top left), 1971–2000 (top right) and 1950–2008 (bottom) (blue = increase of rain; orange = decrease of rain) (after Einfalt *et al.* 2011).

For midwestern stations in the USA, Kunkel *et al.* (1999) show based on the analysis of a 101-year record that the frequencies of 1- to 7-day rainfall extremes of 1-year return period or longer around the turn of the twentieth century (1896–1906) were higher than for other periods of comparable length, except for 1986–96. They warned that the interpretation of the recent upward trends must account for the possibility of significant natural forcing of variability on century timescales. At decadal scales, they found below average frequency of rainfall extremes in the early 1930s for all regions in the USA, above-average frequency in the early 1940s in the west, south and north-central USA, below-average frequency in the early 1950s for most regions of the USA, except the northeast and east-north central regions of the USA. For the 1960s, very low frequency of rainfall extremes was obtained for the northeast USA, but this contrasted with the near-average frequency in the rest of the USA. Below-average frequency in the late 1980s occurred primarily in the western regions, while the frequency of rainfall extremes in the northeast was well above average. During the 1990s, most regions experienced an above-average frequency.

Also Frich *et al.* (2002), but now based on a global dataset, show that global maps of indicators of extreme precipitation during the second half of the 20th century show significant changes from one multi-decadal period to another.

3.3 TRENDS IN URBAN RUNOFF: CHANGES DUE TO CLIMATE AND URBANIZATION

When investigating trends in urban runoffs, which is nearly impossible to do because of the very limited length of available urban drainage flow series, trends might be visible as well. However, these trends would probably reflect changes in urban drainage management (additional or new urban drainage systems built, older systems rehabilitated or renovated, connected to treatment infrastructure), urbanization and increase in pavements and house roofs, rather than climate change.

The impact of climate variability and trends on urban runoff or other urban drainage impacts can be analysed by means of urban drainage models. A long-term rainfall series can be simulated in the model and the runoff results statistically analysed (using similar techniques as discussed in Section 3.1). This has been undertaken by Verworn *et al.* (2008) who studied the impact of trends in extreme rainfall statistics (based on a 15-minute rainfall series within the Emsher-Lippe region) on sewer surcharge frequencies, CSO frequencies and overflow volumes. Sewer surcharge frequencies were studied depending on whether they happen somewhere in the system or for several manholes at the same time. Results showed an increase in the surcharge and overflow frequencies in the last one or two decades. Trends in overflow volumes depend on the trend testing method applied. It was found that, depending on the method applied, trend signs may be opposite. Trends in overflow volumes were found to be positive when total overflow volumes per decade were studied, while they were found to be negative after fitting a linear trend line to the overflow volumes of all events in the time series 1957–2006. The multidecadal oscillations, discussed in the previous section may explain these findings.

Another impact study by Pagliara *et al.* (1998) for Italy made changes in design storms based on the significant rainfall trends they found. The changes were made every four-five years during the period 1955–1993 and the design storms simulated in a sewer system model. Differences of 13% (increase from 1960s to 1990s) in peak discharges were obtained. Changes of 3–5% in pipe diameter are required to bring the peak discharges for the 1990s to the same level as for the 1960s.

3.4 DISCUSSION

The methods discussed in Chapters 2 and 3 for analysing the stochastic/statistical properties of rainfall extremes and properties of variability, trends and non-stationarity are all empirical methods. There is no need to explain how important it is for such methods to be based on good quality observations. It is beyond the scope of this book to present methods for analysing the quality of data, for analysing the temporal consistency of the data and for detecting outliers. For these methods, the reader is referred to Hosking and Wallis (1997) and Tsakalias and Koutsoyiannis (1999) in the first instance. It is worth noting that several studies reported above carried out a data quality analysis. Some adjusted the rainfall measurements, as was done by Vincent and Mekis (2006) for site relocation, changes in observing programs and corrections for known instrument changes or measurement program deficiencies.

For trend testing, urbanization increase might be a possible source of non-stationarity in climatic records, which may influence trend estimates. Dense urban cities may be influenced by the *Urban Heat Island* (UHI) effects discussed in Chapter 5 (e.g. Dettwiller & Changnon, 1976). The reported increases in temperature and induced mesoscale convergence caused by the UHI effect are, however, so small that it seems unlikely to have had a strong effect on the rainfall trends for most cities (Jones *et al.* 1989; Huong & Pathirana, 2011). Karl *et al.* (1988) established a minimum population threshold of 100,000 people for discernible UHI effects. Kishtawal *et al.* (2010) documented for many large cities in India (with population over 5 million) an urbanization signature in the observed frequency of heavy summer rainfall during the June-Sept. monsoon season.

Another possible source of non-stationarity in climatic records is improvement of instrumentation. Around 1990 and around 2000, measurement devices using, respectively, tipping bucket technologies and weighing type instruments were gradually introduced into station networks worldwide. These instruments permit a higher accuracy measurement for short time interval sampling (down to one minute and less) than the traditional paper strip based instruments. Therefore, the results for sampling intervals of 1 hour and below are influenced by this effect, leading to a higher number of events for these sampling intervals than for a 2-hour interval (or longer). Quirmbach *et al.* (2009) reported in their study that the recorded increase in the number of events, which was higher for the $\leq$1 hour intervals than for the longer sampling intervals, might be in part attributed to the instrumentation improvements.

In conclusion, long-term historical trends due to anthropogenic climate change are difficult to quantify and verify because of limited data, instrumental or environmental changes, interannual variations and longer term climate oscillations. The problem of data limitation is very relevant when analyzing trends in extremes. As highlighted by Frei and Schär (2001) and Schmidli and Frei (2005), the signal-to-noise ratio in a trend analysis depends on the record length, the trend magnitude, the "noise" level (e.g. the magnitude of the variations in the extreme values), and the frequency of events under consideration. Unless the trend magnitude is very large, a real trend may not be detectable in a statistical test when we focus on extremes.

In case a historical trend is present and can be detected, extrapolation of this trend to future decades can be made but these will have an even higher degree of uncertainty. For these reasons, as is the case for any modelling application, it is wise to combine the empirical data with the results from physically-based modelling. Physically-based modelling of climate trends is done by means of climate models, which are presented in the next chapter.

Chapter 4

Climate models

Urban drainage system planners need information on future trends, up to 20–100 years from now, depending on the operational lifetime of the system. One option is to analyse historical trends (see previous chapter) and to use these trends for extrapolation. However, one should be very careful in extrapolating these observed trends far into the future. For longer-term projections, a more physically-based approach is required that accounts for the drivers of the change and the temporal projections on these drivers. This can be done using the results from climate projections, simulated by climate models.

In this chapter, a general introduction to climate modelling is given followed by a classification of different types of models (Section 4.1). Also the differences between modelling weather and climate and issues of predictability are discussed. Climate forcing scenarios and GCM simulations for these scenarios are finally presented in Sections 4.2 and 4.3.

4.1 ATMOSPHERIC MODELLING

Weather versus climate modelling

The objective of climate models is to describe long term changes in the atmospheric system, at the global or regional level. They make use of atmospheric models, which are basically the same type of models as used in operational weather forecasting. While weather forecasting attempts to predict the conditions of the atmosphere accurately for a period of about one week, using specific initial conditions given by assimilated observation data, climate models operate for much longer periods, generally from several decades up to the end of the century. For climate models, the initial conditions are less important compared to weather models because the objective is to simulate changes in average or extreme statistical properties of key meteorological phenomena rather than making detailed short-term forecasts.

Historically there were significant differences between the physics in climate and weather models. For example, early weather models ignored radiation in their implementation (Shea *et al.* 1994). However, today the physical processes and their implementations in climate and weather models are nearly identical. In fact recent literature shows many examples of using weather model systems to conduct climate type simulation experiments, for example Lo *et al.* (2008), who used the WRF model for climate downscaling. The main differences are the prediction duration of interest and the "type" of prediction that is being undertaken (Giorgi, 2005):

(1) Prediction of the evolution of the atmospheric state, given an initial condition: this is called predictability of the first kind.
(2) Prediction of the statistics of the response of the atmosphere for some external forcing (e.g. climate forcing by means of scenarios for GHG emissions or atmospheric concentrations): called predictability of the second kind.

Weather prediction is a clear example of predictability of the first kind. The weather prediction rarely shows high accuracy beyond a period of one week. In general, the prediction accuracy within a given system is a function of the temporal scale of the phenomenon that is considered. Typical atmospheric processes leading to rainfall are of relatively short time scales (e.g. cumulus formation: a few hours; synoptic systems: a day or two). Climate projections are essentially of the second kind: they largely "forget" the initial conditions of the simulations and respond to the external climate forcing, behaving much like a boundary value problem. The results of such attempts are inherently probabilistic and it is important to treat them as such in further analyses.

To make a clear distinction between predictions of the first kind and predictions of the second kind, predictions of the second kind are commonly called *projections* rather than predictions.

In between weather predictions (~5–10 days) and climate projections (~30–100 years) comes forecasts on seasonal to decadal time scales. Despite the chaotic nature of the atmosphere, predictability on longer time scales may exist through the large-scale, low-frequency climate oscillations discussed in Section 3.2., such as ENSO, which influences for example seasonal precipitation in most parts of the world (e.g. Trenberth & Caron, 2000). Seasonal to decadal forecasting is performed using dynamical, statistical or a combination of methods. The forecasts generally exhibit good skill in the tropics and for monsoon rainfall. In other regions the skill is generally low but expected to improve with future development of observation systems as well as climate models (e.g., Smith *et al.* 2012).

When the climate change projections are to be made at regional scales, we face further challenges. Historically GCM projections, which cover the entire globe, have been largely limited to coarse resolutions, typically several hundred kilometres, primarily due to computational limitations. While this limitation is continually relaxed, there are other issues that are intrinsic to the atmospheric system that makes it challenging to use local scale projections. First, there are numerous local effects (e.g. geographical features) that modify the atmospheric flow in complex ways. Secondly, small horizontal shifts in storm movement can dramatically alter local precipitation patterns. Further there is evidence that the variability of atmospheric variables (including precipitation) increases with refined spatial scales (Giorgi, 2002).

While a simplification, it is generally true that the smallest space/time scale at which predictions have real information decreases both spatially and temporally with an increasing range of prediction (Figure 4.1). The figure shows that with increasing range of prediction, fine-scale (both time and space) information is lost. Long-range model results should be interpreted as space/time averages over a large range or as statistics, which can be seen as another way of aggregating. Even with sufficiently high grid resolution (which is far from the case today) a GCM will not be able to deterministically predict cumulus formation. However, it may be able to give: 1) rainfall averaged over a large region over a large time period; and 2) statistical information on cumulus formation. This is an intrinsic property of the atmospheric system – not a limitation in current models which can be overcome with better numerical schemes and refined resolutions.

Physical basis

The development of atmospheric models started from the late sixties to the mid-seventies, with the development of the theory of chaos by Lorenz (1975). This theory allowed describing the behaviour of complex systems like the earth's atmosphere. All physically based atmospheric models use some sort of numerical scheme to solve the following five equations in a three-dimensional space:

(1) Conservation of mass
(2) Conservation of momentum
(3) Conservation of heat
(4) Perfect gas law
(5) Conservation (and phase changes) of water

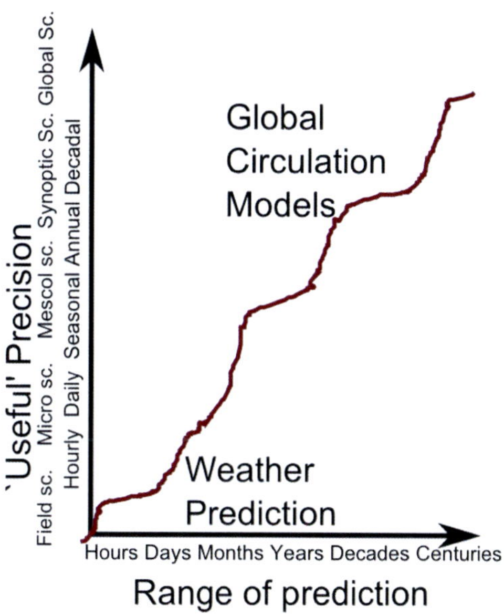

Figure 4.1 A simplified view on predictability of atmospheric conditions.

Note that atmospheric models describe water in a different way than what is done in hydrology. Hydrologists typically model water as an incompressible fluid that does not change its properties significantly within the typical ambient range of temperatures (Chow, 1964; Chow *et al.* 1988). Therefore the motion of water is generally described using the conservation laws of mass and momentum. In the case of gases, it is not realistic to assume incompressibility. They expand significantly with an increase of temperature. In addition to the conservation of mass and momentum, describing atmospheric motion requires a conservation of thermal energy (as described by the first law of thermodynamics) and a relationship between temperature, pressure and volume (as described by the perfect gas law). It is typical to consider water as a separate constituent having its own conservation equation.

Boundary conditions

Like any other type of numerical model, atmospheric models need initial conditions and boundary conditions to operate. The boundary conditions are often divided into top, bottom and lateral types (Figure 4.2).

In most atmospheric models the top boundary conditions use simple parameterizations for important quantities (e.g. radiation exchange). These parameterizations are often simplified and depend on known external processes (e.g. seasonal solar radiation). For example, many GCMs set the model top well above the troposphere making it possible to assume no air exchange over the model top. For instance, the NOAA GISS ModelE (Schimit *et al.* 2006) sets the model top at 0.1 hPa level, which is approximately 64 km above the stratosphere. Modern atmospheric models use a simplified surface model (called Land Surface Model – LSM) to cover the bottom boundary. For example, ModelE uses an improved version of the Rosenzweig and Abramopoulos (1997) LSM, that can represent vegetation, snow and catchment hydrological processes.

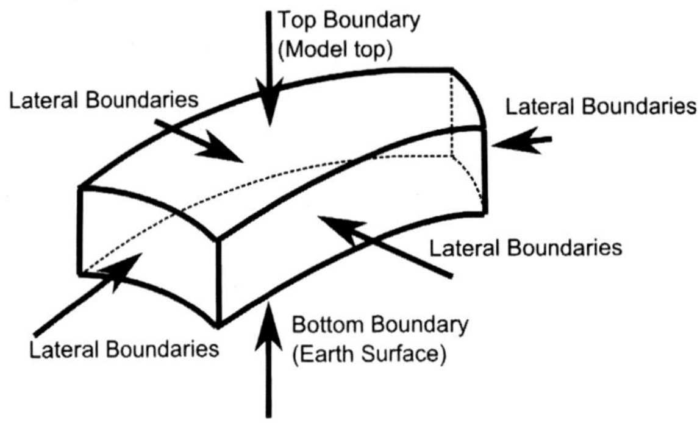

Figure 4.2 Types of boundary conditions in atmospheric models.

When the model coverage is global, as is the case for a GCM in applications of climate change projections and a Global Weather Model (GWM) in weather prediction applications, the model does not have lateral boundaries. Therefore the need for lateral boundary conditions does not arise. This leads to an important ability for these models: given initial conditions they can simulate future atmospheric conditions without any future boundary conditions. Therefore global atmospheric models can make predictions in the real sense of the word.

Regional models

In regional studies, another type of model is used, namely a *Limited Area Model* (LAM). Such as model is called RCM when used for applications of climate change projections and *Regional Weather Model* for numerical weather prediction. As the name implies, regional models have lateral boundaries and therefore depend on the specification of the lateral boundary conditions. These boundary conditions can come from two sources:

- In the case of historical studies they can be based on past observations. However, in order to combine different types of observations with inevitable error sources in a physically consistent way, data assimilation is used with the help of prior knowledge of the state given by a numerical atmospheric model (Wang *et al.* 2000). For historical studies such *reanalysis data* are used as lateral boundary conditions (often also for initial conditions). For more information on the basics of data assimilation and reanalysis the reader is referred to Shea *et al.* (1994).

  There are several publicly available reanalysis datasets:

  - One is the global reanalysis data from the National Centre for Environmental Prediction (NCEP), which are dating back to 1948 and continuously updated (Kistler *et al.* 2001): http://www.esrl.noaa.gov/psd/data/reanalysis/reanalysis.shtml.

  - Interesting is the NCEP FNL Operational Model Global Tropospheric Analyses ds083.2 dataset, a 6-hourly dataset covering the globe at 1 × 1 deg resolution (CISL, 2011) at 26+ atmospheric layers from the earth's surface to 10 hPa (about 30 km) height. They use the Global Telecommunications System (GTS) and other data sources and perform reanalysis using the Global Forecast System (GFS), which is an operational GWM.
  - Other global reanalysis datasets are the 40-year reanalysis data (ERA-40) from the European Centre for Medium-Range Weather Forecasts (ECMWF) at 2.5 deg resolution covering the period 1957–2002 (Uppala *et al.* 2005), and the continuously updated ERA-Interim dataset covering the period from 1979 onwards: http://www.ecmwf.int/products/data/archive/descriptions/e4/index.html.
  - Next to these global datasets, some regional datasets are available, for example for Europe (E-OBS; Haylock *et al.* 2008), India and East Africa (Yatagai *et al.* 2009) and South America (Silva *et al.* 2007). Another useful source of daily rainfall data is the Global Historical Climatology Network (GHCN) – Daily, which contains records from over 75000 stations in 180 countries and territories, with lengths that range from less than 1 year to more than 175 years: http://www.ncdc.noaa.gov/oa/climate/ghcn-daily/
  - One has to note that these reanalysis/historical data are more reliable for more recent periods. A strong increase in the accuracy is expected from the 1980s due to the availability of satellite data. Serious inaccuracies may, however, remain in recent years. For example, Uppala *et al.* (2005) found a serious bias in ERA-40 rainfall data due to the Pinatubo eruption of 1991, which caused significant artificial shifts in some of the assimilated satellite streams and reanalysed rainfall fields.
- For analyses involving future forecasts as in climate modelling but also in operational weather forecasting, the regional models (RCMs or RWMs) depend on predictions made by global models (GCMs or GWMs).

Figure 4.3 shows a taxonomy of atmospheric modelling applications. As outlined above, the two traditional classes of applications are weather modelling and climate modelling. They have little overlap in the temporal range, and strongly differ in spatial scales. Weather prediction modelling typically covers spatial scales from mesoscale (strict technical sense meso-gamma scale, ~10 km) all the way to global scale. Typical examples of models used for such short range predictions are RAMS (Pielke, 1992), MM5 (Grell, 1994) and WRF (Skamarock, 2005). The typical usage of these models is at the scale range from the meso-gamma scale to the synoptic-scale (~1000 km) as they use specified lateral boundary conditions, with a forecast range of several days to weeks. In special research applications the models are sometimes employed at smaller spatial scales (e.g. <1 km, microscale or smaller). They also can be used as global models with special setups. However, the most prevalent use of these models remains within the mesoscale range. This application therefore is also known as *Mesoscale Meteorological Modelling* (MMM) (Pielke, 2002).

For climate models, because they consider long term changes in the atmospheric system, the spatial domains are large compared with the spatial domains covered by the weather prediction models, as also shown in Figure 4.3. They indeed go from global to synoptic-scale level. There also would be the option to use climate models for the smaller mesoscale range, hence to limited-area application domains. This would be useful in applications of urban hydrology. Most useful for these applications would be to restrict the spatial domain to the area of a single city or urban area. Whether this is practically feasible will be discussed in Chapter 5. Atmospheric models bounded to the smaller mesoscale will hereafter be referred to as LAMs. This name refers to the atmospheric model, independent on whether it is used for weather prediction or climate modelling.

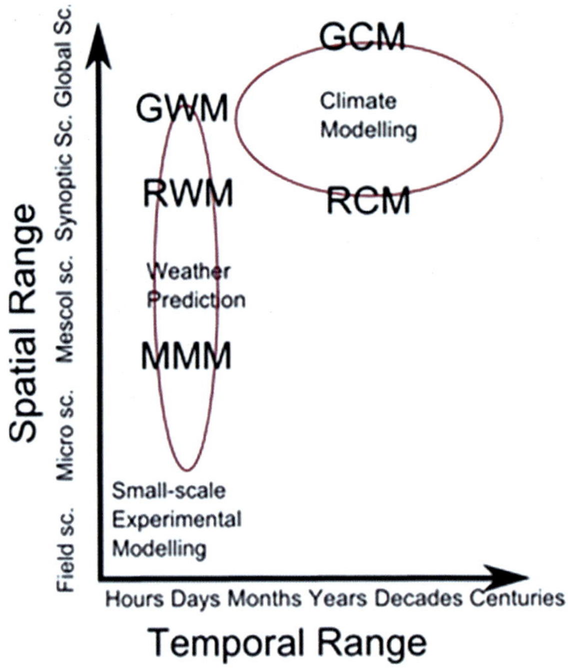

Figure 4.3 A taxonomy of atmospheric modelling applications. Weather modelling applications, while covering micro scale to global scale, rarely attempt to predict more than several weeks into the future. On the other hand, climate models typically cover temporal ranges from days to centuries, but are rarely applied at spatial scales smaller than regional. See the text for the different acronyms used in the figure.

4.2 CLIMATE FORCING SCENARIOS

Climate change impact estimation of variables including extreme rainfall requires prediction of the statistics of the response of the atmosphere for external forcing through changes in GHG emissions or concentrations. In Section 4.1, this type of prediction was called "predictability of the second kind" or projection. It requires scenarios to be defined for the future trends in GHG emissions or concentrations. These are determined by driving functions such as demographic, socio-economic, technological and social development (Nakicenovic & Swart, 2000). Based on various overall scenarios, Nakicenovic and Swart (2000) developed 40 storylines that each describe a possible path. Together the paths span over wide intervals of human population, wealth, GHG concentrations and thus climate. These were summarized in a limited set of scenarios, which are well known as the IPCC Special Report on Emissions Scenarios (SRES) A1, A2, B1, B2, A1B, and so on. (Table 4.1; Nakicenovic *et al.* 2000).

All the SRES scenarios are *baseline scenarios* in the sense that they do not include any explicit climate policy (mitigation), although emission reduction may result from other environmental concerns that are taken into account in some scenarios. The CO_2 emissions from the most frequently used SRES scenarios are shown on Figure 4.4 (coloured lines).

In conclusion, a wide range of possible emission futures remains plausible from a scientific viewpoint. It is important to note that SRES scenarios do not account for the fact that populations might significantly adapt their behaviour due to climate change experiences and/or communication/sensitization. For that reason, there are changes regarding scenarios in preparation for the 5th Assessment Report (AR5) of the IPCC. The new process will be an important change from previous assessments. The move originates from a need to replace the SRES scenarios, and to cover the whole range of published scenarios, including strong mitigation cases. The central concept of this new framework is a set of 4 benchmark scenarios referred to as *Representative Concentration Pathways* – RCPs (Moss *et al.* 2008). By contrast to the SRES emission scenarios, the RCPs are not based on storylines defining the drivers behind the emissions. Rather, the RCPs are defined by selecting concentration pathways and the associated radiative forcing in 2100 so as to cover the full range of scenarios available in the scientific literature. The *radiative forcing* is a measure of the imbalance of incoming and outgoing energy in the earth-atmosphere system, due to climate altering factors. The RCPs are referenced by the radiative forcing reached in 2100, namely RCP8.5 (8.5 W/m^2, representing the largest emissions or high reference position), RCP6, RCP4.5, and RCP3-PD. In the name of the "RCP3-PD" scenario, PD stands for Peak-and-Decline: rather than increasing then stabilizing to a certain value, the radiative forcing is passing through a peak (at 3 W/m^2), then declining and eventually stabilising (the radiative forcing in 2100 was set to 2.6 W/m^2 following an evaluation of the plausibility of such low scenarios). The two lower scenarios are in the range of concentrations typical for mitigation scenarios, and the lowest one is representative of emissions that would follow from substantial mitigation efforts compatible with a limitation of global warming around 2°C, so that the coverage of possible futures is much more comprehensive than with the non-mitigation SRES scenarios (Figure 4.4).

Table 4.1 SRES scenario summary.

Scenario	Description
A1	Fast growing economy, new/efficient technologies, population peak around mid-century and decline thereafter. Three storyline subgroups: fossil intensive (A1FI), fossil energy sources (A1T) and balanced use of all sources (A1B).
A2	Heterogeneous world, preservation of local identities, continuous population growth. Economic/technological progress is more fragmented and slower than in other scenarios.
B1	Global population as in A1, services and information society, clean and resource efficient technologies.
B2	Global population as in A2 but slower evolution, intermediate economic development, more diverse evolution in technology than in the A1 and B1 storylines.

A key idea is that this set of pathways can be used to run climate models while new socio-economic scenarios are simultaneously developed. This parallel process is illustrated in Figure 4.5. When new socio-economic and emission scenarios will be ready, it is expected that it will be possible to link these to the RCPs so as to obtain climate change information from the climate runs based on the RCPs, thus avoiding a need for new climate simulations. A practical consequence for impact and adaptation studies is that they do not only need to wait for the climate simulation results, but they may also need to wait for the availability of consistent socio-economic information from fully defined new scenarios with associated storylines. The RCP process helped to start this process more quickly than would the previously used "linear" approach (Figure 4.5) but it should be clear that the RCPs themselves do not provide complete socio-economic information so that further development is still needed in this area.

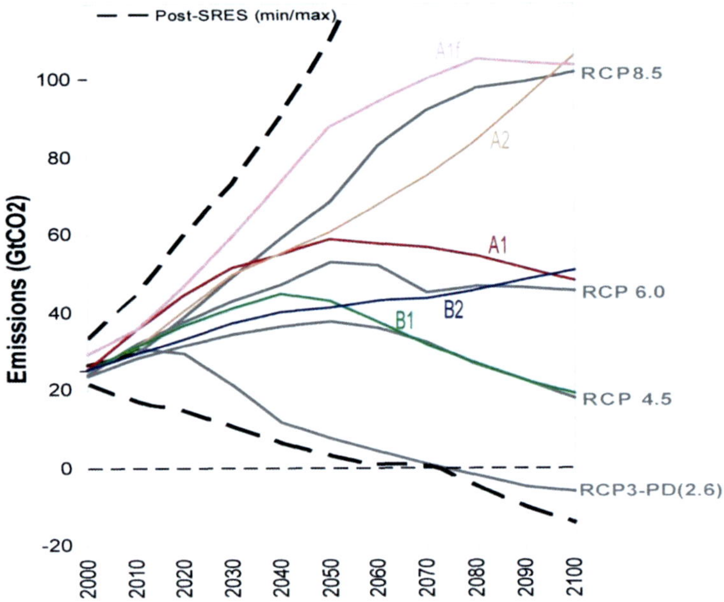

Figure 4.4 Coloured lines: Total carbon dioxide emissions for the SRES scenarios (A1, A2, B1, B2 and A1f: A1 Fossil Intensive scenario (IPCC, 2007). Grey lines: illustrative carbon dioxide emissions for each of the representative concentration pathways (Moss *et al.* 2008).

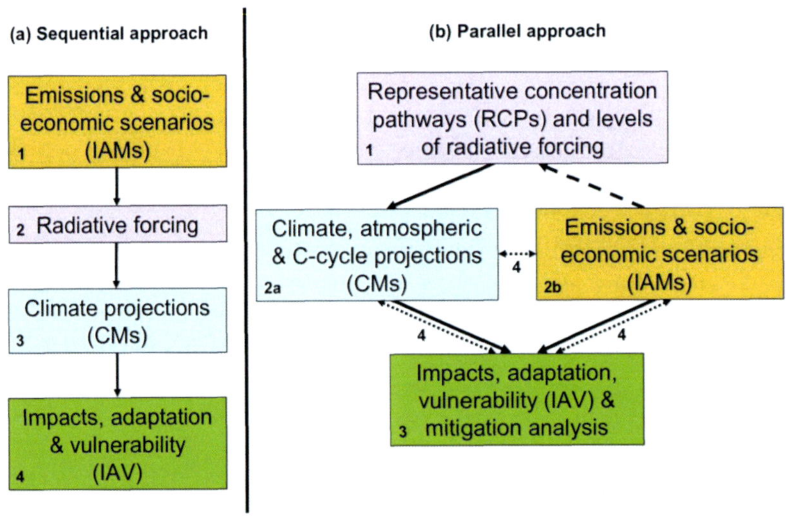

Figure 4.5 Approaches to the development of climate forcing scenarios: (a) previous sequential approach for the SRES emission scenarios; (b) parallel approach of the RCP based scenarios. Numbers indicate analytical steps (2a and 2b proceed concurrently). Arrows indicate transfers of information (solid), selection of RCPs (dashed), and integration of information and feedbacks (dotted) (from Moss *et al.* 2008).

Figure 4.6 shows the main differences in the processes involved when applying SRES emission scenarios versus AR5 RCP based scenarios. The figure is based on the main stages in developing a model of the hydrological impacts from climate change as described by Ward *et al.* (2011). The SRES scenarios worked "forward" from socioeconomic projections to radiative forcings (sequential approach; Figure 4.5). This made it easy to get bogged down in questioning the socioeconomic, technological, and physical assumptions of the scenarios. In contrast, the RCPs are intended to work backwards from assuming forcings of magnitude to the wide range of circumstances that might result in such forcings. This means that the RCPs are "agnostic" to the specifics of the socioeconomic projections; no matter how socioeconomic, politics, and technology are going to evolve during the 21st century. The higher steps in Figure 4.6 of emission scenario definition and carbon cycle modelling thus are eliminated from the AR5 scenario definition. In this book, *climate forcing scenarios* is used as a common term for both the SRES emission and AR5 RCP based scenarios.

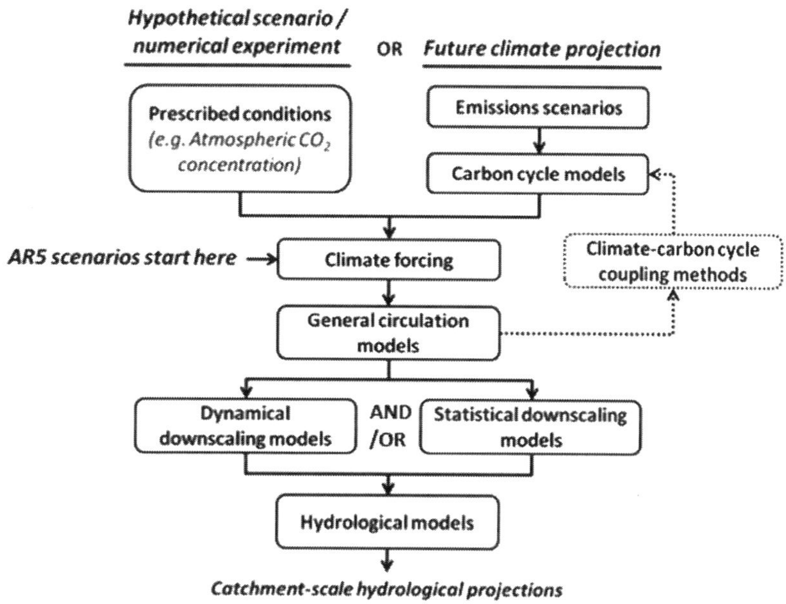

Figure 4.6 Simplified chart of the main processes involved in modelling hydrological impacts from climate change. Note: Dashed lines around climate-carbon cycle coupling methods indicate that not all models are coupled (after Ward *et al.* 2011).

4.3 GCM SIMULATIONS

GCM simulations have/are carried out by several climate modelling groups around the world. Twenty of these groups are involved in the Coupled Model Intercomparison Project (CMIP) of the World Climate Research Programme (WCRP). They carry out coordinated GCM simulations in support of the Assessment Reports of the IPCC. CMIP4, the Coupled Model Intercomparison Project Phase 4 contributed to the 4th Assessment Report (AR4) of the IPCC (IPCC, 2007a). The GCM simulations they produced are for the greenhouse gasses increasing as observed through the 20th century (called 20C3M scenario experiment) and for the 21th century IPCC SRES scenarios. These simulations are provided by the IPCC Distribution Data Centre:

http://www.mad.zmaw.de/IPCC_DDC/html/SRES_AR4/index.html;
http://www.mad.zmaw.de/IPCC_DDC/html/ddc_gcmdata.html.

Currently (2012) the CMIP5, Coupled Model Intercomparison Project Phase 5, is ongoing. This project will deliver many more simulations for the 21st century RCP scenarios as a contribution to the 5th Assessment Report (AR5) of the IPCC. These simulations are done using GCMs which are coupled to a global ocean circulation (atmosphere-ocean general circulation models; AOGCMs).

The simulations cover given periods, both historical periods (called control simulations or baseline periods; e.g. 1961–1990) and future periods (called scenario simulations or periods; e.g. 2071–2100), or are transient for long continuous periods, for example from 1950 till 2100. The control and scenario periods are most often about 30 years or longer, because a 30-year period is likely to contain wet, dry, warm and cool periods (see Section 3.2) and thus is generally considered sufficiently long to define the climate for a given location. A 30-year "normal" period as defined by the World Meteorological Organization (WMO), for example 1961–1990, is recommended by the IPCC for use as a baseline period.

GCM modelling was once strictly limited to the specialized research groups who had access to supercomputers. However, today there are situations where the GCM simulations can be performed by enthusiasts from disciplines outside climate science. For example, EdGCM is a research-grade GCM created by Columbia University and based on the Goddard Institute for Space Studies' General Circulation Model II, which is computationally efficient enough for use on personal computers. EdGCM has a friendly user interface, making it suitable for educational use (Chandler, 2005).

However, for the standard IPCC climate scenarios, the GCM results that are readily available from the IPCC AR4 and AR5 archives make it unnecessary to run GCM experiments for the sole purpose of obtaining the results of standard runs.

The outputs from these climate model simulations are, however, subject to significant uncertainties, due to scaling problems and to the limited knowledge of the physical processes and the obvious high uncertainties in the climate forcing scenarios. Even given the wide range of demographic, socio-economic, technological and social development scenarios considered by the IPCC, the actual future developments may differ significantly from the scenario projections. For example, the SRES scenarios that currently are most often considered do not account for the fact that populations might significantly adapt their behaviour due to climate change experiences and/or communication/sensitization. In terms of climate processes, the increase in GHGs and aerosols will lead to radiative warming, but also to cooling (the so-called global dimming effect, as discussed in Pathirana, 2008). Depending on the specific processes included in the models and their specific descriptions, the results from different GCMs differ, despite their common basis (the fundamental equations). Generally models tend to agree on the direction (i.e. sign) of the change, but differ with respect to the magnitude and/or speed of the change. This again emphasizes the importance of uncertainties in climate projections.

Another potential source of uncertainty are the initial conditions of the climate model. A GCM starts from an unknown and arbitrary initial condition in the pre-industrial era (mid-1800s). Simulations with the same type of GCM, differing only in the initial conditions, will correspond better or worse with the historical climate evolution and will give also different results for the (especially near) future. For general climate features, the uncertainty related to initial conditions of the same GCM may be as large as the uncertainty related to the type of GCM (e.g. Kjellström *et al.* 2010). The impact of initial conditions on precipitation extremes in the Rhine basin for different time scales down to daily was studied by Kew *et al.* (2011), using a set of 17 runs with the ECHAM5 GCM, forced by emission scenario A1B, differing only in initial conditions. Future changes in daily extremes were found to differ by up to 25% for individual members. Similar conclusion was obtained by Willems and Vrac (2011), as will be shown in Chapters 7 and 8.

GCM simulations based on the RCPs were not publicly available at the time this book was prepared. They are envisaged to be ready soon in order to enter the AR5 writing process (to be finalized in 2013/2014). As soon as sufficient climate model runs based on the RCPs become available, it is clear that additional research will be needed to study the effect of these changes in the scenarios (e.g. the effect of mitigation) on the rainfall extremes and urban drainage systems. Questions that may need to be answered are (among others): Is the range of GCM simulations, currently provided sufficiently complete, and are climate simulations available to widen the range if needed? Could impact studies based on the SRES scenarios be somehow "connected" to the new RCP process, and if not, what would be necessary to allow this? It will take time before a full evaluation becomes possible, including linking of the new scenarios selected for the AR5 to climate simulations, detailed assessment of extremes for a set of models, and a range of impact studies. The treatment of uncertainty from scenarios and models, and the possible inclusion of lower emission scenarios in future work are further discussed in the next chapter.

4.4 DISCUSSION

This chapter has discussed climate models as systems of differential equations based on the basic laws of physics, fluid motion, and chemistry. These numerical models represent and couple atmosphere, ocean, land surface and sea ice processes based on assumptions of GHG concentrations in the atmosphere, land use and other critical variables that determine the rate of physical processes in the atmosphere. Future trends in these variables are in turn estimated based on climate forcing scenarios. The changes imposed by these changes in forcing (due to changes in anthropogenic emissions) should be compared to the inherent natural variation of precipitation.

GCMs typically have poor accuracy in simulating precipitation extremes. Moreover, they produce results at spatial and temporal scales that are far too coarse for urban drainage applications. This can be done through the use of high resolution RCMs nested in GCMs. This method is called dynamical downscaling which is presented in the next chapter.

Chapter 5

Dynamical approach to downscaling of rainfall

The previous chapter explained that various types of climate models exist and are broadly classified as either GCMs or RCMs. RCMs account for the sub-GCM grid scale topographical features and land cover heterogeneities. They use initial and boundary conditions from the output of GCMs for selected time periods of the global run. This is commonly known as the nested regional climate modelling technique or dynamical downscaling. Up to now, this approach has been one-way, with no feedback mechanisms from the RCM simulation to the driving GCM. In this simulation scheme, the role of the GCM is to simulate the response of the global circulation to large scale forcing (i.e. GHG concentrations). The RCM accounts for finer scale forcing, like topographic features, in a physical manner, and enhances the simulation of the climatic variables at these finer spatial scales. Although the GCM boundaries in general have a very strong influence on the RCM results (including the effect of changes in GHG forcing on the global climate), RCMs may have a significant role during periods of convective precipitation due to the local convection effects (Rummukainen, 2010). Apart from precipitation, RCMs can provide host of other hydrometeorological variables, on the surface (e.g. solar radiation, sensible and latent heat flux) and three dimensional atmospheric space (e.g. humidity, cloud densities, temperature). These outputs are generated from the physical simulation and hence physically consistent with each other.

When using climate models in studies on climate change, sensitivity studies can give the modeller a thorough insight into the sensitivity of the model output to changes in one or more input parameters or their drivers. Such sensitivity analysis is often undertaken through controlled numerical experiments. Some examples are impact of sea-surface temperature on storm generation over coastal areas, impact of urban growth (as a driver of land use change) on changes of urban microclimates leading to altered rainfall patterns, impact of atmospheric aerosols on rainfall processes and impact of mountains on extreme rainfall.

In this chapter, a brief overview is given of the dynamic downscaling approach (Section 5.1), followed by a discussion of the main tools used in that type of dynamic downscaling by means of RCMs (Section 5.2). Specific attention is given to the method of calculating precipitation (extremes) in these models. Also the concepts of nesting, how local data can be used to improve dynamic downscaling with RCMs, and sensitivity studies, are discussed. Several case studies on fine-scale rainfall simulation from recent literature are presented (Section 5.5) that may help the reader to better understand the concepts discussed in this chapter.

5.1 DYNAMICAL DOWNSCALING

GCMs often run at coarse resolution and are not capable of producing local scale features of rainfall (among other meteorological variables). Therefore, techniques are used to interpolate these results at smaller scales with varying degree of success in reproducing or mimicking local scale features. Dynamic downscaling is where a RCM is used with GCM results as boundary conditions to produce more detailed and locally representative meteorological fields for a specific (limited) domain.

Whereas GCM results are for standard climate forcing scenarios (see Chapter 4), the situation regarding regionally downscaled data is quite different. While several data centres provide a number of sets of regional model results for large regional domains (see Section 5.2), such as Europe, North America, Australia and Japan this is not for every region in the world the case. The resolution of the RCMs obtained for large regional domains, moreover, do not have the required resolution for urban drainage impact studies (see further Section 5.2). This is especially the case for urban rainfall applications where the required resolutions are well below meso-gamma scales. Further, there are often many specific local-forcing conditions (e.g. land use changes) with which the urban scientists would like to model. In this context, being able to perform limited area atmospheric simulations under specific conditions is an important asset for the urban hydrometeorologist.

Historically numerical modelling of the atmosphere even at the regional scale was an exclusive field that was available only for highly specialized research groups and operational forecasters who have access to expensive computational facilities. In addition to the computational expense, early atmospheric models were not readily accessible to the non-specialized user (e.g. urban hydrologist who is interested in the models for the purpose of rainfall downscaling). However, the situation was largely changed through the pioneering work of the National Center for Atmospheric Research, Colorado, USA, who started supporting MM4 (which is the predecessor of MM5) as a community model, providing training workshops and tutorials. This type of models do allow hydrologists to run their own regional climate simulations, not only at the synoptic scale but also at the smaller meso-scale area (or even at the scale of a single city). Today a model like WRF provides a convenient point of entry to limited area modelling for the new user. In the following section, the specific subject of RCMs is discussed. Further in Appendix C, the technical details on how to use WRF (a popular state-of-the-art local atmospheric model) on personal computers are presented.

5.2 REGIONAL CLIMATE MODELS (RCMs)

RCMs in general

Regional climate models, as the name implies, cover a specific horizontal extent on the atmosphere and therefore depend on the specification of lateral boundary conditions (Figure 4.2) in addition to the initial conditions (Pielke, 2002). They numerically solve the differential equations representing the five conservation relationships described in Section 4.1, within the three dimensional space of the model domain to predict the state in the atmosphere in the future. There are a large number of atmospheric models that have been developed during the last several decades that differ in their spatial representation, in the simplifications used in the numerical representation of the conservation equations and in the representation of specialized processes such as boundary layers, cloud processes, and so on. Without attempting to be comprehensive, in this section the discussion here is limited to the widely used class of regional models represented by RAMS (Pielke, 1992), MM5 (Grell, 1994) and WRF (Skamarock, 2005).

The horizontal coordinate system of these models consists of square grids (there are also RCMs that use advanced spectral coordinate systems) on a projected coordinate system where distances are specified in kilometres. The vertical coordinates are typically based on a curvilinear system that follows the surface of the earth (see Figure 5.1 for the vertical coordinate system used in the WRF model).

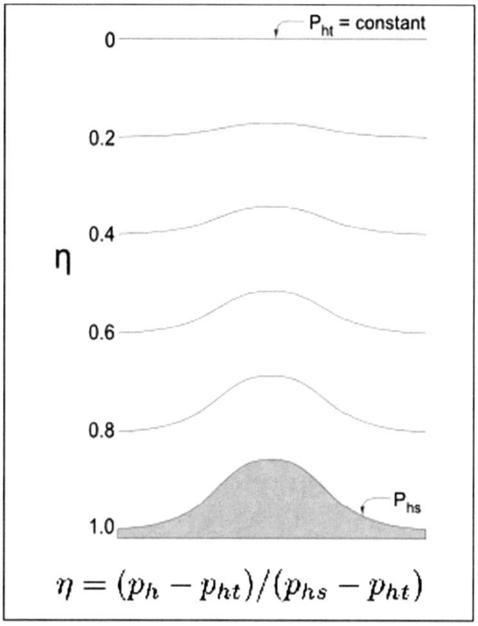

Figure 5.1 Terrain-following η coordinate system used in the WRF model for the vertical representation (after Klemp, 2008).

These models typically use explicit finite difference schemes to numerically solve the governing equations in the three dimensional space. This makes these schemes conditionally stable (Chapra, 2006), with the computational time step being smaller than some function of the grid dimension (known as the Courant – Friedrichs – Lewy condition or CFL condition). For example, in the case of the WRF model, a computational time step in seconds should typically be less than six times the horizontal grid size in km, for stability reasons. This means that for a 10-km grid resolution the computational time step should be less than one minute.

Precipitation computation in RCMs

As stated before, atmospheric models simulate the physical processes in the atmosphere by numerically implementing the governing equations of atmospheric motion. However, in the case of cloud processes, the above statement is in fact a simplification. Important atmospheric processes lead to cloud evolution at relatively small spatial scales. For example, cumulus formation, an important cloud process leading to rain storms, occurs at around 0.1 km – 10 km horizontal scales. It is unrealistic to represent such small scale models at low grid resolutions (e.g. ~100 km) used particularly in global models. Therefore atmospheric models routinely use a scheme called cumulus parameterization where the task is to represent the collective influence of clouds (e.g. rainfall, radiation budget) within a larger area (single grid) parameterized in terms of the large (grid) scale environmental variables. This enables the models to simulate convective rainfall. However, it is important to remember that the primary purpose of cumulus parameterization is not to produce accurate rainfall, but to release model instabilities.

On the other hand, when model grid sizes are small enough, it is possible to represent the cumulus process explicitly by the model physics. This is called explicit rainfall. It is generally agreed that model grid sizes larger than 10 km need some sort of cumulus parameterization for reasonable rainfall outcome as well as model stability. It is generally agreed that scales finer than 3 km do not need cumulus schemes.

While cumulus parameterization has improved greatly over the last few decades (see Arakawa, 2010, for a comprehensive review), still it is a parameterization method as opposed to a primarily physically based computation. It is not advisable to depend on cumulus parameterization in an atmospheric model to produce rainfall magnitude or location accurately. Precipitation focused RCM studies can benefit from grid sizes small enough to safely avoid using convective parameterization (typically <5 km). Examples of such fine-scale precipitation experiments are reported by Nakamura *et al.* (2008) using a 1-km non-hydrostatic regional model for western Japan that explicitly resolves cloud processes, and Beaulant *et al.* (2011) applying a 2.5-km non-hydrostatic model for enabling generation of Mediterranean heavy precipitation.

Nesting in RCMs

GCM results, that RCMs use for lateral boundary conditions, are at quite coarse resolutions. In order to achieve high grid resolutions for the local area of interest, RCMs use nested grid systems. Figure 5.2 shows an example of a telescopic grid (36 km, 12 km and 4 km) to cover the central mountain region of the island of Sri Lanka (Pathirana *et al.* 2006). Cumulus parameterization was used for the 36 km and 12 km grids, but not for the 4 km grid. Only the results of the 4 km (innermost) grid were used for the purpose of analysis, which was to understand the influence of aerosol forcing on rainfall. For this simulation FNL data (CISL, 2011) at 1 degr. grids were used as boundary conditions.

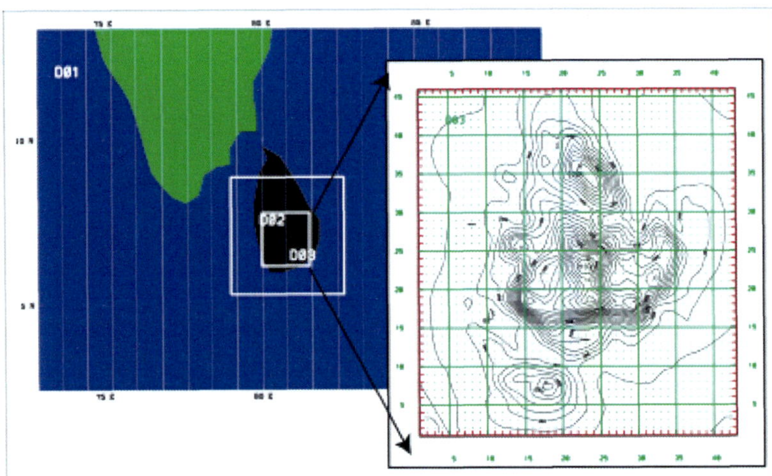

Figure 5.2 Three level telescopic nesting used to resolve the innermost model domain covering the central mountainous area of Sri Lanka at 4 km grid size, while the outermost domain covers an area of approximately 1500 km × 1100 km.

Local data for RCM simulations

Two types of data are needed to complete a RCM simulation setup, namely, static or geographic data and dynamic, initial and boundary condition data. For example, the WRF model needs the following static data:

(1) Surface elevation;
(2) Land use;
(3) Soil properties (type, annual mean temperature);
(4) Monthly albedo variation including maximum snow albedo.

It is possible to obtain most of these static data from global datasets. WRF can use, for example, the global datasets from the USA Geological Survey and the American Geological Institute: the USGS Global GIS database: http://webgis.wr.usgs.gov/globalgis/datasets.htm, and the global databases provided by the Moderate-resolution Imaging Spectroradiometer (MODIS) on NASA satellites: http://geospatialdatawiki.wikidot.com/modis-data-sets.

It is, however, important to supplement these data with local information, particularly at the fine scales involved in urban rainfall studies. Having accurate elevation information, for example, can have a major influence on rainfall quantity and distribution.

Almost always, the initial and boundary condition data primarily consist of some sort of global model result (e.g. GFS) or a reanalysis dataset (e.g. FNL). However, operational weather forecasters routinely supplement these data with local observations such as weather balloon soundings. Incorporation of such data in a physically consistent way to the "main" dynamic datasets is done by *data assimilation*. Data assimilation is the technique by which observations are combined with a weather forecasting product (the first guess or background forecast) and their respective error statistics to provide an improved estimate (the analysis) of the atmospheric state. Variational data assimilation achieves this through the iterative minimization of a cost function. Differences between the analysis and observations/first guess are penalized according to their perceived error. Three-dimensional variational assimilation (3D-Var) uses observations made at a specific time to iteratively nudge the model results (first-guess field). Four-dimensional variational assimilation (4D-Var) can use observations made at different times, and hence involves time as a fourth dimension. Details about data assimilation are beyond the scope of this chapter; the reader is referred instead to Bouttier and Courtier (1999) for an accessible discussion and Daley (1991) for a comprehensive treatment. In Section 5.5 case studies are given with examples of using 3D-Var (or 3D nudging) to prevent RCM results from drifting off from large-scale forcing (coarse global model results).

5.3 RCM SIMULATIONS

RCM simulations for the IPCC SRES scenarios are available for several regions in the world. They are driven by a set of AOGCMs and are provided by several data centres.

For Europe and northern Africa, RCM simulations can be obtained through the EU projects PRUDENCE (http://prudence.dmi.dk/data; Räisänen *et al.* 2004; Christensen *et al.* 2007; Christensen & Christensen, 2007) and ENSEMBLES (http://www.ensembles-eu.org/; van der Linden & Mitchell, 2009). The PRUDENCE RCMs (spatial resolution of approximately 50 km) are mainly driven by the same GCM (HadAM3H) forced with the SRES A2 and B2 scenarios (2071–2100 vs. 1961–1990). The ENSEMBLES RCMs (approximately 25 km) are driven by the ERA-40 boundary conditions as well as different GCMs forced with the SRES A1B scenario (1951–2100).

For North America, RCM simulations are provided by the North American Regional Climate Change Assessment Program (NARCCAP) (http://www.narccap.ucar.edu; Mearns *et al.* 2009). These cover the domain of conterminous United States and most of Canada. The NARCCAP RCMs are run at a 50 km resolution driven by NCEP reanalysis data and the SRES A2 scenario (2041–2070 vs. 1971–2004). Some of these data are however only provided to the partners of the project and not to the public yet.

More limited RCM databases have been set up for regions as Australia and Japan. For Australia, RCM databases are provided by the Commonwealth Scientific and Industrial Research Organisation (CSIRO): http://www.marine.csiro.au/marq/edd_search.Browse_Citation4?txtSession=8180. These are available with restricted access (usually for CSIRO staff). They cover data for the period 1971–2095, created with the CSIRO CCAM (Cubic Conformal Atmospheric Model 805) RCM by dynamically downscaling the GFDLSST GCM output to 60-km resolution over a target area centered over Australia, for the SRES A2 scenario.

For Japan, simulations by means of three RCMs are available from the Japanese downscaling project called S-5-3 (Ishizaki *et al.* 2012). The RCM boundaries were limited to one GCM: MIROC-H. Simulations are provided for the 20th century (20C3M scenario) and the SRES A1B scenario.

Another international ongoing downscaling project is CORDEX (COordinated Regional climate Downscaling EXperiment) of the WCRP. One of its aims is to provide a quality-controlled data set of RCM simulations for the recent past and 21st century projections, covering the majority of populated land regions on the globe. They are based on GCM projections produced within the CMIP5 (see Section 4.3). Their data archive can be found on: http://cordex.dmi.dk/.

There are also several public databases that contain model simulations by one specific institute or data centre or by means of one specific model. One example is the database of the University Corporation for Atmospheric Research (UCAR), which contains results with the NMC Eta model (NCEP, 2005) at 40 km horizontal resolution: http://dss.ucar.edu/datasets/ds068.0/. Another example from Oceania is the CSIRO stretched-grid model (McGregor & Dix, 2001): http://www.ipcc.ch/publications_and_data/ar4/wg1/en/ch11s11-4-3-3-southeast-asia.html.

So far, most RCM simulations are available at daily time scales and at 25 to 50 km spatial scales. Some RCM outputs are also available at hourly time scales and at 10 km spatial scales. While earlier climate model simulations were undertaken for limited periods (time slices) in the past (e.g. 1961–1990) and the future (e.g. 2071–2100), nowadays continuous, transient simulations are often available. In the ENSEMBLES project, transient simulations are made for the period 1950–2100. Up until 2000, the external forcing of the climate system comes from the observed GHG concentrations, while beyond that year it is based on climate forcing scenarios (SRES A1B in the case of the ENSEMBLES project; van der Linden & Mitchell, 2009).

5.4 LIMITED AREA MODELS (LAMs)

The databases reported in previous section provide RCM results at synoptic scale, mostly for entire continents (e.g. PRUDENCE and ENSEMBLES for Europe, NARCCAP for North-America). As explained before, further downscaling by means of more local scale RCM simulations (LAM simulations) might be useful for given areas (e.g. individual countries or even cities), especially in the urban drainage context. It is, however, unlikely that such downscaled data are available for a specific region with the required resolution. As explained in Chapter 4, for urban rainfall applications this resolution is well below the meso-gamma scale (~10 km).

Computational times allow single mesoscale or sub-mesoscale LAM simulations to be conducted for a given local area (see Box 5.1), but it is not practically feasible to conduct several runs for long simulation periods (30 years or more) with different models. Therefore, for urban drainage impact investigations we currently rely on the synoptic-scale RCM simulations available in public databases such as the ones outlined in Section 5.3. The fine-scale LAM simulations provide detailed information that can be used in the current practise but only complementary to the set of synoptic-scale RCM runs. The fine-scale LAM-based information can for instance be used as information for better understanding the quality of the RCM results on extreme rainfall. RCM data quality is discussed in Chapter 6.

Box 5.1 Running LAMs on a personal computer

Until recently, simulation with atmospheric models, including LAMs, could only be undertaken with dedicated super-computer facilities that were accessible for specialized research groups. However, today researchers routinely run RCMs or LAMs on personal computers to perform downscaling and sensitivity studies. A survey of recent literature on the simulations with meso-gamma scale LAMs shows that almost always the period of analysis of these studies are much less than the norm in climate studies, which is at least 30 years. The period of analysis sometimes is a few years, often less. For example Shem and Shepherd (2009) in their study on sensitivity of rainfall to urban growth used two storm events. Pathirana *et al.* (2007) used a series of five day simulations covering a period of six months to study the impact of aerosols on rainfall over the central mountains of Sri Lanka. Lo *et al.* (2008) used a series of 7 day simulations covering one year to compare three approaches to dynamic downscaling. While it could be argued that such studies are not representative enough to be considered as studies on "climate" in the conventional sense, they nevertheless are useful in understanding the sensitivity of various influencing parameters on high resolution hydrometeorological outcomes. Dynamical downscaling studies covering periods comparable with the 30 years standard are rare in the recent literature.

In many situations it is possible to setup quasi-2D idealized model domains (e.g. with little domain width) which are very economical in simulations. These therefore provide an economical way of running a large number of simulations; see Colle (2004), Pathirana *et al.* (2005, 2007), among others, for examples. The WRF model provides a convenient facility to perform idealized (sensitivity) studies (see Skamarock, 2007).

An interesting point of discussion is whether researchers involved in surface studies (e.g. urban drainage engineers, hydrologists, etc.) should attempt to study and master the use of LAMs for dynamic downscaling and sensitivity studies or should it be left in the domain of the atmospheric scientist. Our opinion on this is as follows: today, there is no need for such experts to master GCMs or RCMs for synoptic scale simulations. Results from such models are routinely made available and could be meaningfully used in other studies. However, when it comes to mesoscale LAMs, the situation is quite different. There are numerous applications of mesoscale or sub-mesoscale LAMs to which the "surface expert" can meaningfully contribute and derive benefit. Therefore mastering LAM simulations can be recommended as a useful goal. In Appendix C, instructions are provided on running a WRF model on personal computers. This appendix does not attempt to be comprehensive, but only to provide a "quick-start" guide for the topic and pointers to numerous excellent resources available in the public domain which can be used to develop the expertise further.

5.5 FINE-SCALE RAINFALL RESULTS IN CASE STUDIES

As explained in previous section, it is useful in the context of dynamic downscaling to understand the impact of global and local changes on the model projections, including rainfall, at the local scale. This can be done by sensitivity studies. Sensitivity analysis in the context of dynamic downscaling is an emerging field in global change studies. It deals with the analysis of the sensitivity of an input variable of the atmospheric model, which is the magnitude of influence of a change in such a variable on the model output. Traditionally the field of sensitivity analysis is routinely used to understand the uncertainty in the model result. Another important use of sensitivity analysis is to play what-if analyses exploring the impact of varying input assumptions and scenarios. In fact, most climate model results available today can be effectively considered as *sensitivity studies* (Jones, 2001), that is what will be the sensitivity of the atmospheric system to a particular climate forcing, rather than predictions in a strict sense (Giorgi, 2005).

There are a number of examples of the use of numerical experiments with RCMs in the context of sensitivity studies to understand the influence of various atmospheric and geographical parameters on local-scale rainfall, from which the following are cited here: Buzzi (1998), Pathirana (2005), Singleton and Reason (2006). These studies among many others examined the effect of mountains on local-scale rainfall distribution using sensitivity studies. Deb *et al.* (2008) examined the sensitivity of sea surface temperature on extreme rainfall over Mumbai city, India, focusing on the 2005 July torrential rainfall event, which resulted in more than 1000 mm of rainfall in the vicinity of the city (Vihar Lake). Thielen (2000), Shem and Shepherd (2008), Huong and Pathirana (2011) examined the sensitivity of urban rainfall on the urban conglomeration size and urban growth.

Several illustrative case studies of using atmospheric models for deriving fine-scale rainfall are presented hereafter.

Orography and rainfall

Mountains are largely responsible for high local rainfall in some of the world's most rainy places. For example, the precipitation of Mawsynram village in North-Eastern India (the location with the world's highest annual rainfall, of approximately 11,000 mm), is largely due to the interaction of nearby Khasi hills with the moist monsoon air from the Bay of Bengal (Thapliyal & Kulshrestha, 1992; Murata *et al.* 2007). Traditionally high rainfall associated with mountains was largely explained by upslope condensation due to upward forcing of moist air due to the topographic barrier. However, there are a number of different mechanisms for orographic enhancement of precipitation (Houze, 1994), many of which involve complex interaction of radiative heating with mountain topography, triggering convective break-up. In the recent literature many studies describe using dynamic atmospheric models to understand the sensitivity of topography and other influencing parameters on orographic rainfall.

Many early RCM simulation studies on the dynamics of winds and mountain interactions focused on dry-air flow (Durran & Klemp, 1982, 1983; Nance & Durran, 1997, 1998; Doyle & Durrran, 2001; Pathirana *et al.* 2003). The first numerical experiments involving the rainfall aspects of orographic interaction include Colle (2004) and Pathirana (2005). Both studies employed the MM5 model with special idealized model initiations. The initial and boundary conditions were ideally specified rather than using global model projections or reanalysis data. The topography conditions were also idealized. Figure 5.3 shows an example of such an ideal model setup. Figure 5.4 shows the details of rainfall and related meteorological parameters that can be obtained from those simulations.

Better results from dynamic downscaling

A typical approach to dynamical climate downscaling employs a continuous simulation of a RCM with a single initialization of the atmospheric fields and frequent updates of lateral boundary conditions based on GCM outputs or reanalysis data sets. Due to the typical behaviour of RCMs, the simulations could drift away from the large-scale forcing (the coarse pattern provided by the GCM output, which is assumed to be correct) as the duration of the simulation increases. One solution to this problem is to run the RCM only for a few days, after which the model is reinitialized with the large-scale data. This controls the downscaling field tightly with the large-scale forcing and prevents excessive drifting.

However, there is another problem that prevents frequent model re-initialization: RCMs need considerable spin-up times to bring the atmospheric fields to dynamic equilibrium which can typically be from a few hours to a few days. Therefore, frequent re-initialization can be quite wasteful in terms of computation effort (Figure 5.5). For example if the model spin-up time is 2 days and the time with useful results is 3 days, covering a period 300 days, 100 model runs are needed, each running for a period of five days. Therefore 500 days equivalent run time would be needed out of which 200 days (40%) will be "wasted" as spin-up time.

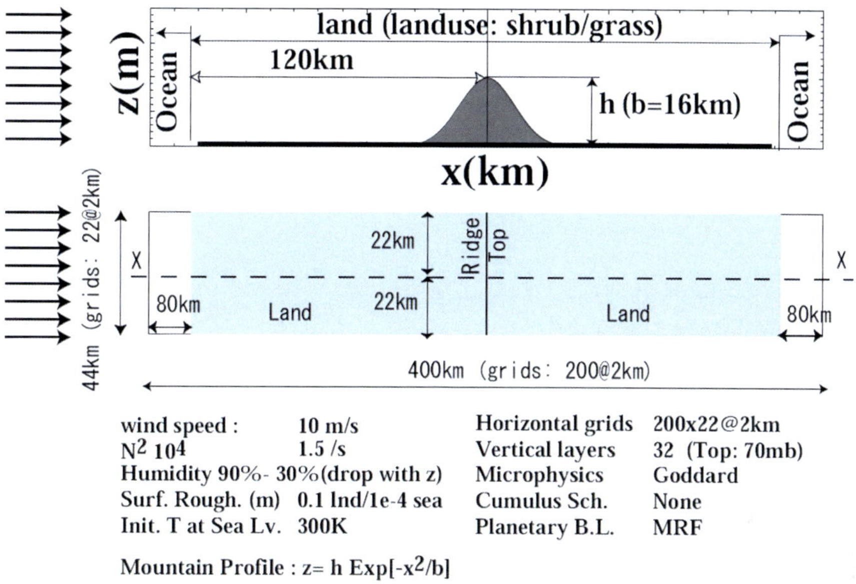

Figure 5.3 An idealized model setup for a sensitivity study on orographic rainfall (after Pathirana *et al.* 2005). The lateral boundary conditions were constant speed moist wind profiles.

Another approach that can be used to reduce model drift is to use a data assimilation technique to nudge the RCM projections with large-scale forcing fields. This technique is called analysis nudging.

Lo *et al.* (2008) examined the relative performance of the above three approaches to dynamic downscaling, namely 1) continuous integration, 2) re-initialization of the model after a short run and 3) continuous integration with 3D analysis nudging. They used the WRF model setup for North America using 178 × 144 grids with 36-km resolution. They used the FNL data (CISL, 2011) as lateral boundary data for the simulation period covering the year 2000. The model spin-up time was set at 1 day. Therefore in case 2 above, each reinitialized model run continued for 7 days and the first day's results were ignored. Figure 5.6 shows the results for daily precipitation over continental USA. The model results were compared with observed daily rainfall series and the correlations of the two series at each point were computed and plotted. As shown in Figure 5.6, the worst performance (due to model drift) is the case of continuous integration without nudging. Re-initialization at every 6th day improved the model performance greatly. However, the best results were obtained by continuous integration with 3D analysis nudging.

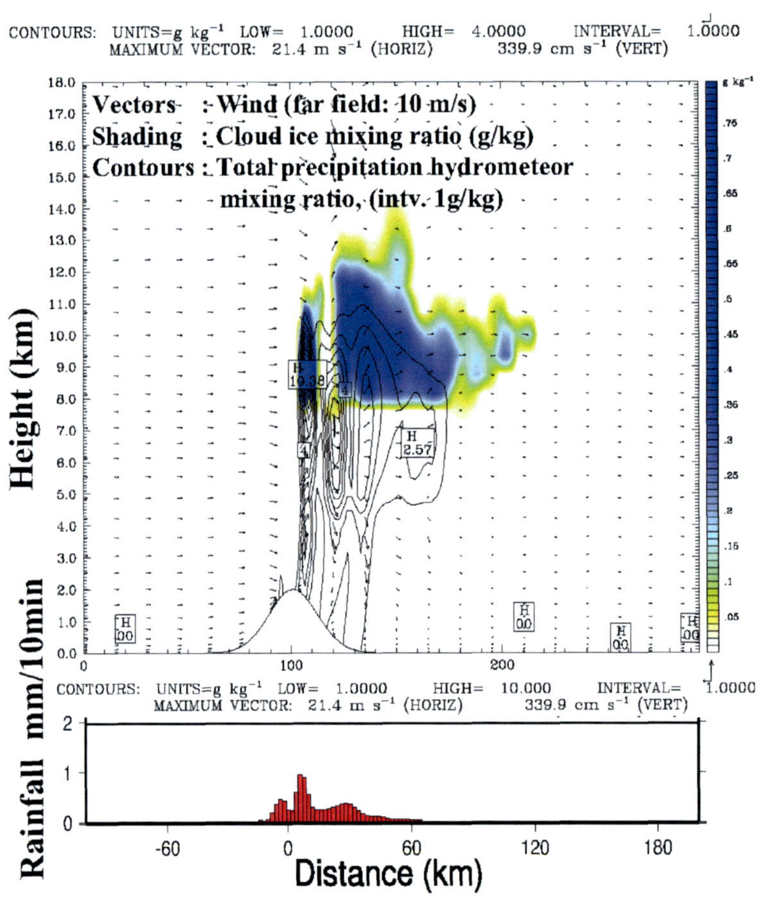

Figure 5.4 Typical output of RCM simulation of orographic rainfall (after Pathirana, 2005). Top: Cloud and rainfall on longitudinal cross section (200 min after model start). Bottom: Rainfall accumulated in the next 10 min.

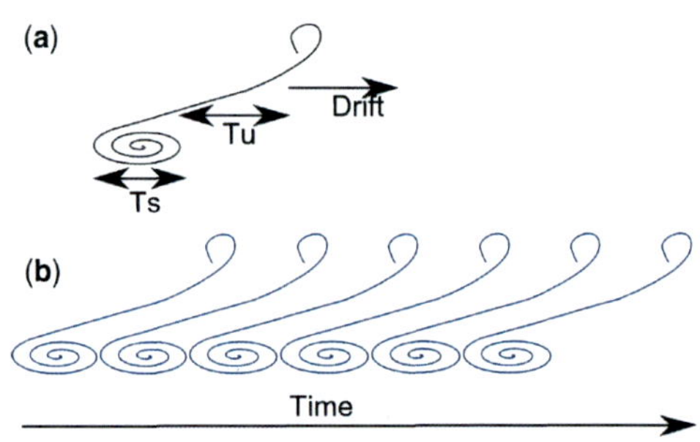

Figure 5.5 An illustration of spin-up time, model-drift and reinitialization. (a) Ts – Model spin-up time, Tu – Time with useful results (before starting to drift significantly). (b) How a series of model runs (re-initialized) can be sequenced to cover a continuous time period. Optimal simulation period for each model run should be Ts + Tu. The results of the spin-up period Ts should be neglected.

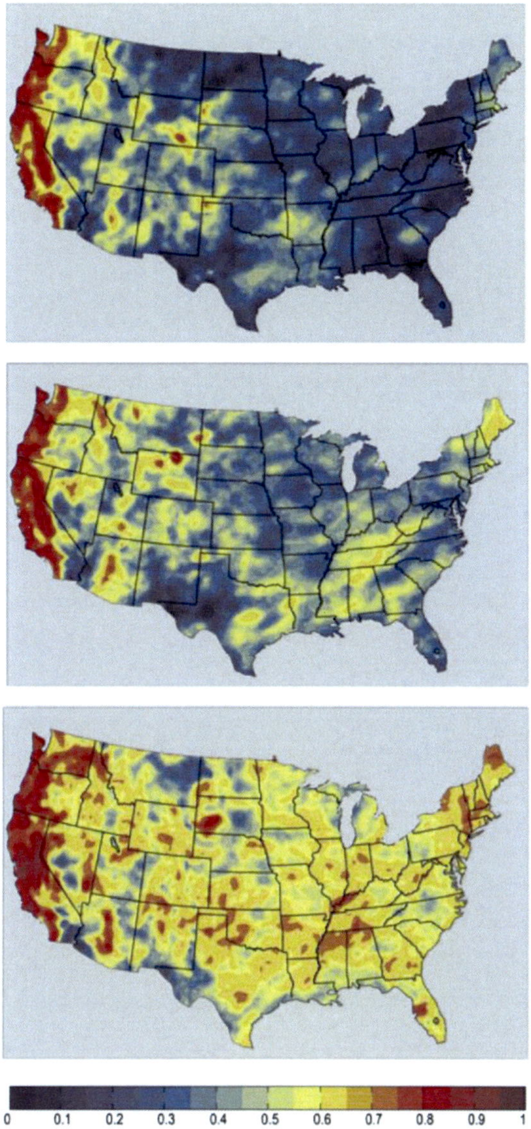

Figure 5.6 Correlations of observed and simulated daily rainfall time series for continuous model integration (top), re-initialization every 6th day (middle) and continuous model integration with 3D nudging (adapted from Lo *et al.* 2008).

The 3D nudging was done for the whole atmospheric column except the planetary boundary layer (PBL). When the PBL was included in the nudging the results somewhat deteriorated (not shown in Figure 5.6). The deterioration in precipitation simulation due to nudging toward surface variables within the model PBL is possible because of large differences in topography between the coarse-resolution FNL and the WRF downscaling domain. Ignoring the unrealistic surface variables in the large-scale field in 3D analysis nudging produces a better result in precipitation simulation.

Urban land use change and local rainfall

Urbanization changes land use properties. It is well understood that these changes generally decrease the flow resistance of the land-surface and reduce the infiltration capacity. These well-understood changes increase urban runoff volumes and flow rates. The question whether there is any significant change in the urban microclimate due to these changes, leading to significant impacts on urban rainfall, has been examined since the 1960s. Today, there is increasing evidence that the changes in the radiation and heat balance affected by changes in surface albedo and vegetation cover on the urban micro-climate can have significant impacts on the precipitation patterns over urban centres and their surroundings. These hydro-meteorological effects can be due to a) microphysical changes resulting from urban pollution, b) increased boundary-layer roughness due to urban structures and c) heat anomalies resulting from changes in albedo and latent heat flux. This was termed in Chapter 3 the UHI (Urban Heat Island) (Sagan *et al.* 1979). While the UHI effect on radiation, temperature and wind has been documented relatively early (Taha *et al.* 1988; Landsberg, 1981), modelling investigations on the impact on rainfall appeared later in the literature. A review of early attempts was summarized by Lowry (1998).

Numerical modelling experiments are extremely important in understanding and quantifying the possible effect of UHI on rainfall, as this is probably the only way to conduct controlled studies at city and regional scales to investigate the sensitivity of various influencing parameters. Shepherd (2005) noted that there had been relatively few studies in this field. Since then there have been a number of reports on such experiments.

Lin *et al.* (2008) used the MM5 model to conduct numerical experiments on the sensitivity of the size of the urban area on the hydrometeorological processes over the western plains of Taiwan. They introduced artificial urban land use clusters of varying sizes to the land use map used by the MM5 simulations. The case study with an urban area of 15 km × 15 km showed a sensible heat flux of about 500 W/m^2, which is about three times that of the land use without such artificial urban areas. They found that the UHI plays a significant role in initiating thunderstorm activity in the eastern plain of Taiwan when there is a significantly sized urban cluster. This was largely due to the uplifting of air masses by convective break-up in the urban area. However, when the urban area is absent, the mountains on the downwind area played a role in uplifting the air masses, causing thunderstorm activity.

Lei *et al.* (2008) studied the effect of explicit representation of the urban energy balance in RCMs in reproducing extreme rainfall results. They simulated the 26 July 2005 exceptional heavy rain event in Mumbai, which caused 944 mm rainfall in a single day, using the RAMS model coupled with an explicit urban energy balance scheme. They concluded that the extreme rainfall event was triggered by the urban processes over India interacting with winds influenced by sea-surface temperature gradients in the Indian Ocean.

Shem and Shepherd (2009) conducted sensitivity experiments on three land use scenarios for Atlanta, USA. They used the WRF model combined with the Noah land surface model, in a telescopic three-level nesting system to simulate a region around Atlanta at a 3.3 km grid resolution. Cumulus parameterization was used for the outer domains, but was switched off for the innermost domain. They conducted simulations with three land use scenarios: a) Atlanta city's current land use, b) a future land use scenario with high urbanization and c) urban land use removed from the model domain. They observed that the model simulations with urban land use produce 10% to 13% percent more rainfall than the scenario with no urban land use.

Pathirana *et al.* (2012) used Dynamica-EGO, a dynamic land use simulator to "project" the urban growth of the city of Mumbai up to the year 2035 (Figure 5.7). Using these projections to provide land use data, they used the WRF-Noah model to investigate the influence of urban growth on extreme rainfall intensities. Figure 5.8 shows the rainfall accumulations over the Mumbai area for the simulation of the extreme rainfall event of July 2007 (using FNL data as initial and boundary conditions), with current and future land use conditions. Based on the simulations of several historical rainfall events, they concluded that if the land use change due to urbanization takes place as modelled, the current 100 year return period rainfall (of duration 1 h) would have a return period of approximately 40 years in the future.

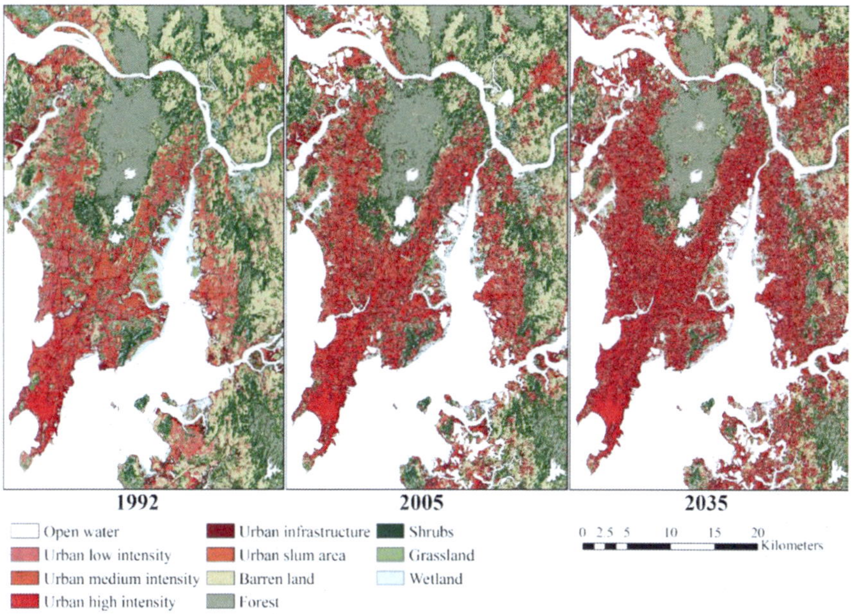

Figure 5.7 Urban growth of Mumbai city by 2035 (right), based on past growth patterns (1992, right and 2005, middle) (after Pathirana *et al.* 2012).

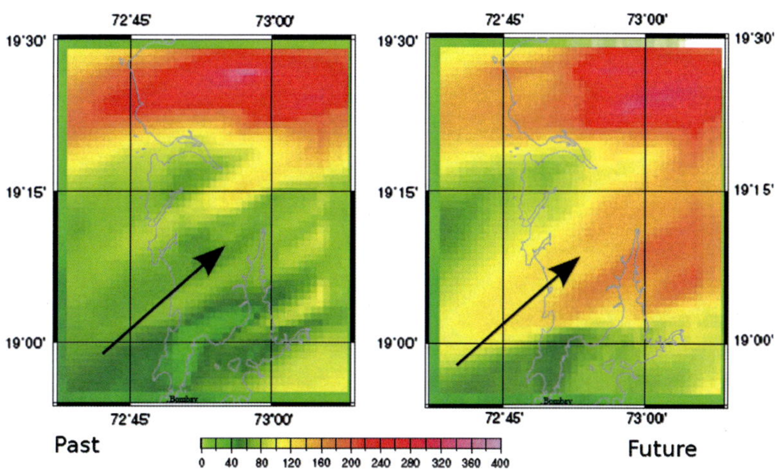

Figure 5.8 Accumulated spatial rainfall from the simulation of the 2007 July rainfall event over Mumbai city. Left: Simulation with current land use. Right: With future land use. The arrows show the prevailing wind direction during the event.

5.6 DISCUSSION

In this chapter we discussed the use of regional meteorological models (RCMs and LAMs) to study the sensitivity of local rainfall and other hydrometeorological variables to the changes in model input. When the changed parameters are the boundary conditions for the RCM, given by a GCM, we call this technique dynamical downscaling. However, many other types of sensitivity studies are also undertaken, including investigations of the effects of changes in land use or temperature (e.g. urban heat island effect).

There are several points that are useful to reiterate regarding RCM simulations. First and foremost is that cumulus parameterization should not be relied upon to provide reliable rainfall estimates. Therefore it is essential to use a small enough grid size (typically <5 km) in the analysis to be able to switch off the cumulus scheme. Most of the accessible GCM projections today are at resolutions of around 100 km if not more. To have a model domain that is large enough to be sensitive to this coarse resolution and at the same time fine enough to be useful at urban scales (and to safely switch off cumulus schemes), it is almost always useful to use a telescopic nesting scheme. Finally, most of the modern RCM simulations use a detailed LSM scheme instead of a simple surface model. This is particularly important in studies where surface changes play a central role. Modern RCMs/LAMs provide convenient ways of linking with a LSM. For example WRF can be easily coupled with Noah-LSM.

The results shown above demonstrate that mesoscale and sub-mesoscale LAMs such as city scale LAMs have their importance. This importance will further increase in the future with better knowledge on fine-scale rainfall generating processes and improved model parameterizations. In the current practise of urban drainage impact analysis, we have to rely on the synoptic scale RCM simulations provided by public databases, given that only these allow to obtain in a reasonable time span sufficiently long simulation runs (30 years or more) for a set of models and simulations. However, shorter, fine-scale LAM simulations are useful as a complementary source of information to get insight in how local scale features affect fine-scale rainfall intensities. The fine-scale LAM not only introduces a structure in the climate change signal, for example due to the fine-scale topographical variations. Gao *et al.* (2008) demonstrated that the stronger topographic forcing of fine-scale climate model may also simulate significantly different climate change patterns compared to the driving RCM or GCM. They emphasized the importance not only of using high resolution models to realistically capture the topographical forcing, but also the importance of using large domains to include relevant forcings and allow the regional model to more freely develop regional scale circulations.

In any case, before using the results from RCMs or LAMs for climate change impact studies, their reliability needs to be assessed. This can be done using several approaches, as discussed in the next chapter. For the study of urban drainage impacts, the primary question is: how reliable are the RCM results for rainfall extremes at fine spatial and temporal scales?

Chapter 6

Evaluation of dynamically downscaled rainfall

The dynamical downscaling technique presented in Chapter 5, uses physically-based regional climate models (RCMs) with GCM output as boundary conditions to simulate the atmospheric processes at a higher resolution than that of the GCM output. Before the results of such climate model simulations are used for climate change impact investigations, there is a need for evaluating the accuracy of the rainfall results. While comparisons of RCM results with observations have in many cases demonstrated good agreement, especially at large spatial scales (e.g. Räisänen, 2007), this is not necessarily the case for the smaller spatial scales. Moreover, results for extreme rainfall are less reliable than for mean rainfall conditions or for variables such as temperature (Räisänen, 2007; Hawkins & Sutton, 2010). Section 6.1 presents different approaches for evaluating the accuracy of RCM rainfall results due to the reasons explained in Section 6.2, and reviews some typical results. Section 6.3 outlines how uncertainty in RCM simulation results can be addressed. The focus obviously is on urban-scale rainfall extremes. Focus in this chapter, however, lies on the sets of RCM simulations available in the public databases (see previous chapter), which provide sets of simulations for 30 years or more. As previous chapter explained, smaller scale LAMs may be useful as a complementary source of information to the urban hydrologist. Such LAMs can be validated using the same methods as discussed in this chapter.

6.1 RELIABILITY OF CLIMATE SIMULATIONS BY GCMs AND RCMs

Evaluation of the accuracy of the rainfall results of climate model simulations can be undertaken using several approaches. The standard approach is to evaluate the climate model's ability to accurately simulate the current climate. It is a standard practice of climatologists to initialize the models with pre-industrial control simulations, to impose the natural and anthropogenic forcing thought to be important in climate evolution for the past decades, and to verify the climate model outputs in particular with respect to their consistency with the statistical characteristics of the present climate. This requires the control simulations to be compared with historical rainfall observations, through comparison of mean and extreme intensities at the temporal and spatial scales of the climate models. Accurate modelling of long-term cumulative rainfall effects is important for modelling of river catchment water balances, including groundwater. In the urban drainage context, the event-based occurrence of high rainfall intensities is more important.

Validation of climate model simulations has to be undertaken for periods of several decades. As explained in Section 4.3, a 30-year control period is most often considered to sufficiently capture the natural variability. Data from several decades are also required when one wants to compare statistical characteristics between historical and future climatic conditions. Such comparison indeed needs a sufficiently long period to get representative statistics. Another reason is related to the numerical computations: climate model runs start from arbitrary initial conditions, which are not necessarily "in phase" with the real climate. The effect of these phase errors is reduced when considering longer periods.

Given that rainfall intensities strongly differ depending on time and space scale, comparison of climate model results with observations should be done for identical scales. This can be easily done for the temporal scale, but is more difficult for the spatial scale, as explained in Section 2.7. Accurate rainfall observations are often only available at the point scale, while climate models produce results at much coarser scales. Spatial disaggregation methods or areal reduction factors could be applied to the results of climate models or upscaling methods to the observations. These methods were presented in Chapter 2 (Sections 2.3 and 2.7). Surprisingly, these methods are seldom applied in the current practise of climate model validations, possibly because hydrology and climate science are two separate worlds. Another reason is that application of these down- and upscaling methods is time-consuming, while their added value is rather limited because the scale difference will be covered by the statistical downscaling steps that are applied next (after the model validation) when it comes to climate change impact investigations.

A first example shown on the validation of RCM results is taken from the ENSEMBLES project (van der Linden & Mitchell, 2009). The RCMs participating in that project were evaluated using different performance measures for their ability to simulate current climate conditions (Christensen *et al.* 2010). Six measures were analysed, including large-scale circulation patterns (F1), seasonal mean temperature and precipitation (F2), distributions of daily and monthly temperature and precipitation (F3), extreme daily precipitation and daily minimum and maximum temperature (F4), long term trends in temperature (F5), and annual cycle of temperature and precipitation (F6). The results are shown in Figure 6.1. No model performs best with respect to all measures, and the differences between the models are largest for simulation of seasonal patterns and extremes (F2 and F4). These results emphasise the general conclusions from performance studies of climate models, that there is no single best model, and the use of a set of climate models, called *ensemble*, is recommended to provide more reliable conclusions on climate change impacts (see Section 6.3).

When the climate model simulated rainfall characteristics (e.g. mean intensities, variances or quantiles) deviate significantly from the historical values, the climate model simulation may be considered inaccurate and removed from further analysis. Evaluation of the significance of the deviations is typically based on hypothesis testing or estimation of confidence intervals. In the methods based on hypothesis testing, the null hypothesis of no difference is tested and climate model simulations rejected when the null hypothesis is rejected at a given significance level (such as 5%). The confidence intervals on the observed statistics are most often based on resampling methods. For instance, Dibike *et al.* (2008) applied the Wilcoxon rank-sum test to examine the significance of the median difference, and estimated confidence intervals using bootstrap simulations. Confidence intervals based on the observed extremes can be used to evaluate the significance of the under- and overestimations of the extreme value statistics of the RCM/GCM results (see example in Figure 6.4).

For the Uccle station in central Belgium, Baguis *et al.* (2010) calculated the mean error and the mean squared error in the daily rainfall intensities, as well as the number of days per year with more than 10 mm, 20 mm or 40 mm, or less than 0.1 mm of rainfall, the daily rainfall quantiles and the regional rainfall covariance. They did not only test the current and past climate conditions, but also looked at outliers in the future projected climate change. Outliers were defined as having more than 3 times the interquartile range higher or lower than the upper or lower quartile in comparison with the complete set of climate model simulations considered. From the 31 RCM simulations they analysed, three were rejected.

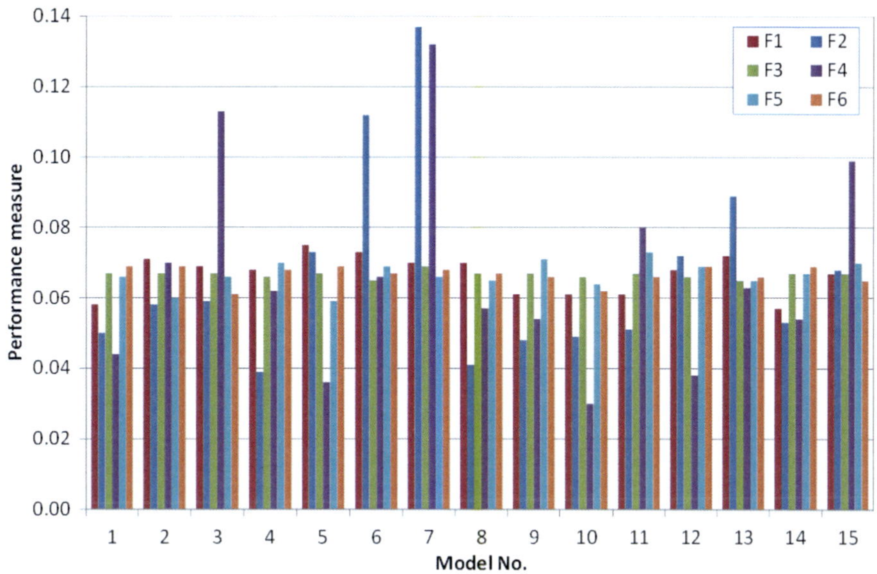

Figure 6.1 Performance measures for the 15 RCM models in the ENSEMBLES project (data taken from Christensen *et al.* 2010). Larger performance measures correspond to better performance. The performance measures have been normalised so that the sum of a measure for the 15 models equals 1.

From the accuracy evaluations available, it appears that some RCMs have a strong bias towards overestimated frequency of wet days, generating very frequent but very small rainfall amounts (Räisänen *et al.* 2004), although this problem of "constant drizzling" also depends on season and region. In some regions/seasons it is the other way round. The bias might be due to the grid scale. Area-averaging over a grid cell means that the grid cells have more days of light rainfall (Frei *et al.* 2003; Barring *et al.* 2006), and also reduces the magnitude of extremes compared with point rainfall.

Although testing of short-term rainfall on the RCM grid scale based on high-resolution gauges has not been undertaken very often, some results can be found in the literature. Figure 6.2 shows a typical result when comparing extreme rainfall statistics from climate model results with rain gauge observations. The quantiles based on rain gauge observations are larger, especially at fine temporal scales. As illustrated in Figure 6.2, this difference might be even larger than the climate change signal itself.

The difference in the rainfall results of GCMs and RCMs and rain gauge observations as plotted in Figure 6.2 has been shown by several researchers. The bias in extreme rainfall characteristics is generally increasing for decreasing temporal scale. For the Uccle station in central Belgium, Willems and Vrac (2011) compared extreme rainfall statistics for durations between 1 and 15 days with the same statistics derived from 17 GCM runs with the ECHAM5 model using the A1B emission scenario (Figure 6.3). The results are presented in the form of IDF relationships (see Section 2.5). In Figure 6.3, the historical

IDF relationships at Uccle (based on 10-minute rainfall intensities for the period 1967–1993, and published in Willems, 2000) are compared with the results from the ECHAM5 runs for daily or longer time scales. Significant differences are observed with the GCM results showing lower rainfall intensities than the historical observations. The differences are greatest for the daily time scale and reduce towards the larger time scales.

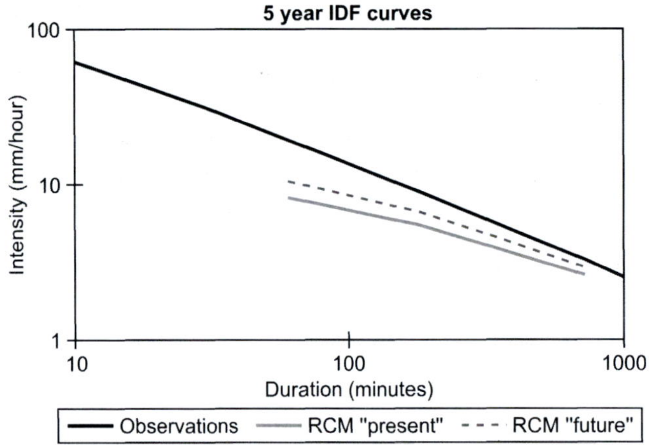

Figure 6.2 Example of estimated IDF curves based on observed and RCM output of extreme rainfall intensities (after Arnbjerg-Nielsen, 2008).

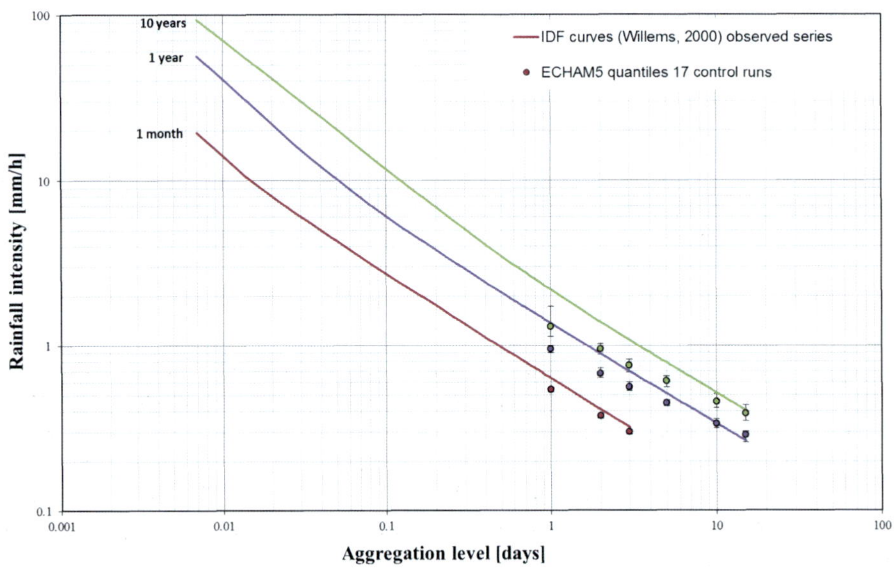

Figure 6.3 Comparison of historical IDF relationships at Uccle, Belgium, with the results of 17 ECHAM5 control runs (mean and range of the 17 runs are shown) (after Willems & Vrac, 2011).

Another comparison between climate model results (two different GCM models with the A2 emission scenario) and historical point rainfall data has been analysed for Dorval station in Quebec (Nguyen *et al.* 2008b) for the period 1961–1990, and for coastline areas in northern Canada (Dibike *et al.* 2008). The differences were in these cases only studied for daily or coarser time scale, but again systematic underestimations (dependent on the return period) were found. See Section 7.3 for more information on these studies and Figure 7.8 for some results.

Olsson *et al.* (2009) found for the grid cell covering a rain gauge at Kalmar, Sweden, that RCM results overestimate the frequency of low rainfall intensities, and therefore the total volume, but that higher intensities are reasonably well reproduced.

Hanel and Buishand (2010) compared 1-day and 1-hour precipitation extremes in five ERA40-driven RCM simulations with radar-observed estimates during a 10-year period. Distinct errors in the 1-hour RCM-simulated extremes were found, whereas the 1-day extremes were simulated quite well.

In a study by Sunyer *et al.* (2012) daily rainfall extremes were extracted from four different RCM/GCM models from the ENSEMBLES database for Denmark. The extreme value statistics from the four models were compared to the statistics based on observations (see Figure 6.4). The extreme value statistics derived from the four climate models vary considerably and provide both higher and lower rainfall depths compared to the observed extreme value statistics.

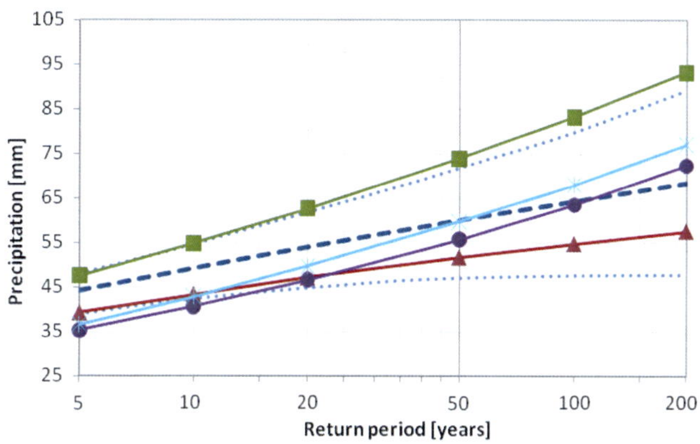

Figure 6.4 Estimated extreme value statistics of daily rainfalls of the observed record (- - - -) and RCMs for the control period: HIRHAM-ARPEGE(—▲—), HIRHAM-ECHAM5 (—■—), REMO-ECHAM5 (—●—) and ALADIN-ARPEGE (—✳—). The 95% confidence intervals for the observed extreme events are shown as dotted lines (·····) (after Sunyer *et al.* 2012).

Mishra *et al.* (2012) recently showed that RCMs with both reanalysis and GCM boundary conditions behave similarly and underestimate 3-hour precipitation maxima across almost the entire USA. The results indicate that RCM simulated 3-hour precipitation maxima at 100-year return period could be considered acceptable for stormwater infrastructure design at less than 12% of the 100 urban areas investigated (regardless of boundary conditions). RCM performance for 24-hour precipitation maxima was slightly better, with performance acceptable for stormwater infrastructure design judged adequate at about 25% of the urban areas.

For Southern South America, Menéndez *et al.* (2010) analysed one GCM and five RCMs and concluded that the models succeed in reproducing the overall observed frequency of daily precipitation, but that most but not all models underpredict rainfall amounts. Their models tend to underestimate the number of events with strong or heavy rainfall. Menéndez *et al.* (2010) moreover have shown for three periods with anomalous climate conditions that the models have regime dependence, performing better for some conditions than others. Marengo *et al.* (2010) compared Mann-Kendall based trend testing results obtained from GCM 20C3M scenario simulations versus observations at 104 stations in South America, covering the period 1960–2000. They found that the models may give strong under- or overestimation of the trends. Independent on this bias, the models they studied exhibit in general a larger frequency of extremes compared to observations.

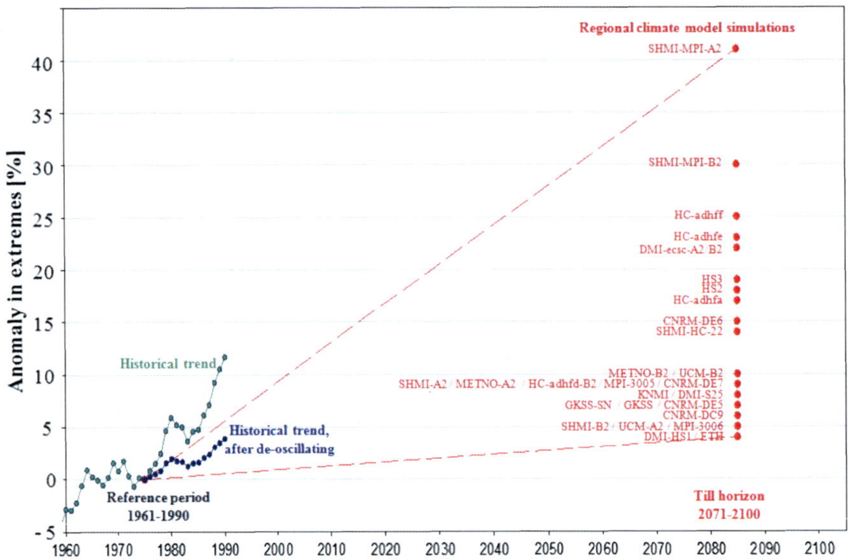

Figure 6.5 Comparison of the change (called anomaly) in quantiles of daily rainfall extremes (of return period larger than 1 year) at Uccle, Brussels, in the winter season Dec.-Feb., for a 30-year window (1961–1990), before and after approximate elimination of the multidecadal oscillation effects on the total historical trend, for the PRUDENCE RCM results for 2071–2100 vs. 1961–1990.

While discussing the need to compare and validate climate model results with observations, there is another important issue to consider. Validation using climate model control runs based on past climate records, does not guarantee that the climate model will be able to reliably simulate the impacts of changes caused by specific climate forcing scenarios. A close and physical consistent match of the RCM simulation results with observations provides insufficient evidence for credible future projections (Gutowski *et al.* 2008). The ability of the model to reliably simulate the impacts of changes caused by specific climate forcing scenarios can be partly tested based on transient historical runs for several decades and then comparing the observed and simulated trends in climate (taking into account the recent historical increases in GHG concentrations). As the latter increases are limited in comparison with what is projected for the future, the skill of the model to project relative changes under changing GHG forcing can never be fully verified. Another option is to compare the magnitude of the recent trends observed in historical climate records with the magnitude of future projected changes. This has been done in Figure 6.5 for daily rainfall extremes in Belgium. In Section 3.2 it was explained how for these data, significant trends were identified for recent years after approximate elimination of the influence of natural variability (multi-decadal climate oscillations). When the historical trend in extreme rainfall quantiles is plotted starting from the 1961–1990 baseline period (which is equal to the climate model control period) up to the most recent year of available observations, for moving time windows of 30 years, then increasing historical trends were found that are consistent (in sign and order of magnitude; see Figure 6.5) with the future projections made by RCMs in the PRUDENCE database (see Section 5.3).

Another study which compared the magnitude of recent trends observed in historical climate records with the magnitude of future projected changes is the one by Gregersen *et al.* (2011). Based on a transient RCM simulation for a period of 151 years (1950–2100), they concluded for Denmark that the increase in the frequency of the extreme events predicted by the model (0.4% pa) was significantly lower than the increase observed during the past 31 years (2.5% pa). This did not come as a surprise if one takes the underestimation of the RCM in describing rainfall extremes (Arnbjerg-Nielsen, 2008) into account. That (future) extreme rainfall projected by models may be underestimated because most models seem to underestimate observed increases in precipitation due to warming was also suggested by Min *et al.* (2011).

6.2 REASON OF DIFFERENCES BETWEEN GCM/RCM RESULTS AND OBSERVATIONS

As explained above, there are two main reasons for the differences between the climate model results and the observations. The first reason is that the present generation of RCMs does not include proper descriptions of physical properties such as convective rain cells and cell clusters (Section 5.2). This leads to a bias in the estimated extreme rainfall intensities from the climate models. The second reason is the difference in spatial scales between the climate model results and the rain gauge observations.

Related to the coarse spatial scale, RCMs use parametric representations of the physical processes describing precipitation. These parameterizations are modules based on observations and physical theory, to represent processes with scales too small to be resolved on the spatial scale of the model grid. They lead to precipitation results of limited accuracy, especially for generation of convective and extreme rainfall. This is, on the one hand, due to our limited understanding of the essential processes involved in precipitation formation. These range from large-scale atmospheric dynamics, mesoscale convective circulations, to the local precipitation microphysics at the smallest spatial and temporal scales (Baker & Peter, 2008). At smaller scales, the natural internal variability or "noise" unrelated to external forcings increases (Hawkins & Sutton, 2009). There are still many important small-scale processes at these scales, which are unresolved or not represented explicitly in the models, such as the representations of clouds, convection and land-surface processes (Randall *et al.* 2007), including the complexity of the feedbacks involved. Extremes may also be impacted by mesoscale circulations that AOGCMs and even current RCMs cannot resolve, such as low-level jets and their coupling with heavy rainfall intensities (Anderson *et al.* 2003; Menéndez *et al.* 2010). Another issue with small-scale processes is the lack of relevant observations.

On the other hand, it is also due to our limited ability to model these processes in global and regional climate models. The spatial resolution of the climate models is coarser than the spatial extent of convective clouds (typically 1–10 km^2; Niemczynowicz, 1988) that generate high rainfall intensities over short durations. Some cloud processes of high importance for urban rainfall generation, like droplet formation, draughts and turbulent mixing, take place on an even smaller scale (Baker & Peter, 2008). Furthermore, local, short-duration precipitation-generating mechanisms are difficult to resolve because of numerical stability and computational efficiency considerations, hence limiting the temporal and spatial scales in the models, although some short time scale convection-resolving experiments are being performed. van Meijlgaard *et al.* (2008) reported that the RCM simulation time step can be as small as a few minutes, but that the results are very uncertain at this temporal scale. The spatial scales can be as small as 1–2 km, the scale at which clouds and convection are resolved, using non-hydrostatic mesoscale models, but simulations are only feasible for short periods of only a few months or a few years at most (Grell *et al.* 2000; Hay *et al.* 2006; Hohenegger *et al.* 2008; Knote *et al.* 2010). For larger spatial scales of less than 10–20 km, double-nesting may be required by embedding the very high-resolution model within a coarser-scale RCM. Less common is the use of stretched grids with variable resolution.

Due to the parameterizations applied to solve the scale problems, climate models simulate the frequency, distribution and intensity of extreme rainfall less well than other variables such as temperature (Räisänen, 2007; Fowler *et al.* 2007; Fowler & Wilby, 2007; Hawkins & Sutton, 2010). For the same reason also the changes in extreme rainfall due to anthropogenic changes are simulated less well (e.g. Randall *et al.* 2007; Alexander & Arblaster, 2009; Maraun *et al.* 2010). This often leads to a wide range of changes or strong inconsistency across climate models (Tebaldi *et al.* 2006). For a given climate forcing scenario, the accuracy of the (relative) changes in extreme rainfall might, however, be higher than the accuracy of the absolute extreme rainfall values themselves. This is because the changes are mainly controlled by the large-scale anthropogenic forcing, while the absolute values are strongly controlled by small-scale processes and local conditions. Variability internal to the climate system plays a larger role on regional scales than on global scales.

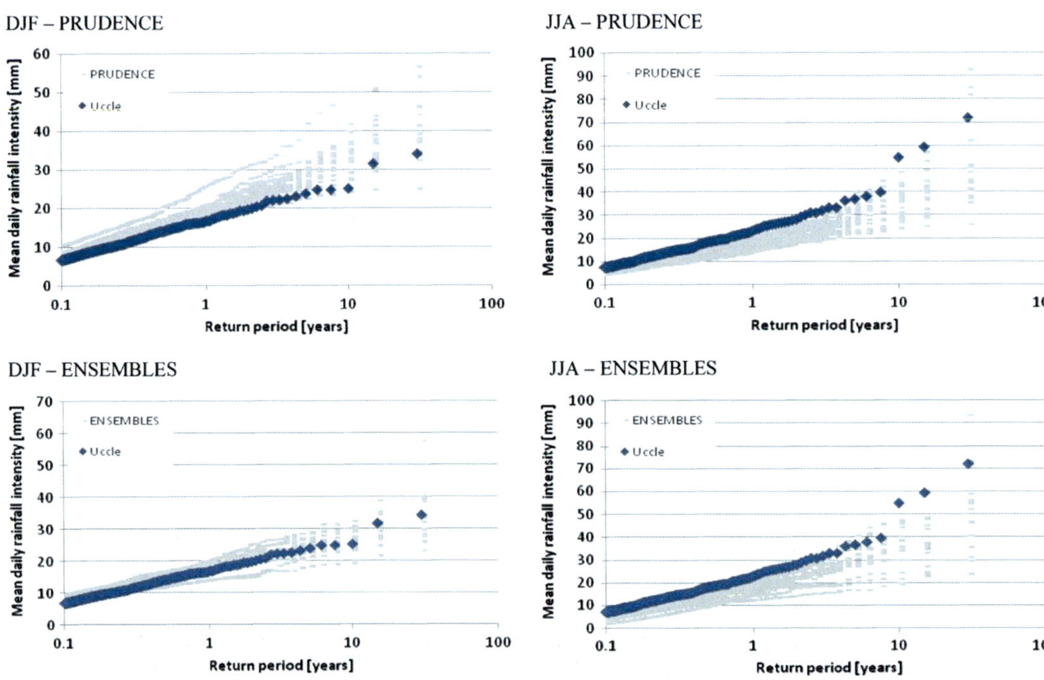

Figure 6.6 Validation of daily precipitation extremes at Uccle, Brussels, in winter (DJF) and summer (JJA) months. Top row: PRUDENCE RCM runs. Bottom row: ENSEMBLES RCM runs (after Staes *et al.* 2011).

Climate science, however, evolves fast, such that newer generation models may have a higher accuracy compared to older generation models. This is illustrated in Figure 6.6 comparing the PRUDENCE and ENSEMBLES results for all RCM runs with available results for the main Belgian meteorological station at Uccle. It is clear from Figure 6.6 that, while the PRUDENCE runs show systematic overestimation of daily precipitation extremes in the European winter (DJF) season, these are strongly reduced in the more recent ENSEMBLES runs. For the summer (JJA) season, the systematic underestimation remains, which may be due to the fact that current RCMs are still too coarse to resolve convective precipitation. Also for the winter season, lower rainfall intensities are expected for the RCM results in comparison with the observed point intensities, because of the spatial scale difference (grid-averaged precipitation versus point precipitation), so that the results suggest that precipitation is better represented in the last simulations performed.

The improvements in the RCM outputs may be due to improvements in the RCMs and/or improvements in the GCMs in which the RCMs are nested. Note that the importance of these two contributing factors can be studied by replacing the GCM based boundary conditions by historical reanalysis data (Section 4.1). Figure 6.7 shows the results from the CLM3.0 model (http://www.clm-community.eu). In order to eliminate the influence of the GCM in which the CLM is nested (i.e. the influence of the boundary conditions), CLM results driven by lateral boundary conditions from reanalysis data (ERA-40 and NCEP) were considered. This focuses on the ability of the RCM to produce extreme precipitations with statistical properties that match observations (the state of the atmosphere at large scales is similar in both models and observations, therefore the model – observation differences resulting from natural variability are strongly reduced). Figure 6.7 shows that CLM3.0 ERA40 results match well the Uccle historical daily rainfall extremes, also in the summer season. This shows that this RCM is able to reproduce rainfall extremes well when the model is forced by historical large-scale atmospheric circulation information at its boundaries. However, there may still be some differences because grid-averaged extremes differ from point-scale values. The rainfall biases in the RCM results (PRUDENCE, ENSEMBLES; see Figure 6.6) thus appear to mainly result from biases in the GCM forcings. This is consistent with other studies that found that the GCM explains a large fraction of the biases, although the regional model may have a larger role in summer due to local effects such as convection (e.g. Rummukainen, 2010).

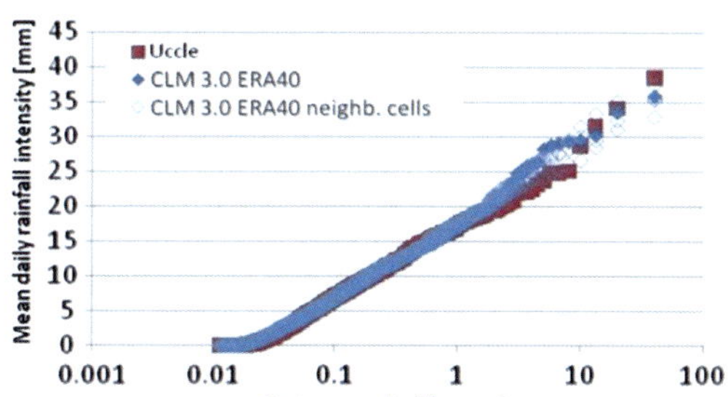

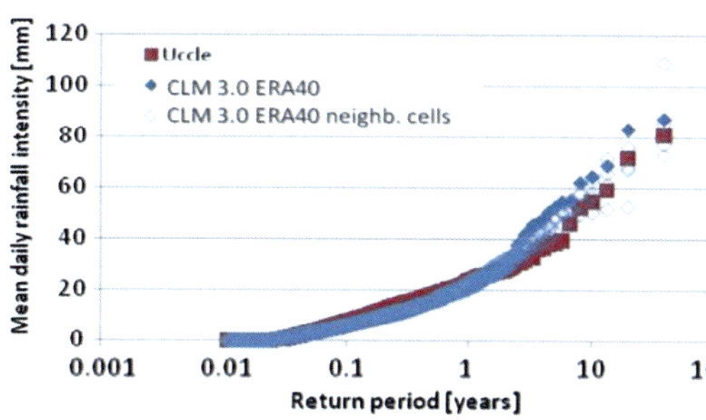

Figure 6.7 Validation of daily rainfall extremes, based on the CLM3.0 ERA40 results for the grid cell covering the Uccle meteo-station, Brussels. Left: DJF (winter); right: JJA (summer) (after Staes *et al.* 2011).

Rainfall intensities aggregated over time scales of one day or longer are shown by IDF curves in Figure 6.8. As for the 1-day time scale, the ERA40 forced CLM3.0 results tend to show smaller biases than the mean value from ENSEMBLES simulations, although some overestimation of the larger events (10-year return period) is noticeable, especially for the shorter time scales shown (1 day).

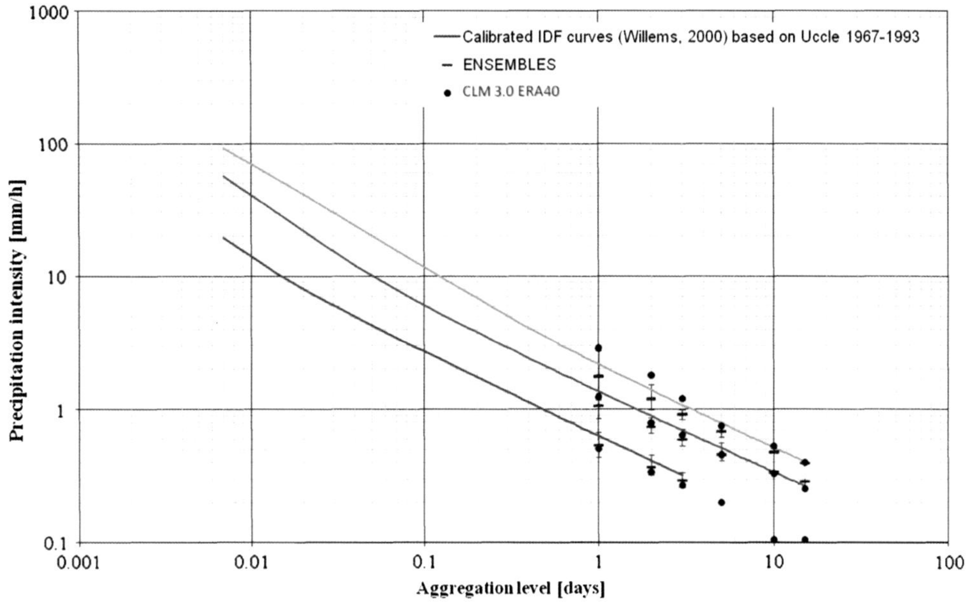

Figure 6.8 Comparison of rainfall IDF relationships for Uccle, Brussels, with ENSEMBLES and CLM3.0 ERA40 results. Three return periods are shown: 0.1 year (red line), 1 year (blue line) and 10 years (green line) (after Staes *et al.* 2011).

6.3 UNCERTAINTY IN CLIMATE IMPACT PROJECTIONS FROM VARIOUS GCMs/RCMs AND DIFFERENT SCENARIOS

Due to the large uncertainties in climate modelling of rainfall extremes (as described in the previous section), future climate change impact projections will also be highly uncertain. The uncertainties in future projections are larger than for the past climate simulations, due to additional uncertainties in the estimation of future climate forcing and unknown feedback mechanisms in the climate system (see Chapter 4).

Given the compounding nature of these uncertainties in modelled climate projections, it is important to understand where (or when) particular factors are dominant. Hawkins and Sutton (2009) gave a useful assessment of the three main sources of uncertainty in current GCM projections of global temperature change and related this to the timescale of the projection. The sources of uncertainty they considered were:

(i) internal variability (natural climatic variability and short-term changes that can cause temporary departures from longer-term trends);
(ii) model uncertainty (differences in numerical schemes, modelling assumptions, coupling processes, parameterization, etc.); and
(iii) scenario uncertainty (differences in projected climate forcing and the resultant forcing).

They showed that in the short term internal variability dominates and in the medium term (several decades) model uncertainty dominates the overall uncertainty in temperature projections. However, the relative influence of scenario uncertainty grows rapidly over time and by 2100, scenario uncertainty dominates overall uncertainty. Results for Europe are shown in Figure 6.9.

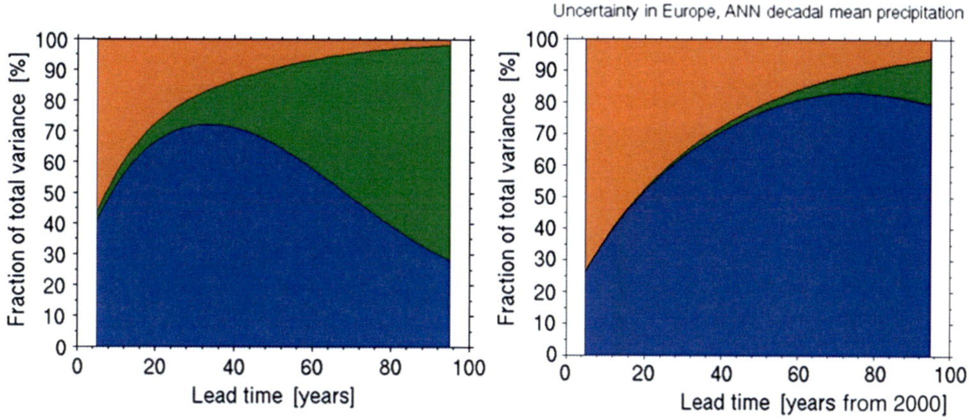

Figure 6.9 Relative contributions to total uncertainty of climate projections for Europe from internal variability (orange), model uncertainty (blue) and scenario uncertainty (green). Left panel: temperature, right panel: precipitation (after Hawkins & Sutton, 2009; 2011: Retrieved 31 March 2012, from http://climate.ncas.ac.uk/research/uncertainty/).

A similar analysis of uncertainty in future precipitation changes was undertaken by Hawkins and Sutton (2011), who found that, unlike temperature, a combination of modelling uncertainty and internal variability heavily dominates projected global and regional changes in precipitation up to 2100 (see Figure 6.9). This supports earlier research by Covey *et al.* (2003) and Arnell (2003) who found that the variability in long-term hydrological projections were dominated by modelling differences rather than by climate forcing trajectories. Déqué *et al.* (2007) concluded that for Europe the uncertainty in mean changes in rainfall relating to the choice of the driving GCM is generally larger than that due to the choice of the RCM, the climate forcing scenario or natural variability. However, for summer rainfall in Europe the choice of the RCM was found to be of equal importance as the choice of the GCM (Déqué *et al.* 2007; de Elía *et al.* 2008). The choice of the RCM is also more important for changes in extremes rather than changes in mean values (Frei *et al.* 2006; Fowler *et al.* 2007).

In short, the current GCMs have large uncertainties in the projection of long-term precipitation changes. This is in part due to the fact that precipitation, unlike temperature, is a secondary variable or output from GCMs. This is unfortunate because while it might be a secondary output from a GCM, precipitation is a primary input to hydrological models.

This means that climate impact projections depend highly on the specific GCM and/or RCM used as well as the specific climate forcing scenario considered. For these reasons, in any climate change impact study several climate models and several forcing scenarios should be considered. This *ensemble modelling* approach allows ensemble-based probabilistic projections, which are more useful than deterministic projections using "best guess" scenarios (Palmer & Räisänen, 2002; Tebaldi *et al.* 2005; Smith *et al.* 2009; Semenov & Stratonovitch, 2010). However, this approach may be difficult in practice for urban impact studies, because of the large computational resources often associated with hydraulic modelling. Moreover, there is always a risk of underestimating the overall uncertainty due to lack of knowledge of key processes (Walker *et al.* 2003). Different models share the same level of process understanding and sometimes even the same parameterization schemes and code (Knutti *et al.* 2010). It therefore must be recognised that the total uncertainty of climate projections is likely to be larger than that exhibited by an ensemble of models.

For central Belgium, Figure 6.10 shows the results of changes in extreme rainfall characteristics based on 31 climate model simulations with 11 different RCMs. Climate factors (i.e. the ratio between future and current statistics; see further Chapter 8) were found to depend strongly on the RCM simulation considered, hence showing that climate change impact investigations based on only one or a few climate model simulations may not be reliable.

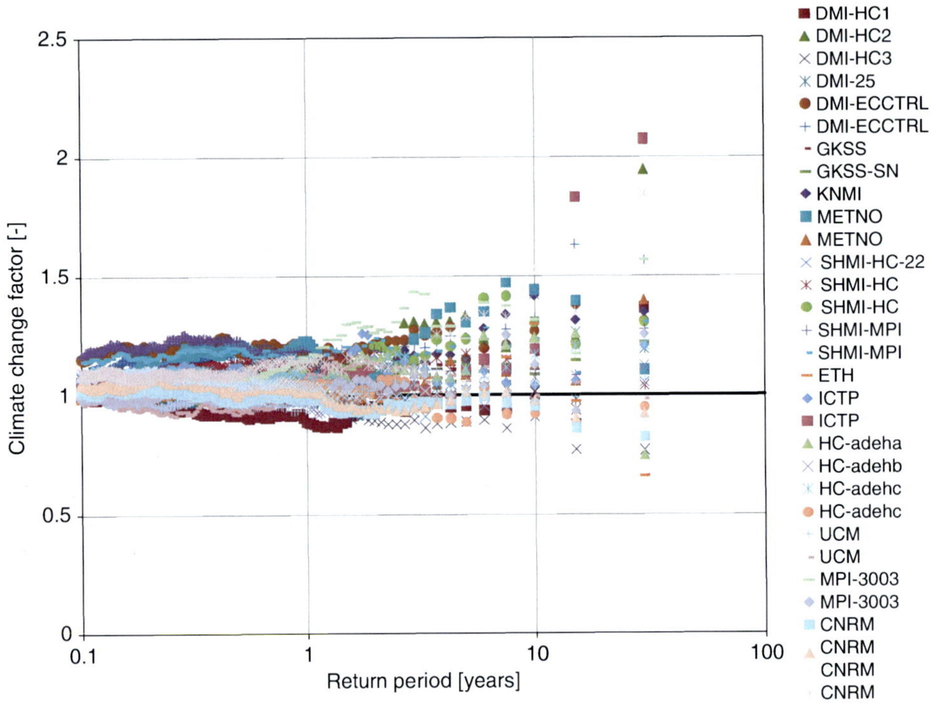

Figure 6.10 Climate factors versus return period for 31 PRUDENCE RCM simulations for daily rainfall as function of the exceedance probability at Uccle, Brussels, during the summer season, 2071–2100 vs. 1961–1990 (adapted from Ntegeka *et al.* 2008).

When analysing Figure 6.10, one can note that the differences between different climate models increase with increasing return period. This is not only due to the higher uncertainty of the models in describing rainfall extremes, but also due to natural variability. The empirical return periods in Figure 6.10 are computed based on rank statistics after sorting the rainfall extremes in the 30-year simulation periods. Rainfall intensities of longer return periods have a higher uncertainty when empirically derived based on rank statistics.

Another factor to be taken into account when analysing rainfall changes based on long-term RCM runs, is the influence of climate oscillations, as discussed in Section 3.2. Figure 6.11 shows the comparison between the multidecadal oscillations derived from the observations at Uccle, Brussels (see Section 3.2) and the ones obtained from the CLM3.0 model with lateral boundary forcing from ERA40 re-analyses data for the period 1961–2000. Despite the limited time period considered, it appears that the CLM results follow the historical observations quite closely, showing part of an oscillation. This confirms that given the appropriate large-scale conditions, RCMs can provide the correct temporal oscillations in extreme rainfall statistics.

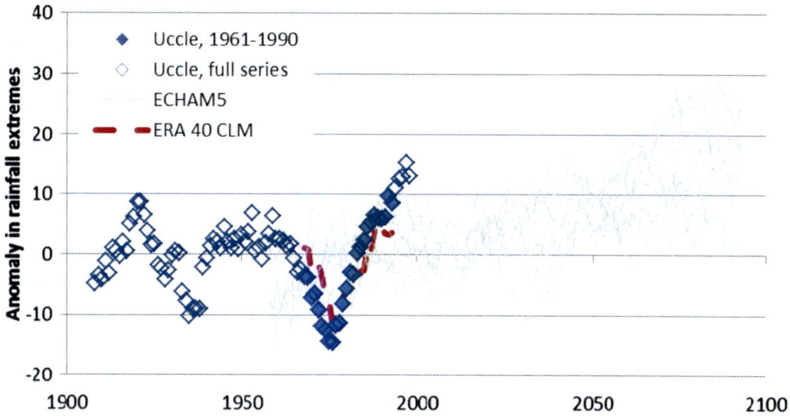

Figure 6.11 Anomalies in daily rainfall extremes based on the CLM3.0 results at Uccle, Brussels, driven by ERA40 data, and comparison with 17 ECHAM5/MPI-OM A1B runs ESSENCE project (after Staes *et al.* 2011).

Figure 6.11 includes another comparison with the results from an ensemble of 17 ECHAM5 GCM simulations driven by the SRES A1B emission scenario, each with slightly different initial conditions. The results also show an oscillatory behaviour in which the period and amplitude characteristics are similar to past observations for at least some of the ensemble members, although the amplitude of the oscillations provided by the model may be smaller, and oscillations may sometimes be faster. This preliminary analysis thus suggests that GCMs may represent the physical processes that are responsible for the oscillations, and could therefore be a useful tool to investigate their origin. This is encouraging but further investigations are necessary to confirm that the oscillations found in the model results are due to the same mechanisms as the ones operating in the real climate system. The climate models may indeed not fully capture the decadal and multi-decadal variations, given that the primary drivers of the oscillations are often largely unknown.

In any case, the oscillations shown by the GCM are not in phase with the observed ones, but this was expected since the oscillations are attributed to (unforced) natural variability (the initialisation of the model does not relate to a particular date in the past, and differs between the ensemble members). The fact that the RCM simulations forced by reanalysis were in phase with the observations confirms that the oscillation signal is found in the large-scale atmospheric circulation (although its origin is likely to involve other climate system components).

These multidecadal oscillations must be taken into account when building climate change projections and conducting impact analyses. Indeed, if the study period is shorter than an oscillation, the impact modeller has to keep in mind that the results may be biased from the long-term averaged climate. Due to the oscillations, an input series period covering an oscillation peak may provide overestimated risk figures (but the maximum will be addressed), while using an input series that covers an oscillation low would result in underestimation of the risk. This is also important for model validation if it is not performed on the basis of reanalysis forcing (e.g. in the case of GCMs), because if the validation period does not cover an integer number of oscillations, then phase differences between the model and the observations may result in model-observations differences, even in the computed average climate statistics, while there is no actual model deficiency (bias) behind these differences.

There is ongoing research to find the drivers of the oscillations such that they can be better taken into account both in the historical trend analysis (Chapter 3) and to include these drivers in the climate models. In Australia, two major research projects of this kind have been started recently. The first is the South Eastern Australian Climate Initiative (SEACI) which aims to improve understanding of the causes and impacts of climate change and variability in south-eastern Australia, and improve projections of future climate. This project extends over the entire Murray-Darling Basin (which alone covers 14% of Australia's landmass), the state of Victoria and southern South Australia (www.seaci.org). A major aim of the SEACI research project is to determine the extent to which climate change and variability are affecting rainfall and runoff in south-eastern Australia. SEACI has investigated the relationship between south-eastern Australian rainfall and the Sub-Tropical Ridge (STR) intensity and position. For example, a range of datasets and methods indicate that the tropics are expanding, pushing the downward descending arm of the Hadley circulation further south. These observed changes are changing the nature of the rainfall across south-eastern Australia: rain bearing systems affecting south-eastern Australia are less often due to mid latitude cyclones and increasingly due to larger systems centred further north. This signal is seasonally dependent and peaks during summer and autumn, providing insight into the observed autumn rainfall deficit.

The second major Australian project is the Goyder Institute for Water Research's Climate Projections Project (http://goyderinstitute.org/index.php?id=31). In terms of climate drivers, this project examines the influence of the ENSO cycle, the PDO, the IOD, and the SAM on rainfall and other hydro-climatic variables. It also investigates how these modes of variability interact and potentially moderate each other and examines their influence on local synoptic systems (such as cut-off lows, atmospheric blocking and tracks of low and high pressure systems).

Further insight into these issues of natural climate variability and climate drivers may possibly come also from ongoing research efforts regarding predictability of seasonal-to-decadal scale climate change, which will be addressed in IPCC's AR5.

6.4 DISCUSSION

Validation of GCM or RCM results for local conditions is required before these models are used as input for local climate change impact studies. This is commonly done by comparison with statistics obtained from long-term historical rainfall series. The review of GCM and RCM validation results presented in this chapter for rainfall extremes has shown that climate models may strongly deviate from observations. This is partly due to differences in spatial scales. Accurate rainfall observations are often only available at the point scale, while climate models produce results at much coarser scales. Second reason of the strong deviations is the poor accuracy of the climate models in simulating rainfall extremes.

Statistical tests can be applied to the initial set of available climate model runs and after testing, rejected runs could be removed, which are the ones that are inconsistent with the current or past climate observations. However, it became clear from the discussions in the chapter that one has to be careful with such rejections for several reasons: natural variability and limited length of the available time series, difference in spatial scale, influence of climate oscillations. Moreover, results in this chapter have shown that for rainfall extremes of short duration, including convective storms, deviations from historical values might be systematic for most if not all climate models and bias correction is needed to avoid rejection of the majority of models.

The spatial scale differences can be resolved by statistical downscaling methods, as discussed in next chapter. They can be combined with bias correction, as also described in next chapter. The discussion in next chapter will make clear that bias correction can be implemented without explicit quantification of the bias, when the assumption is made that the bias in current and future climates is the same. This is another reason why care is required before rejecting climate model runs based on control run validations. It explains why some experts (e.g. Baguis *et al.* 2010) rejected climate model runs only when they have both a high bias for the control period, and at the same time predict future climatic changes that deviate from the range given by other climate models or runs.

This chapter has moreover shown that climate change impact analysis does not translate into the simple task of focusing on one single impact scenario (e.g. design storms to be increased by a single percentage change in magnitude). The need to cover a broad range of scenario changes is due to our current lack of knowledge regarding the exact nature and magnitude of future climate change impacts, although these impacts are likely to be significant. This is explained further in Chapters 8 and 9.

It is clear that there is a need to adapt the results of climate model simulations (at spatial scales of hundreds or tens of kilometres and at monthly, daily or hourly temporal scales) so that they can be meaningfully used at the more detailed scales (spatial scales of a few kilometres and temporal scales as low as 5 or 10 minutes) of urban drainage models. This can be done using statistical methods, which are also a potential way to tackle the uncertainties and biases in dynamically downscaled urban rainfall extremes, as explained in the next chapter.

Chapter 7

Statistical approach to downscaling of urban rainfall extremes

Previous chapter concluded that the output from GCMs and RCMs cannot be used directly to assess the influence on rainfall extremes relevant for urban drainage, and announced that the problem can be overcome by statistical downscaling techniques. These techniques aim to scale the outputs from climate models, both in space and time, down to the scale of urban hydrological impact modelling. The downscaling can even be done furtherdown to the point scale in order to provide comparable values with historical rainfall series; this is needed for the verification task discussed in Chapter 6.

Statistical downscaling commonly refers to both types of adjustments of GCM/RCM results. Section 7.1 briefly outlines some of the available techniques for statistical downscaling and includes a comparison with dynamical downscaling. Different types of methods are presented in the subsequent Sections 7.2, 7.3, 7.4 and 7.5. These provide literature overviews and discussions on the underlying assumptions and their applicability. Regardless of the method used, verification of the downscaling results is needed (Section 7.6).

7.1 MOTIVATION FOR STATISTICAL DOWNSCALING AS COMPARED TO DYNAMICAL DOWNSCALING

It should be noted from Chapter 5 that although the term dynamical downscaling is used to describe the application of RCMs or LAMs to produce rainfall at local scales, based on GCM output, what is done is in fact a full physical simulation of the atmospheric system within the model domain. The GCM output fields are used as lateral boundary conditions and are not limited to precipitation. For dynamical downscaling it is typically necessary to specify pressure, lateral velocity, temperature, moisture content and elevation (geopotential height) as lateral boundary conditions. Clouds and rainfall are generated within the model using the governing equations. While it is possible to specify hydrometeorological variables like cloud density at the lateral boundaries, these are optional and often unnecessary to get good rainfall simulation within the model domain. In this sense, dynamical downscaling is markedly different from statistical downscaling techniques in that it "generates" rainfall based on atmospheric physics (Figure 7.1). Note that both techniques should not be applied separately as the figure suggests, but can be combined, which is most often done.

A primary advantage of dynamic downscaling is the ability to produce a complete set of meteorological variables besides rainfall (e.g. cloud densities in a three-dimensional space, surface temperature on a two-dimensional surface). Since the process of generating rainfall in dynamical downscaling is based on the atmospheric physics and calculated by solving the governing equations, it is guaranteed that the output quantities are physically consistent with each other. For example, when heavy rainfall happens in a location the cloud formation is always consistent with such a storm activity, so that the expected reduction of shortwave radiation on the earth surface (due to cloud cover) is automatically ensured. While some elementary relationships between various meteorological variables can be embedded in statistical downscaling, it is impossible to achieve complete consistency among all variables.

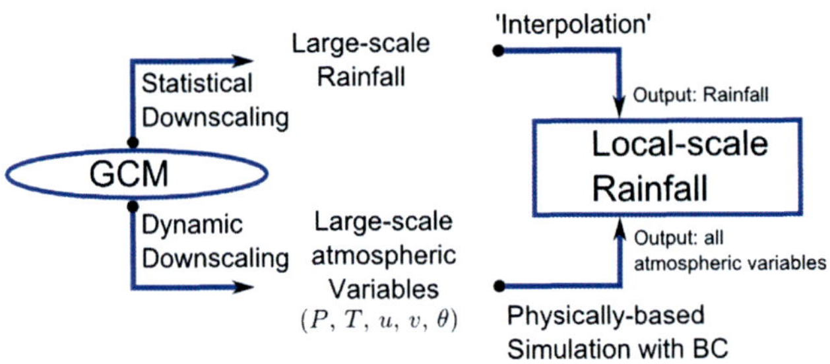

Figure 7.1 The two paths from GCM to local scale rainfall. Statistical downscaling primarily uses "large-scale" GCM rainfall projections and interpolates these values. Dynamical downscaling uses primary atmospheric variables predicted by GCMs and runs local-area models to physically simulate the atmosphere. The latter provides the output of all the atmospheric variables at the finer scale (*P*: pressure; *T*: temperature; (*u*,*v*): horizontal velocity components; θ: moisture content; BC: boundary conditions).

The dynamically downscaled rainfall is also consistent with local geographical features, for example, topographic variations (Pathirana *et al.* 2005) and landuse (Shepherd, 2010). This is an important advantage when generating rainfall products for local applications such as urban storm drainage modelling.

On the other hand, dynamically downscaled rainfall does not necessarily show consistency with the large-scale rainfall generated by the GCM that provided the boundary conditions. Suppose a GCM with 100 km × 100 km horizontal resolution was used to provide boundary conditions to a RCM with 1 km × 1 km resolution. Aggregation of rainfall in the 10000 grid points of the RCM corresponding to a grid point in the GCM would not result in rainfall volumes equivalent to those predicted by the GCM grid. However, it is possible to use GCM rainfall to calibrate and validate the performance of the RCM.

Perhaps the most severe constraint in using dynamical downscaling is the computational expense. Atmospheric models have three-dimensional grid systems. Therefore, the increase of resolution by a factor of 10 (say 10 km to 1 km) will increase the number of computational cells roughly by a factor of between 100 and 1000 (It is not always necessary to increase the number of vertical levels proportional to the increase of horizontal grids). Due to the nature of numerical implementation of typical RCMs, the computation time-step needs also to be reduced by the same factor to prevent numerical instabilities (say from 10 s to 1 s). Therefore the total computational effort will increase by a factor of 1000 to 10000 when the grid resolution is increased by a factor of 10. This poses considerable challenges in applying dynamical downscaling with the objective of producing a high resolution rainfall product to be used at urban scales. However, the last two decades have seen an explosive growth of computational power. Today, RCM simulations are routinely performed on consumer grade personal computers. Of course, as explained in Section 5.2, next to the increase in grid resolution and related numerical issues, physical parameterisations have to be changed as well. Below grid sizes of about 3 km, cumulus and convective parameterization are no longer needed and the processes can be modelled explicitly by the model physics, although some problems with non-stationary behaviour remain. Another problem is that the high resolution model is nested in a coarser driving model that involves different parameterization schemes. The RCMs moreover most currently do not include coupling between ocean and atmosphere (Wang *et al.* 2004). Questions also remain about the extent to which RCM biases in rainfall results are inherited from the driving model.

Climate model users most often apply dynamical downscaling up to the level of the highest resolution RCMs available, after which point further statistical downscaling may be required. This is not always a good approach. The previous chapter discussed how high resolution RCMs may fail to adequately describe the local surface processes over heterogeneous regions. In these cases, one could argue that a better approach would be to make use of the lower resolution climate model results and an extra statistical downscaling step (Dibike *et al.* 2008). Other authors have shown that in some cases the RCM based dynamical downscaling method slightly outperforms statistical methods in projecting daily rainfall extremes, as was the case in the test made by Vrac *et al.* (2007a). The question is whether, due to the parametric representation of the precipitation processes, simulation results from current RCMs should be extracted at very fine scales. Below which scale the climate models should not be used is, however, a debatable question (about 30 min or 1 hour and about 3 km?). It depends on the spatial scale at which cloud formation and convection processes can be simulated explicitly without the need to use physical parameterisations (see Section 5.2).

Less debatable is that below the scales of available RCM runs, further downscaling needs to be done using statistical methods. The basic principle of statistical downscaling is to use empirically-based relationships to convert the coarse-scale climate model outputs to rainfall data at the finer urban drainage scales. In this approach, the coarse-scale climate model based information is used as basis for the prediction. These are called *predictors*, while the fine-scale rainfall results are termed *predictands*.

Figure 7.2 shows that, while for other hydrological applications such as water supply, irrigation, river hydrology and rainwater harvesting, no temporal downscaling is required, urban hydrology should involve both spatial and temporal downscaling. The rainfall generating processes indeed occur over temporal scales ranging from multi-decadal to sub-minute resolutions with corresponding changes in spatial scales. For urban drainage one has to go down to the minute resolution.

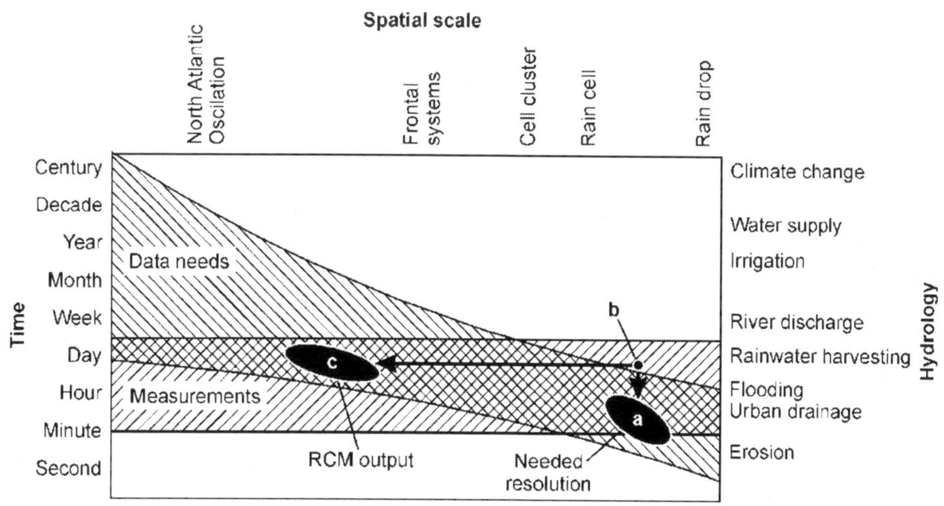

Figure 7.2 Statistical downscaling of RCM outputs down to the scale required for urban hydrological impact studies requires both temporal dowscaling (a – b) and spatial downscaling (b – c) (adapted from Arnbjerg-Nielsen, 2008).

The statistical model can only be based on historical data, thus assuming that the transfer from the predictors to the predictands will not significantly change under changing climatic conditions.

Statistical downscaling methods have the advantage of being computationally inexpensive, and are potentially able to produce finer spatial scales than dynamical methods that use RCMs. Different types of methods exist, each with very specific underlying assumptions. These are presented in the following sections.

Section 7.2 starts with the *delta change* method, being a fundamental statistical downscaling principle. Other statistical downscaling methods can broadly be classified into the following three types (Hewitson & Crane, 1996; Wilby *et al.* 1998; Nguyen *et al.* 2006; Fowler *et al.* 2007; Vrac & Naveau, 2007):

- Empirical transfer function based methods (Section 7.3);
- Re-sampling methods or weather typing (Section 7.4);
- Conditional probability-based or stochastic modelling methods (Section 7.5).

7.2 DELTA CHANGE AND CLIMATE FACTORS

At the interface between climate science and urban drainage, hence when transferring climate model outputs as predictors to inputs in urban drainage models as predictands (Figure 1.1), either the climate model rainfall results are used directly, or only the information regarding the rainfall changes. The rainfall changes can be represented in the form of *change factors*, which are also called *climate factors* or *perturbation factors*. The climate factor is calculated as the ratio of a statistical (extreme) rainfall intensity characteristic (e.g. mean, variance or a certain quantile) in the climate model scenario simulation to the statistical characteristic in the climate model control simulation. Climate factors for extreme rainfalls corresponding to a certain return period can be calculated by applying extreme value analysis to the control and future scenario results using the methods described in Section 2.4. Climate factors can also be calculated for the number of rain storm events, proportion of wet days, dry spell length, dry-wet transition probability, and so on. To account for seasonal differences in the changes, monthly or quarterly change factors are typically estimated and applied in the statistical downscaling.

The climate factor represents the percentage difference in the rainfall characteristic due to climate change (i.e. a climate factor of 1.2 represents 20% increase in the considered rainfall characteristic). In the urban drainage context, the climate factor is investigated for rainfall intensities at the characteristic time scales of urban drainage systems. These should correspond to the times of concentration (response times to rainfall) for different locations in the system. For urban drainage flood risk studies, these may be limited to the most critical locations for urban flooding and sewer surcharging along the urban drainage network. For most systems, the relevant times of concentration range between a few minutes and a few hours (Butler & Davies, 2010).

The future rainfall characteristics are then estimated by superimposing the changes on the observed statistics. This method is called the *delta change approach* or in short *delta approach* (Lettenmaier *et al.* 1999). Another name used is method based on *Model Output Statistics* (MOS) (Wilks, 1995). The delta factors can indeed be seen as MOS values. When the observed values are considered at the scale required for the urban drainage impact analysis or at the point scale, the method derives the changes at the scale of the climate model outputs and applies the same changes to the finer scale observations. Hence, when the delta approach is applied in downscaling, the assumption is made that the changes are scale independent.

An implicit assumption of the delta approach is that climate model bias in the considered rainfall statistic is unchanged in a future climate. Thus, the delta approach provides bias correction without explicit quantification of the bias. By using the differences between the current and future climate from the climate model, there is no need to apply the absolute values of the climate model results in the climate change impact study. This means that a model that has poor performance (i.e. a model with a high bias in certain rainfall statistics), still might be useful for quantifying changes caused by increases in anthropogenic forcing.

The climate factor may depend on the temporal scales (rainfall aggregation level), but the assumption is often made that this dependence is small or can be neglected for the range of scales involved in the interfacing between the climate model outputs and the urban drainage impact model (Figure 1.1).

In traditional delta change applications (e.g. Lettenmaier *et al.* 1999), only changes in the mean rainfall intensity are considered, for example, applying change factors per month. While this approach is suitable when considering impacts that are driven by changes in average rainfall conditions, it may be inappropriate when changes in variability or extremes are important. The proportion of wet and dry days and the temporal variability relative to the mean will remain the same as in the observed time series. Therefore, more advanced methods have been developed that superimpose changes on variability, wet-dry spell properties and other rainfall characteristics of the rainfall series.

Sunyer *et al.* (2012) applied a method that considers changes in both the mean and the variance of daily rainfall. The observed time series is corrected by assuming a certain functional relationship between future (P^{fut}) and observed (P^{obs}) rainfall (Leander & Buishand, 2007). Eq. (6) is used to estimate the time series for the future. It depends on two parameters a and b. The parameters are estimated using the monthly mean rainfall ($\bar{P}$) and coefficient of variation $CV(P)$ for the future scenario, respectively, $\overline{P^{fut}}$ and $CV(P^{\text{fut}})$, which are calculated from the change factors of the mean and the variance. Parameter b is first estimated, using a simplex optimization algorithm to solve Eq. (7). The value of a is afterwards estimated using Eq. (8).

$$P^{\text{fut}} = a(P^{\text{obs}})^b \tag{6}$$

$$\overline{P^{\text{fut}}} = a\,\overline{(P^{\text{obs}})^b} \tag{7}$$

$$CV(P^{\text{fut}}) = CV((P^{\text{obs}})^b) \tag{8}$$

The time series for the future using this method will have a monthly mean equal to the mean of the observed time series multiplied by the change factor of the mean and a monthly variance equal to the variance of the observed time series multiplied by the change factor of the variance. As in the case of the basic delta change method where only the mean is changed, the proportion of dry days will remain the same as in the observed time series.

Ntegeka *et al.* (2008) applied a time series perturbation approach where both perturbations in the number of events and in the probability distribution of rainfall intensities were applied at the daily time scale. For each month of the year, the number of wet days in the rainfall series was calculated and perturbed by removing or adding wet days. Given that the future climate for Belgium tends towards a smaller number of wet days, some wet days had to be removed. A removal operation was tested and compared with methods where isolated wet days that are situated in dry periods are converted into dry days. After the wet day frequencies have been changed, the rainfall intensities of these wet days were changed following a *quantile perturbation based method*. Rainfall intensity changes were applied depending on the cumulative probability or return period of the daily intensity. Different delta changes are thus applied to different days, depending on the probability of the rainfall intensity on that day (delta changes to quantiles). It is the same methods as described in Box 3.2 for testing anomalies and trends in rainfall quantiles, but now applied to climate model results for future and historical periods instead of historical periods and sub-periods. Similar approaches, particularly aimed at high-resolution tipping-bucket data, have been made by Olsson *et al.* (2009, 2012b). In the latter, an event-based analysis is used to analyse changes in the number of events. Based on the results, selected observed events are removed or duplicated to represent the expected future changes.

When calculating change factors for probabilities (e.g. probability of wet or dry days, transition probabilities, or autocorrelation), Burlando and Rosso (1991) proposed to apply the logit-function transformation ($x/(1-|x|)$) on the probability or correlation (x) before calculating the change factor, to ensure that the estimated values for the future period will remain in the interval $[-1, 1]$.

Removal/addition of wet/dry periods in a time series, or changes to the series of rainfall intensities to match statistical properties such as wet/dry day frequencies, wet/dry spell lengths, distribution variance, skewness, and so on can be performed using either deterministic or stochastic approaches. The stochastic approaches will be presented in Section 7.5.

7.3 EMPIRICAL TRANSFER FUNCTIONS

General empirical downscaling

Empirical transfer function based methods make use of empirical relationships or transfer functions between statistics of the predictands and the predictors. These are derived from historical (observed or estimated) values of both the predictands and the predictors, and afterwards applied to the climate model based future simulations of the predictors to generate small scale values. Generally an explicit function is used to describe the cross-scale relationship between the coarse and small scale statistics. The predictor variables should ideally be highly correlated to the predictand variables. The transfer function can take many forms such as any type of regression relation, equations based on rainfall time scaling laws, artificial neural network (ANN) models, GLMs, and so on. The predictor and predictand variables can be considered as time series, such that the value in each time step can be downscaled to obtain a time series of rainfall. This involves an approach to transfer the changes in the statistics to changes in the full time series. This is often done using a method that is similar to the delta change approach, but now the changes are obtained from the observed series and applied to the climate model results whereas in the delta change approach they are obtained from the climate model results and applied to the observed series. The generated series can then be used in urban drainage continuous simulation models for assessing impacts such as flooding and CSO overflows; see Chapter 9. Another approach is to pre-process the climate model output series and obtain statistics (e.g. empirical frequency distributions or estimated probability distributions at specific temporal and spatial scales), and to apply the transfer functions directly on these statistics or distributions to obtain the small-scale results. From the discussion in Chapter 2, it is clear that the latter approach is most useful in combination with an impact method based on design storms (e.g. impacts on sewer surcharges or floods). This is because such a method will obtain rainfall statistics or frequency distributions rather than full time series.

A typical example of a time series version is the popular regression-based statistical downscaling method (SDSM) proposed by Wilby *et al.* (2002). It allows to consider several predictor variables: 2 m daily mean temperature, near surface specific humidity, near surface relative humidity, mean sea level pressure, zonal component of geostrophic airflow, meridional component of geostrophic airflow, geostrophic airflow, vorticity, 500 hPa geopotential height, and also the predictand value from the previous day, and wether the day is wet or dry. Another example is provided by Dibike *et al.* (2008), who used multiple regression relations between daily precipitation predictand and climate predictors from the NCEP reanalysis data set for the periods 1961–1990 (calibration period) and 1991–2000 (validation period). They considered the commonly used predictors for downscaling of precipitation, which are (mean) sea level pressure, geopotential height, zonal wind velocity, and specific or relative humidity (at 500 and 850 hPa). Once the downscaling model was calibrated and validated using the NCEP predictors, the corresponding GCM predictors were used to downscale to daily precipitation.

Vrac *et al.* (2007b) used a non-linear and non-parametric technique considering same predictor variables (this time taken at the ground, or vertically integrated for specific humidity), and extended with temperature, dew point temperature, dew point temperature depression (representing the degree of saturation in water vapour in the atmosphere) and wind direction. Daily precipitation was considered as predictand. The dew point temperature and dew point temperature depression have shown good explanatory power for downscaling precipitation by Charles *et al.* (1999), Vrac *et al.* (2007a) and Vrac and Naveau (2007). Vrac *et al.* (2007b) concluded that these atmospheric variables provide more realistic downscaling results when they are combined with some geographical predictors. They suggested the following four geographical variables: elevation, diffusive continentality (which represents the shortest distance to the coast), advective continentality (which represents the degree at which incoming air mass paths travels over land versus over the ocean) and W-slope (which represents the degree of air mass uplifting, hence potentially cooling and precipitating, due to the presence of mountains).

Olsson *et al.* (2004) used two serially coupled ANN models to downscale 12-hour catchment precipitation from a gridded 20 × 20 km meteorological analysis, using as predictors wind speeds at 850 hPa and precipitable water. Kang and Ramirez (2009) used ANN modelling based on GCM results on precipitation, sea level pressure, temperature and surface upward latent heat flux to obtain daily rainfall predictand results.

Also Coulibaly *et al.* (2005) and Sharma *et al.* (2011) applied an ANN, more specifically a time-lagged feedforward neural network. This is an ANN that includes a memory structure in the inputs. The major assumption is that the local weather is not only conditioned by the present large-scale atmospheric state, but also by the past states (Coulibaly *et al.* 2005). They explained that ANNs are highly adaptable and are capable of modeling complex nonlinear processes. However, ANNs appear to have difficulty downscaling rainfall owing to their inability to reproduce some of the two key features of a high-resolution rainfall time series: intermittency and variability. Coulibaly *et al.* (2005) concluded that their ANN model tends to generate too many small intensity rainfall days and consequently underestimate dry spell lengths.

Olsson *et al.* (2012a) have shown that further advancements could be made by making the transfer function depending on RCM process variables characterizing the current weather situation such as cloud cover and precipitation type. They found that the RCM rainfall intensity results were lower than the rain gauge intensities, and that the underestimation was more severe for convective precipitation events than for stratiform events. They explained the differences by the typical spatial size for these two types of precipitation events. A process-based approach was developed in which 30-min values of different cloud cover variables were used to estimate the wet fraction corresponding to the different precipitation types. These fractions were used to convert the grid average rainfall into a local intensity, with a corresponding probability of occurrence in an arbitrary point inside the grid. It should be emphasized that RCM-simulated precipitation types and cloud cover are highly uncertain, as are therefore the estimated local intensity. Evaluation in Stockholm, Sweden, however showed a reasonable agreement with observations and theoretical considerations, which supports the approach (Olsson *et al.* 2012a).

It is important to note that while the rainfall variable can be used as a predictor in any of the above-mentioned downscaling methods, some modellers prefer to exclude the rainfall climate model output and instead they use other climatic variables for increased accuracy. Downscaled rainfall results then can be obtained in a next step based on another downscaling method.

Separation of downscaling and bias correction steps

The empirical downscaling approach presented above does not include bias correction. That is why these methods also are referred to as *Perfect Prognosis* (PP) methods. The empirical transfer function performs the downscaling, by applying it to the climate model results, assuming that the latter does not have a bias at the scale of the predictor. This is different from the delta approach, where the downscaling step and the bias correction step are combined.

Other methods separate these two steps. As a first step, empirical correction factors or functions are applied to obtain bias corrected climate model outputs (for rainfall or other variables). This is followed by a second step where the bias-corrected climate model outputs (rainfall and/or other variables) are transferred to fine scale rainfall (downscaling). As is the case for all downscaling methods, the transformation in the second step can account for differences in both temporal scale (e.g. from daily to sub-daily rainfall) and spatial scale (e.g. from grid averaged rainfall to point rainfall). The first bias correction step can obviously only be done in the case where the predictor and predictand variables are at the same temporal and spatial scales. This would require observations to be available at the same scales as the climate model predictor variable(s). In most cases, this does not pose a problem for the temporal scale. When daily climate model outputs are considered as predictors (i.e. rainfall), daily observations are required, which are available in most regions of the world. The main obstacle for separating the downscaling and the bias correction step, is the spatial scale. Accurate ground observations of rainfall so far can only be obtained by rain gauges (Sevruk, 1989; Groisman & Legates, 1994; WMO, 2008a), which are point measurements, while the climate model predictors are gridded. A separate bias correction then requires ground station observations to be interpolated in space, which in most practical cases can introduce quite significant interpolation errors (see Section 2.7). Further, data from sufficiently dense station networks are often not available, particularly at sub-daily time scales.

Instead of using rain gauges, radar data can also be used to support the spatial interpolations. Preferably, radar data are combined with rain gauge observations because of their higher accuracy in measuring rainfall intensities (O'Connel & Todini, 1996; Collier, 1996; Grimes *et al.* 1999; Harrison *et al.* 2000; Sokol, 2003; Einfalt *et al.* 2004). Recent advances in the measurement of rainfall at small urban scales are new radar technologies and processing techniques, which allow a reliable means of obtaining rainfall data with a spatial scale of 1 km^2 or less and a temporal resolution of 5 minutes or less (Einfalt *et al.* 2004; Michelson *et al.* 2005). Radar technologies that are currently extensively tested for measuring rainfall with high spatial resolution are the Local Area Weather Radars (LAWRs) or X-band polarimetric radars (e.g. Thorndahl & Rasmussen 2011). These can be seen as an alternative to C-band and S-band radar for local urban areas. Another promising technology under development is microwave technology in commercial wireless links (e.g. Leijnse *et al.* 2010).

It may be considered to use freely available, continental-scale data bases of spatial precipitation observations (e.g. E-OBS in Europe; Haylock *et al.* 2008) or meteorological reanalyses (e.g. ERA40; Uppala *et al.* 2005), but their accuracy differs between regions and for a particular location careful analysis is always required.

This *upscaling* of rainfall observations by areal rainfall interpolations can be undertaken for the full time series, or for rainfall statistics (e.g. rainfall distributions or quantiles). ARFs are commonly used in hydrology to downscale grid averaged rainfall to point rainfall, or to upscale from point rainfall to grid averaged rainfall (see Section 2.7 for more details). Due to the many difficulties in the upscaling of point rainfall, the bias correction and downscaling steps are most often combined.

Some modellers separate the temporal and spatial downscaling aspects. In the separation between the downscaling and bias correction steps, the temporal downscaling step can be easily separated from the other steps of spatial downscaling and bias correction. The latter two steps can be combined by comparing the climate model outputs with observations at the temporal scale of the climate model outputs, for example, comparing gridded rainfall outputs with point rainfall observations. The temporal downscaling step then remains.

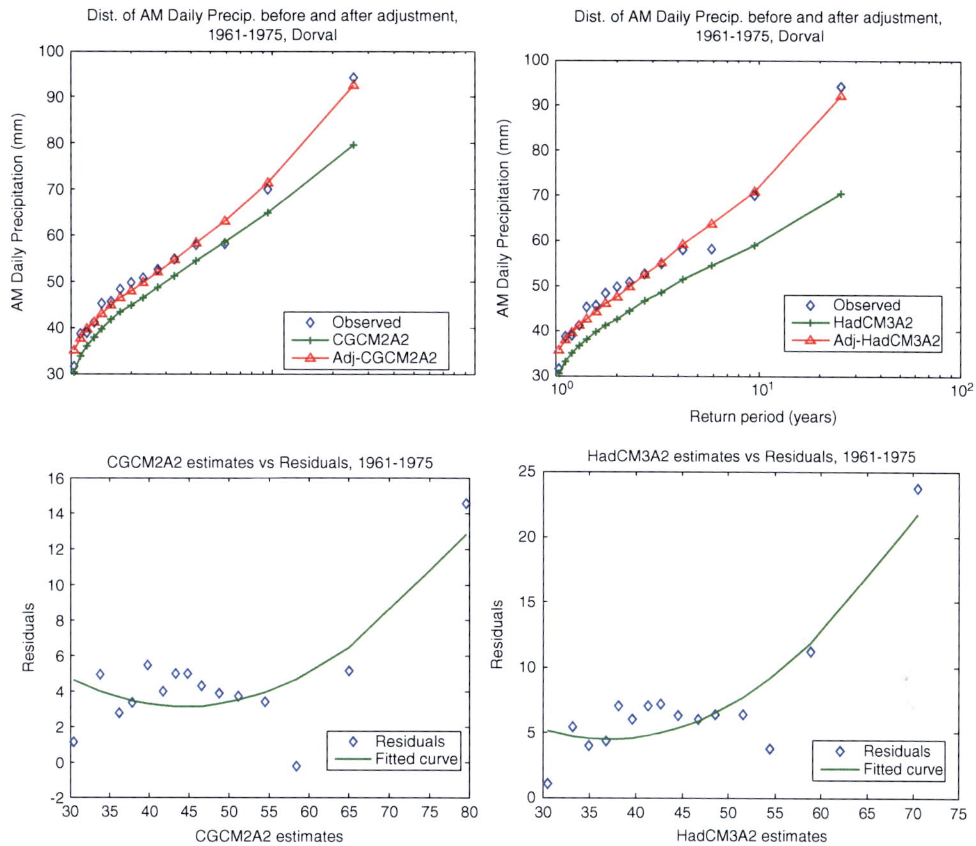

Figure 7.3 Daily rainfall quantile plots before and after bias correction and spatial downscaling based on a second order polynomial correction function (Dorval station, Quebec): (top) correction of CGCM2 and HadCM3 A2 climate model results for calibration period 1961–1975, (bottom) calibration of correction function for period 1961–1975 (after Nguyen *et al.* 2008b).

Temporal-spatial empirical downscaling was applied by Nguyen *et al.* (2008a, 2008b) to describe the linkage between large-scale climate variables as provided by GCM simulations with daily extreme precipitations at a local site. The SDSM approach of Wilby *et al.* (2002) was used together with a temporal downscaling procedure to describe the relationships between daily extreme precipitations with sub-daily extreme precipitations using the scaling General Extreme Value (GEV) distribution (Nguyen *et al.* 2002; see Section 2.5 and Figure 7.3). Nguyen *et al.* (2008a, 2008b) demonstrated the feasibility of this downscaling method using GCM climate simulation outputs, NCEP reanalysis data, and daily and sub-daily rainfall data available at a number of rain gauge stations in Quebec (Canada). Combined bias correction and spatial downscaling of the GCM-downscaled annual maximum daily rainfalls, based on a second-order polynomial function, was required to achieve a good agreement with the observed at-site daily values (see Figure 7.3). After obtaining the bias-corrected downscaled annual maximum daily rainfalls at a given site, further temporal downscaling to sub-daily maximum rainfall intensities was obtained by means of scaling the GEV distribution (Nguyen *et al.* 2002). Based on the concept of scale-invariance, where moments of the rainfall distribution (GEV in this case) are a function of the time scale, which has scaling properties (Nguyen *et al.* 2007), probability distributions of sub-daily rainfall intensities were accurately obtained from the distribution of daily rainfall intensities.

Quantile mapping

The bias correction step of Nguyen *et al.* (2002, 2008a, 2008b) is based on matching rainfall quantiles. This can be undertaken based on empirical probability distributions or after calibration of theoretical distribution functions to the data. Similar method can be applied to the downscaling step, to match large with small scales quantiles. The method is often referred to as the *quantile mapping* approach (Box 7.1). When applied to the SOM approach, it can be viewed as the delta approach applied to quantiles (see also Section 7.2). When the focus in urban drainage is on rainfall extremes, the correction could be limited to quantiles above a given threshold.

Box 7.1 Bias correction of RCM rainfall by quantile-quantile mapping based on the gamma distribution

It is well known that RCM rainfall data are generally biased. A common situation is to have an overestimated frequency of wet periods and an inaccurate frequency distribution of non-zero intensities (e.g. underestimated extremes), as compared with observations. These biases may strongly affect the results from impact models, not least the ones focused on hydrological consequences. Therefore different methods for bias correction have been developed and applied.

One common approach is quantile mapping, which uses cumulative distribution functions (CDFs) for observed and simulated rainfall to remove biases. Essentially, this approach replaces the simulated rainfall value with the observed value that has the same non-exceedance probability. Both empirical and fitted theoretical CDFs have been used. In the latter case, for daily (and possibly also sub-daily) rainfall the gamma distribution often provides a good fit. The gamma distribution has the probability density function:

$$f(x, \alpha, \beta) = \frac{1}{\Gamma(\alpha)\beta^{\alpha}} x^{\alpha-1} e^{\frac{x}{\beta}}$$

where α is a shape parameter, β is a scale parameter and $\Gamma(\alpha)$ is the gamma function. The CDF of the gamma distribution has the form:

$$F(x, \alpha, \beta) = \frac{1}{\Gamma(\alpha)} \gamma\left(\alpha, \frac{x}{\beta}\right)$$

where γ is the lower incomplete gamma function defined as:

$$\gamma(s, x) = \int_0^x t^{s-1} e^{-t} dt$$

By construction, the gamma CDF has an inverse which we denote $F^{-1}(x, \alpha, \beta)$. The ML estimate of the shape parameter α has a closed form formula while β has not. However, both equations for the ML estimators are numerically stable, so methods such as Newton's method can be applied to find these parameter values.

Let X_{OBS} be a time series of observed non-zero rainfall in a reference period and X_{SIM}^{ORG} the corresponding original (raw) data from an RCM simulation in the same period. Fitting gamma distributions to the time series by the ML method yields parameters α_{OBS}, β_{OBS} and α_{SIM}, β_{SIM}. Corrected RCM precipitation X_{SIM}^{COR} is then generated by the transformation:

$$x_{SIM}^{COR} = F^{-1}[F(x_{SIM}^{ORG}, \alpha_{SIM}, \beta_{SIM}), \alpha_{OBS}, \beta_{OBS}]$$

Assuming that the bias is independent of the period considered, the transformation may be applied to adjust future scenarios.

Some common issues associated with quantile mapping of daily (and sub-daily) rainfall are worth emphasising. A common RCM bias is an overestimated frequency of wet days with very low intensities. This may erroneously bias the distribution fitting towards low intensities and a common approach is therefore to use a cut-off threshold that adjusts the simulated frequency of wet days to agree with the observations. Further, it may be considered to use other distributions than the gamma distribution and also to fit different theoretical distributions to different parts of the empirical distribution. For example, extreme value distributions (e.g. Gumbel) may be suitable for values above some high precentile (e.g. 95%).

Examples of recent applications of the type of quantile mapping outlined here include Piani *et al.* (2010), Themeßl *et al.* (2010) and Yang *et al.* (2010).

Yang *et al.* (2010) applied a distribution-based quantile mapping approach for adjusting daily RCM outputs prior to hydrological climate effect simulations. In their approach the RCM outputs were compared with gridded 4 × 4 km fields of interpolated precipitation observations in a historical control period (typically 1961–1990). For precipitation, a cut-off was used to remove spurious low precipitation amounts and a double gamma distribution was used to rescale the remaining intensities (one distribution for "normal" intensities and one for extreme intensities, above the 95% quantile). In that approach, bias correction and spatial downscaling were combined.

Rosenberg *et al.* (2010) applied the quantile mapping approach for bias correction of hourly RCM series. The quantile mapping was, however, applied on the monthly values. The hourly values within the month were rescaled correspondingly, but after the RCM results were truncated so that each month had the same number of nonzero hourly values as the observed series.

7.4 RE-SAMPLING METHODS OR WEATHER TYPING

In the re-sampling or weather typing methods the downscaling is still performed based on relations between the coarse scale climate model based predictors and the fine scale rainfall predictand, but they are not put in the form of a regression relation or transfer function. The time series results are used such that for each future event (e.g. day) in the climate model output, a similar situation is sought in the historical fine scale rainfall series. The fine scale rainfall observation for that event is copied as a future rainfall event (Zorita & von-Storch, 1998; Benestad *et al.* 2008). Predictions for the future are thus obtained by copying most similar events (analogue events) from the past. Pressure fields from climate models are most often used as predictors. Using a classification scheme, different weather types can be identified based on the pressure fields and rainfall observations taken for days with the same weather type. Weather types are selected for which relationships exist with the fine scale rainfall predictand. The weather typing approach therefore is based on synoptic similarity. Bárdossy and Caspary (1990) and Benestad *et al.* (2008)

describe how to find robust weather types. A disadvantage of the technique is that no rainfall intensities that are more extreme than the most extreme event in the past will be considered; only the sequence and frequency of the events will be changed.

Several procedures exist for classifying weather types. They are based on either professional knowledge (subjective classification) or statistical characteristics derived from the observations (objective classification). They aim to differentiate long-term historical observations into several representative climate patterns or weather types (WTs) to describe certain climate conditions in the study area. In contrast to other downscaling approaches, multi-grid points rather than a single grid point are used to capture the local climate circulations over a large area.

The subjective classifications are conducted with long-term collecting data and experts' knowledge of the local region. They are normally based on air pressure distribution over a certain window. One well-known subjective classification is the one by Hess and Brezowsky (1969), used by the German Weather Service. The Hess and Brezowsky WTs are classified using air pressure distribution over Western Europe and the North Atlantic sector. Atmospheric circulations are differentiated according to the movement of frontal zones, the location of high and low pressure centres, and the cyclonic and the anti-cyclonic rotations. Hess and Brezowsky (1969) finally generated 29 WTs and one unclassified WT. The other important subjective WTs include the Schüpp's WTs for Switzerland (Schüpp, 1968) and the Lamb WTs (Lamb, 1972; Jones *et al.* 1993) for the British Isles. Generating subjective classification requires long observation series and a good understanding of local climatic phenomena, which is only feasible in a few countries.

The objective classification is a semi-automated or automated technique that pertains to mathematical approaches, for example, hierarchical methods (Johnson, 1967), k-means methods (MacQueen, 1967; D'onofrio *et al.* 2010), cluster analysis (Kyselý & Huth, 2006), singular value decomposition (Gelati *et al.* 2010) and correlation methods (Yarnal, 1984). Jenkinson and Collison (1977) developed an automated classification technique, considering the 28 WTs of Lamb (1972), which were originally determined by subjective classification. The Jenkinson-Collison classification method is automated based on sea level pressure values at 16 locations in the North-West Atlantic region centred around the studied location. From the 16 pressure values, pressure gradients and vorticity indices are computed and the WTs determined (Jones *et al.* 1993; Demuzere *et al.* 2009). Comparison between the subjective and objective Lamb WT classifications can be found in Jones *et al.* (1993). Automated classification methods have become more and more important in the field of statistical downscaling due to their flexible application for individual sites where no subjective classification exists.

For use in statistical downscaling, the WTs are most often defined for daily conditions. The daily WT sequence is then formulated as a given condition for obtaining downscaled rainfall intensities. How the conditioning is done, what WTs are used or which atmospheric variables are considered for defining the WTs, differs strongly from one method or study to the other.

Yiou *et al.* (2008) determined dependencies between the parameters of precipitation extreme value distributions for France and weather patterns obtained from pressure data over the North Atlantic. The dependencies were found to be different for mean and extreme precipitation. A clustering algorithm applied to geopotential height data over the North Atlantic was used to obtain the weather patterns. The most extreme precipitation storms in summer coincided with the blocking weather regime, which tended to favour dry summers and was characterized by an anticyclonic atmospheric circulation pattern. This was explained by the fact that intense convective episodes, bringing flash precipitation, often occurred at the end of heatwaves, which bear dry conditions over Europe. Vrac and Naveau (2007) made use of precipitation pattern types (based on the application of cluster techniques to spatial rainfall data) rather than weather types. They related the temporal changes in these types of precipitation patterns to large scale atmospheric variables (using a Markov chain model), and related parameters of precipitation probability distributions to the precipitation pattern types. Based on daily rainfall data from Illinois, USA, they found that such use of precipitation patterns is more efficient than the use of weather types based on large-scale atmospheric circulation patterns.

Gelati *et al.* (2010) obtained WTs for downscaling of daily precipitation in Denmark and southern Sweden using singular value decomposition based on geopotential height and relative humidity at different pressure levels.

Willems and Vrac (2011) compared several types of weather typing based downscaling methods with the direct use of precipitation results from climate models. This was undertaken specifically for urban drainage applications, thus for downscaling down to 10-minute rainfall intensities. They found that the classical weather typing based method, in which rainfall changes are being solely explained by changes in atmospheric circulation, underestimates changes in short-duration rainfall extremes in comparison with other statistical downscaling methods. They have further advanced the weather typing methods accounting for the fact that precipitation change not only depends on change in atmospheric circulation, but also on temperature rise. For the different weather types, rainfall probability distributions were changed as a function of the change in temperature distribution. This was undertaken based on the concept that the intensity of heavy rainfall events may increase with temperature at a rate proportional to the Clausius-Clapeyron (C-C) relationship. The C-C relationship implies an increase in the atmospheric moisture holding capacity of approximately 7% per deg. C of warming. The C-C dependence is especially strong for extreme rainfall events, which can substantially deplete the atmospheric moisture column (Trenberth *et al.* 2003; Lenderink & Van Meijgaard, 2008; Haerter & Berg, 2009; Berg *et al.* 2009; Lenderink *et al.* 2011). Rainfall events are indeed fed by atmospheric moisture transport in convergence areas surrounding synoptic and convective storm systems (Trenberth *et al.* 2003).

Figure 7.4 shows some results by Willems and Vrac (2011) of the change in precipitation quantiles (scenario versus control period) for time scales between 1 and 15 days for the direct precipitation results of the climate model (mean of 17 ECHAM5 A1B runs), and between 10 minutes and 15 days after statistical downscaling of these ECHAM5 runs. When the advanced weather typing based method is compared with a quantile perturbation based downscaling method (see Section 7.2), where changes in sub-daily rainfall intensities are assumed identical to the changes in the daily intensity quantiles (control versus scenario period), similar results are obtained.

Lall *et al.* (1996) and Rajagopalan and Lall (1999) applied resampling using the K-nearest neighbour (K-NN) resampling approach, proposed by Young (1994). The K-nearest neighbours were selected in terms of a weighted Euclidean distance between a number of circulation properties. Higher weights were given to the closer neighbours. To incorporate autocorrelation, resampling was done dependent on the simulated values for the previous day. Also Buishand and Brandsma (2001) used K-NN resampling, but extended the approach for rainfall generation at multiple sites, accounting for the interstation correlations. Sharif *et al.* (2007) applied this approach to generate daily rainfall, followed by a cascade model type disaggregation step to obtain hourly rainfall series. Only data for events with a substantial amount of rainfall were disaggregated because of the focus of their study on extremes. To overcome the major limitation or resampling methods that they cannot produce more extreme values than the historical ones, Sharif and Burn (2007) added a random component to the individual resampled values.

Instead of using climate analogues from the past, climate analogues from other locations can also be considered. In such approach, locations are selected for which it is expected, based on the results of climate models, that future climate conditions will be similar. The method is also called *climate matching*. This method has been tested by Arnbjerg-Nielsen (2012), assuming the future climate in Denmark in 2100 might become similar to the present climate in Northern France and Germany. Similar changes in precipitation quantiles were obtained using the precipitation results from a high resolution RCM.

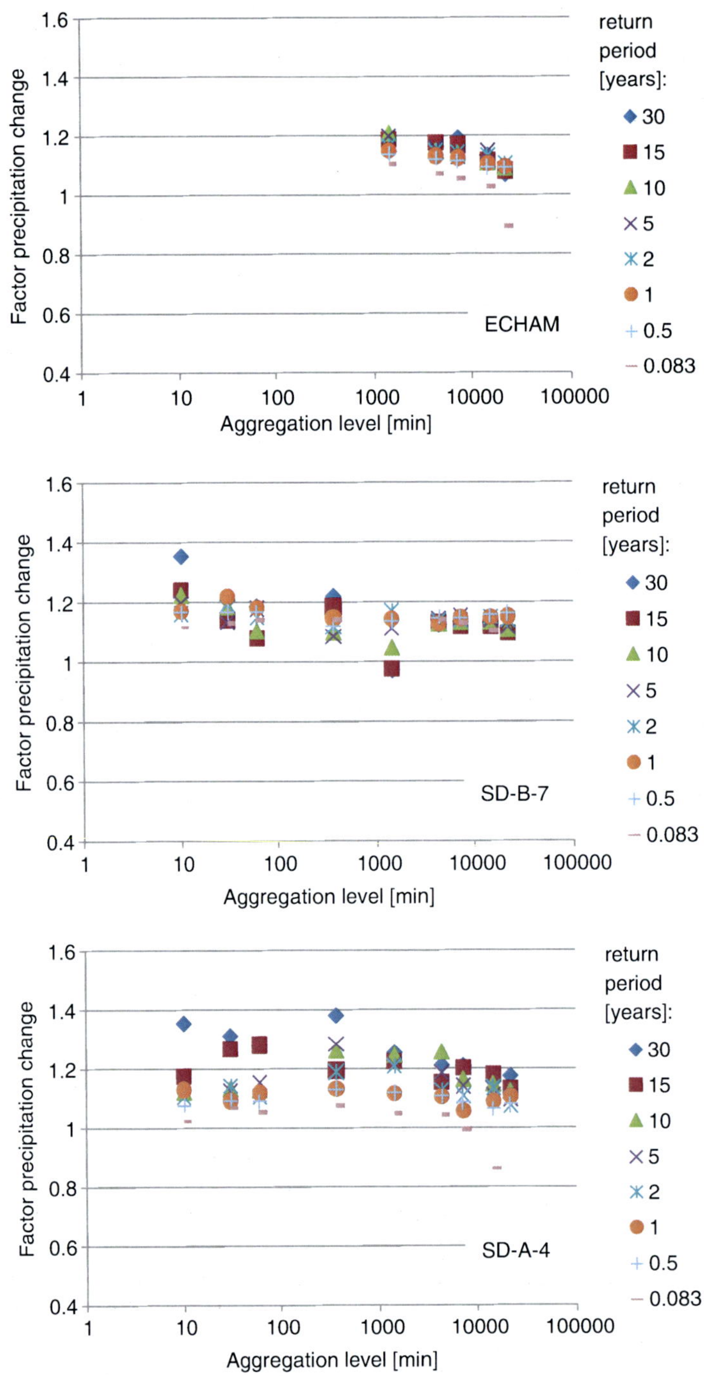

Figure 7.4 Climate factor versus time scale (from 10 minutes to 15 days) for various return periods (based on mean of 17 ECHAM5 A1B runs, 2071–2100 vs. 1961–1990): direct climate model results (ECHAM); weather typing based downscaling method (SD-B-7); quantile perturbation based method (SD-A-4) (adapted from Willems & Vrac, 2011).

7.5 CONDITIONAL PROBABILITY-BASED OR STOCHASTIC MODELLING

Conditional probability-based or stochastic modelling methods use stochastic rainfall models with parameters conditioned on the coarse scale climatic predictors. A stochastic rainfall model is a model for rainfall that takes into consideration the presence of some randomness in one or more of its parameters or variables (rain storm intensity, size, velocity, direction, etc.). The stochastic rainfall generators discussed in Section 2.1 are commonly applied. As discussed in that section, the stochastic models can be pure rainfall generators (Schreider *et al.* 2000) or can be conditioned on global climatic variables such as atmospheric circulation or weather types (Wilks & Wilby, 1999; Katz *et al.* 2003; Fowler *et al.* 2000, 2005; Burton *et al.* 2008). In the generator by Fowler *et al.* (2000, 2005), the NSRP model parameters can be conditioned, for instance, on the Lamb weather types.

Generators based on stochastic modelling are also called *weather generators*, given that they can be extended to describe other weather variables next to rainfall. The parameters of the stochastic model have to be obtained from statistical analysis of time series (observed or climate model based), and can be altered in accordance with future climate model simulation results. For the Markov chain based approaches, for instance, the parameters are the transition probabilities and the wet time step intensity distribution. For the stochastic pulses generators (e.g. NSRP or BLRP), it are the distributions of the rain storm properties as defined in Section 2.1. However, often the distributions are not directly changed but the statistics to which the parameters of the generator are traditionally calibrated by optimization (see Section 2.1). For the LARS weather generator (LARS-WG), the distributions of wet and dry spell lengths and precipitation amount, which are defined as histograms, are changed. This is done by multiplying the intervals of the histogram from the observed period using change factors derived from the climate model results. Other approaches consider conditioning of the stochastic model parameters based on large scale climate predictors or weather types, such that the parameter changes are obtained through the changes in the predictors or weather types. Chun *et al.* (2009) made use of GLMs to describe the wet day occurrence using a Markov based approach and the rainfall amount using a gamma distribution, based on sea level pressure, temperature and relative humidity as predictors. They made use of the GLIMCLIM software by Chandler (2011) (see Section 2.1). Gelati *et al.* (2010) applied a non-homogeneous hidden Markov model to downscale atmospheric patterns to daily precipitation in Denmark and southern Sweden, conditioned on WTs.

Changes in the stochastic model parameters and distributions or statistics can, directly or indirectly through changes in the predictors or weather types, be derived from comparing the climate model control and scenario runs. Sunyer & Madsen (2009), for instance, calculated change factors for the transition probabilities (dry–wet and wet–wet) in the Markov chain model by Brissette *et al.* (2007) they applied and for the statistics used to define the wet day rainfall intensity distribution, in their case limited to the mean. The change factors were used to estimate the statistics for the future which were then used as input in the weather generator. They also tested the NSRP generator by Cowpertwait *et al.* (1996) and Burton *et al.* (2008). For this generator, they applied change factors to the following statistics that were used to calibrate the parameters of the generator by means of optimization: mean, variance and skewness of daily rainfall amounts, dry day probability, and lag-one autocorrelation. Next to the Markov chain and NSRP generators, Sunyer and Madsen (2009) also tested the LARS-WG generator. The histogram distributions in LARS-WG were changed but only for the mean wet and dry spell lengths and the change in the mean rainfall amount. The frequency in each interval was not modified.

Sunyer and Madsen (2009) compared the three stochastic rainfall generators (Markov chain model, LARS-WG and NSRP model) for downscaling of rainfall. It was found that all three models represented well the increase in the number of extreme events, but only the NSRP model reflected well the change in extreme rainfall amounts. Sunyer *et al.* (2012) applied the three rainfall generators and two delta change methods (see Section 7.2) based on, respectively, correction of the mean and a mean and variance correction procedure for downscaling rainfall of four different RCM/GCM projections from the ENSEMBLES data base. The estimated quantiles of the 20-ensemble (four RCMs times five downscaling methods) projected daily extreme rainfall by 2100 are compared with the current extreme value statistics in Figure 7.5. The applied downscaling methods resulted in very different extreme rainfall estimates. The NSRP rainfall generator provided, in general, the largest increase in quantile estimates. The NSRP was the only downscaling method of the five applied methods that explicitly accounted for changes in the skewness, which is an important statistic for the simulation of extreme events.

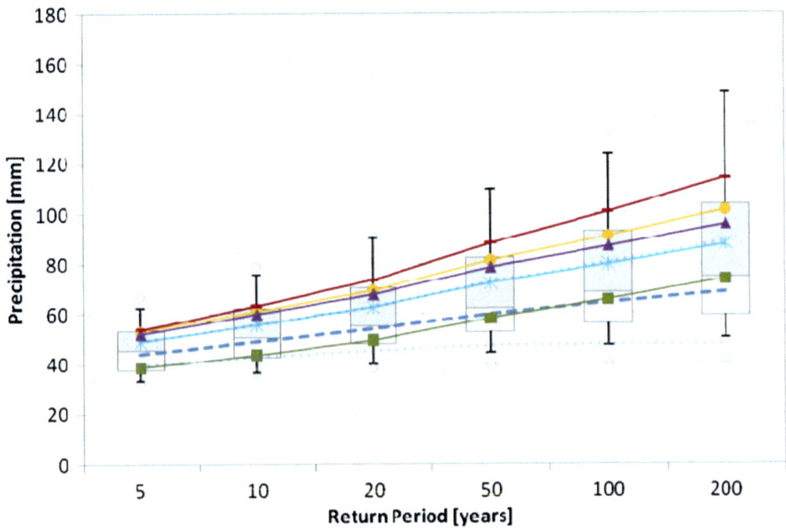

Figure 7.5 Estimated quantiles of POT rainfall extremes for the observed (- - - -) time series 1979–2007 and for all generated time series for the future period 2071–2100 (box plot). Results of the five statistical downscaling methods (change in mean (✳), change in mean and variance (●), Markov chain (■), LARS (▲) and NSRP (–) applied to the HIRHAM-ECHAM5 A1B results. The 95% confidence intervals for the observed extremes are shown as dotted lines (·····) (after Sunyer *et al.* 2012).

Onof and Arnbjerg-Nielsen (2009) made use of the BLRP generator, to generate rainfall at the hourly time scale. They followed a two-step approach where the rainfall generator captured the storm structure at the hourly time scale, and where in a second step a disaggregator was used (based on multi-scaling) to bring hourly rainfalls down to 5-minute scales. The BLRP model was also applied by Segond *et al.* (2007) to downscale daily rainfall, obtained with a GLM, to hourly rainfall for an urban catchment in the UK. The hourly rainfall patterns were then interpolated (between rain gauges) linearly in space using inverse distance weighting to obtain hourly rainfall estimates over the whole catchment. Such applications are often based on delta change methods, that is identification of properties/variables that are assumed to be scale invariant from the regional climate model scale to the urban catchment scale.

Another option for this type of statistical downscaling is to directly use the coarse scale rainfall estimates from the climate models as predictors of fine scale rainfall. The statistical downscaling techniques then are similar to the ones that are commonly used in statistical hydrology for temporal and spatial disaggregation of rainfall; see Section 2.3 for an overview of these methods.

Recently, stochastic downscaling methods based on random cascade models (Section 2.2) have been developed and applied in the context of climate modelling. Sharma *et al.* (2007) downscaled bias-corrected GCM rainfall and applied it in hydrological impact modelling. Kang and Ramirez (2010) developed a downscaling model based on random cascades intended for application to GCM-simulated rainfall. In the model, a reference scale is identified above which the rainfall fluctuations are assumed scale-dependent and below which they are assumed self-similar.

7.6 VERIFICATION OF STATISTICALLY DOWNSCALED CLIMATE MODEL RESULTS

Regardless of the downscaling method employed, verification of the downscaled climate model results under the present climate is always needed. This can be achieved in a similar manner to the verification of the climate model outputs (based on hypothesis testing or confidence intervals); see Chapter 6. However, in this case the verification is undertaken at the scales relevant for the urban drainage impact modelling rather than at the scale of the climate model outputs. The most common method applied in the urban drainage field is the comparison of quantiles or IDF relationships (e.g. Verworn *et al.* 2008; Olsson *et al.* 2009; Willems & Vrac, 2011) that combine both frequency and intensity of the rainfall events (see Section 2.5). Given that urban drainage design and planning is largely based on probabilities of reaching certain severities, it is important that rainfall exceedance probabilities and related probabilistic results are captured well by the downscaled climate model results.

Willems and Vrac (2010) compared historical IDF relationships with statistically downscaled RCM and GCM outputs for central Belgium, see Figure 7.6. The comparison covers a range of durations from 1 to 15 days for 17 GCM runs with the ECHAM5 model forced with the A1B emission scenario. The historical IDF relationships are based on 10-minute observations of rainfall intensities for the period 1967–1993 (Willems, 2000). Figure 6.3 has shown that without statistical downscaling significant differences were observed: the GCM results displayed lower rainfall intensities than the historical observations, with the greatest differences occurring at the daily time scale. Using quantile perturbation based statistical downscaling, the climate model results were downscaled from daily grid-averaged precipitation to 10-minute point precipitation (Figure 7.6). After downscaling, the bias is greatly reduced.

Dibike *et al.* (2008) compared their downscaled and observed daily rainfall intensity distributions by means of box plots of the distributions, and plots of observed versus downscaled quantiles; Figure 7.7 and Figure 7.8. They also compared the mean and standard deviation of these distributions and tested the deviation in the mean based on the Wilcoxon rank-sum test, and the 90% bootstrapping based confidence intervals.

Figure 7.9 shows the validation results of the downscaling method by Nguyen *et al.* (2008b) by means of quantile plots. These validation results show similar accuracy as the calibration results shown before in Figure 7.3.

Rosenberg *et al.* (2010) tested their bias correction method based on quantile mapping for two RCM runs and found for the State of Washington that the bias in the average 1-hour annual maxima for 1970–2000 reduced from −33.2/−19.2% to −7.3%/−13.4%.

An interesting question addressed by Sharma *et al.* (2011) is whether statistical downscaling is beneficial in comparison with dynamical downscaling, hence whether raw RCM data should be further downscaled before use in impact models. They tested the SDSM approach and the ANN model of Coulibaly*et al.* (2005) for one RCM run and the daily rainfall results at Chute-des-Passes and Chute-du-Diable in

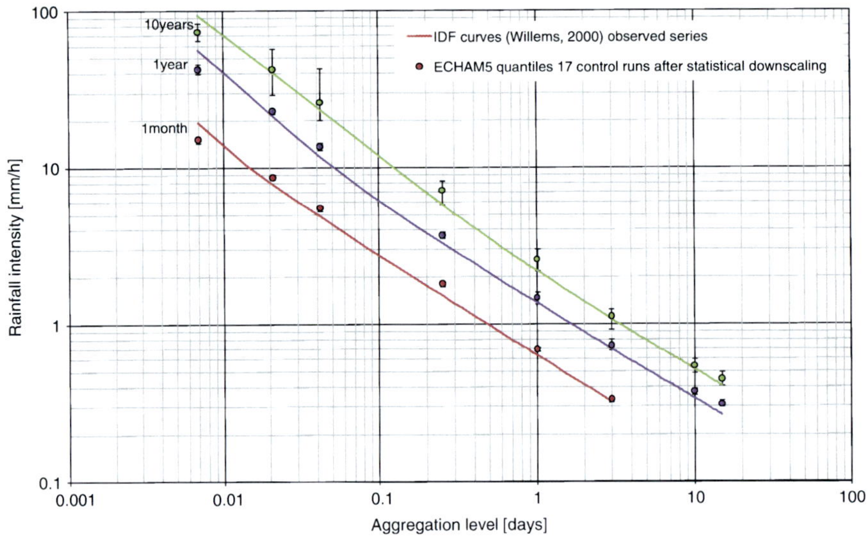

Figure 7.6 Comparison of historical IDF relationships for different return periods (1 month, 1 year and 10 years) at Uccle, Belgium, with the results of a quantile perturbation downscaling method based on 17 ECHAM5 A1B runs (the mean, highest and lowest of the 17 runs are shown for the control period 1961–1990) (adapted from Willems & Vrac, 2011; also in Willems *et al.* 2011).

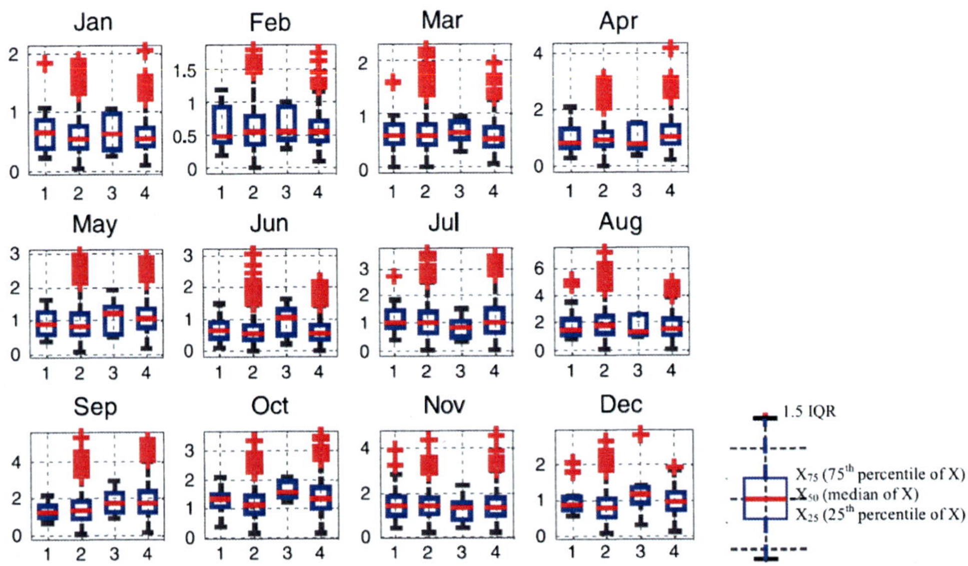

Figure 7.7 Box-plots of daily rainfall values of 1: observed series calibration period 1961–1990, 2: downscaling results calibration period, 3: observed series validation period 1991–2000, 4: downscaling results validation; statistical downscaling by SDSM using NCEP reanalysis data predictors, for Cape Dorset station, northern Canada (after Dibike *et al.* 2008).

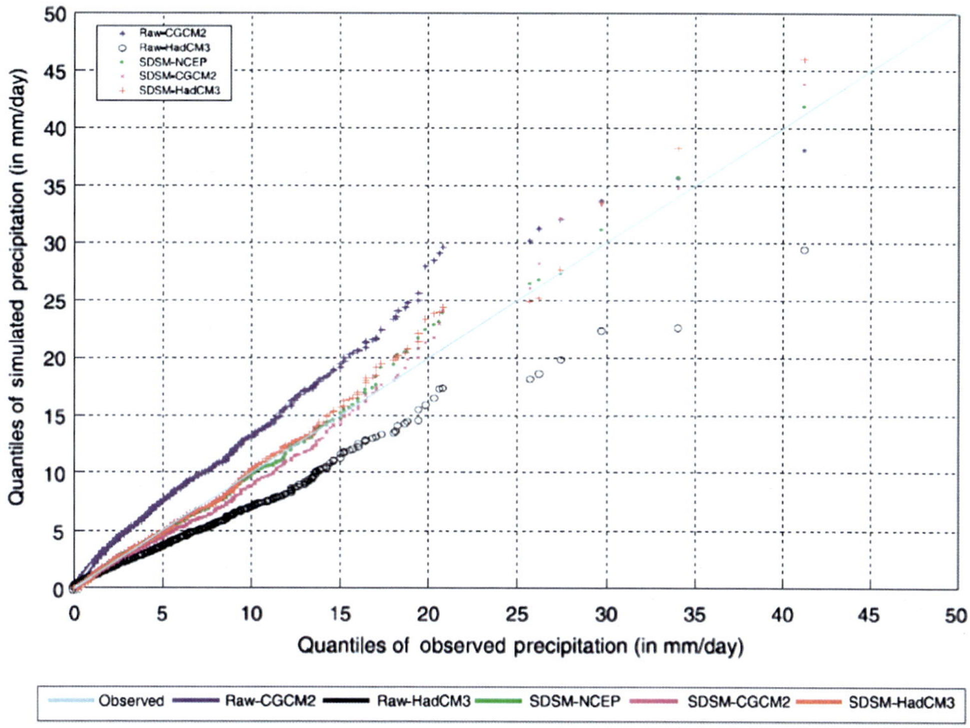

Figure 7.8 Empirical quantile-quantile plots for the observed and GCM-based daily rainfall of the CGCM2 and HadCM3 A2 results before and after statistical downscaling based on the SDSM model for 1961–1990 (Cape Dorset station, northern Canada), for validation period 1961–1990 (after Dibike *et al.* 2008).

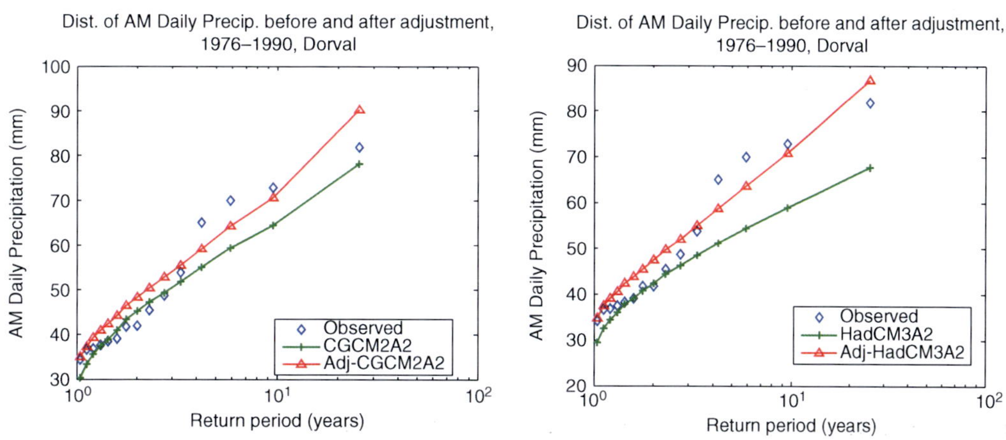

Figure 7.9 Comparison in daily rainfall quantile plots of the CGCM2 and HadCM3 A2 results before and after bias correction and spatial downscaling based on a second order polynomial correction function calibrated for 1961–1990 (Dorval station, Quebec), for validation period 1976–1990 (after Nguyen *et al.* 2008b).

northeastern Canada. The data for 1961–1980 were used for calibration of the downscaling models and the remaining period 1981–1990 for validation. SDSM-based downscaled RCM data were found to have lower root-mean square error (8.3 versus 15.8 and 7.8 versus 13.6 for the two stations), mean absolute error and mean relative error values for daily rainfall than raw RCM data. The ANN-based downscaled data also had lower root-mean square error, but the mean absolute and relative error values were higher because of the inability of the ANN to properly account for days with no precipitation (see also Section 7.3).

7.7 DISCUSSION

The statistical downscaling methods discussed in this chapter attempt to overcome the problem of scale and high biases in rainfall characteristics at small scales in two different ways: by direct bias correction and downscaling, or by estimating rainfall intensities using other climatic outputs (e.g. atmospheric or sea level pressure, temperature, wind speed, humidity) as predictors. At the same time, they should not be seen as the perfect solution. As Hundecha and Bárdossy (2008) concluded, for the same reason as why small-scale rainfall intensities are not well described by climate models, the statistical downscaling are less reliable during periods when the local climate is determined by local convective processes than during seasons when the local climate is determined by large-scale circulation.

A key aspect of quantile mapping is the access to good observations. These data should cover a long period, a period of 30 years is often used as historical reference period (see Section 4.3). The observations should have the same temporal resolution and preferably also the same spatial resolution as the RCM data. Generating accurate spatial precipitation observations is however not trivial. Interpolation from a sufficiently dense gauge network, taking into account meteorological and geographical conditions, potentially supported by radar observations, conceivably provides the most accurate estimates. Also point observations are sometimes used in quantile mapping, which then becomes a mixture of bias correction and downscaling from RCM grid scale to local scale. This may work well for temporal resolutions down to ~1 day, where the statistical properties of precipitation at the RCM grid scale and the local scale, respectively, are generally rather similar. For gradually higher temporal resolutions the statistical difference between spatially averaged and local precipitation may rapidly increase, reflecting different dominant rainfall processes, which complicates bias correction and downscaling.

Regardless of the statistical downscaling method employed, it has to be assumed that the relations between local scale precipitation and larger scale climatic variables remain valid under future climate conditions. These assumptions so far cannot be verified or supported by reference to any climate modelling. Limited support can be found by verifying that the relationships are applicable for a wide range of rain storm conditions and climatic regimes. Some researchers try to tackle this issue by dividing the total calibration period into sub-periods representing for example wet or warm years, expected to represent future climate. They then calibrate on the remaining years and verify the performance on future years. Using this approach, Timbal (2004) and Timbal and Jones (2008) succeeded to reproduce the observed rainfall decline in the late 1960s in the southwest of Australia and in the mid-1990s in the southeast of Australia.

As pointed out previously, good agreement of the statistically downscaled results with observations for the past and current climate, does not, however, guarantee that the present-day downscaling relationships will continue to be valid under future climate change. There is a way to partly test the latter assumption by checking whether the statistical downscaling methods produce future projections that are consistent with the RCM projections (driven by the same GCM simulations) (Vrac *et al.* 2007a). This validation is, however, incomplete because it ignores the shortcomings/biases of RCMs. The method incompletely assumes that these biases are already entirely generated by the GCM, such that it makes sense to compare dynamical and statistical downscaling results for the same GCM run. Another approach is to compare the statistical downscaling results for the most recent decades with observations, after the downscaling relationships have been calibrated to older records (that were less affected by anthropogenic climate change). This was done by Dibike *et al.* (2008) considering the period 1961–1990 for calibration of the SDSM downscaling method and the period 1991–2000 for validation.

Given the limited possibility to validate the statistical downscaling assumptions, good practice would involve assessment of the uncertainties associated with the downscaling step. Given that future climate conditions are highly uncertain, it is clear that such quantification is very difficult (which probably also explains why few past studies have attempted to quantify downscaling uncertainty). Instead of quantifying statistical uncertainties, it would, however, be possible to deal with scenario uncertainties. In the same way as we can use an ensemble modelling approach using several climate model runs (several climate models, climate forcing scenarios and initial conditions; see Chapter 6), it would be good practice to apply several downscaling approaches (as e.g. considered in Sunyer *et al.* 2012). This would require an ensemble of statistical downscaling techniques and scaling assumptions to be considered.

The dynamical and statistical downscaling methods discussed in Chapter 5 and in this chapter can be applied separately or combined. Downscaling for hydrological impact investigations through the use of RCM results basically involves both steps of downscaling: dynamical and statistical. It is based on the view that the regional climate is conditioned by two factors: the large scale (global) state and the local physiographic features. The large scale state (the predictors) is represented by the climate models (GCMs; or dynamically downscaled by RCMs), while the local physiographic features are so variable in space and time that they cannot be accurately covered at the moment (mainly due to numerical limitations, but also due to lack of knowledge; see Section 6.1) by the climate models. Statistical downscaling based on historical data is therefore often used instead.

In light of the dynamical and statistical methods described in Chapters 5 and 7 to downscale and verify the (changes in) rainfall extremes for the small urban scales, the key question addressed in the next chapter is: what do downscaled climate scenarios say about the future urban rainfall extremes?

Chapter 8

Future changes in rainfall extremes

GCM model runs have been downscaled by a number of researchers to investigate climate change impacts on local scale precipitation. Through intercomparison of the simulations with RCMs/GCMs for both current and potential future climate conditions after forcing with climate scenarios (scenario simulations), and possibly further statistical downscaling, changes in extreme rainfall conditions have been investigated. Often the changes in precipitation extremes in the future are calculated as climate factors or changes in IDF relationships (Section 8.1), which can vary strongly spatially (Section 8.2). Section 8.3 finally addresses the issues of uncertainty, confidence and likelihood assessment in the rainfall changes.

8.1 AT-SITE CHANGES IN RAINFALL EXTREMES

Climate factors of extreme rainfall events were introduced in Section 7.2 as the ratio of rainfall intensity in the scenario simulation to the intensity in the control simulation for the same exceedance probability or return period. Such quantile-based climate factors are the result of both a change in the intensity of rain storms and a change in the number of storms (Figure 8.1). Investigation of the dependence of the climate factor on both aggregation level and return period allows climate change to be formulated in terms of probability distributions of rainfall intensities, rain storm peak intensities, rain storm cumulative volumes, wet/dry spell lengths, wet/dry day frequencies, and so on, or combined in terms of changes in the rainfall IDF relationships (Figure 8.1). The calculation of quantile-based climate factors does not only originate from the tradition in urban drainage to use design storms based on rainfall quantiles. It also originates from the knowledge that it is very difficult to attribute an individual historical event to specific causes (anthropogenic climate change in this case), whereas it may be possible to estimate the influence of the climate change on the likelihood of such an event (Hegerl *et al.* 2007).

Figure 6.10 has shown an example of quantile-based climate factors derived by Ntegeka *et al.* (2008) for 31 RCM simulations for Belgium. They calculated these climate factors for the daily, weekly, monthly and seasonal time scales. The calculations were undertaken both separately for each month of the year, and for the winter (Oct. – Mar.) and summer (Apr. – Sep.) seasons. For each month or season, return period and time scale, an ensemble set of 31 climate factors was thus obtained. Ntegeka *et al.* (2008) processed these ensemble sets to obtain the highest, median and lowest factors (among the 31 factors). They also calculated (for each climate model simulation and time scale) the mean climate factor for all extreme rain storms having return periods of 0.1 year or higher. The climate factors for the summer rainfall extremes were found to decrease with increasing time scale, as shown in Figure 8.2. This decrease was due to a decrease in the number of extreme rain storm events in summer.

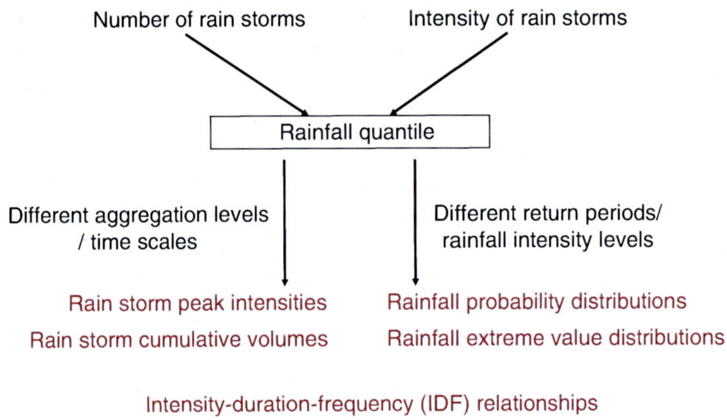

Figure 8.1 Variables to be considered in urban drainage related to statistical downscaling.

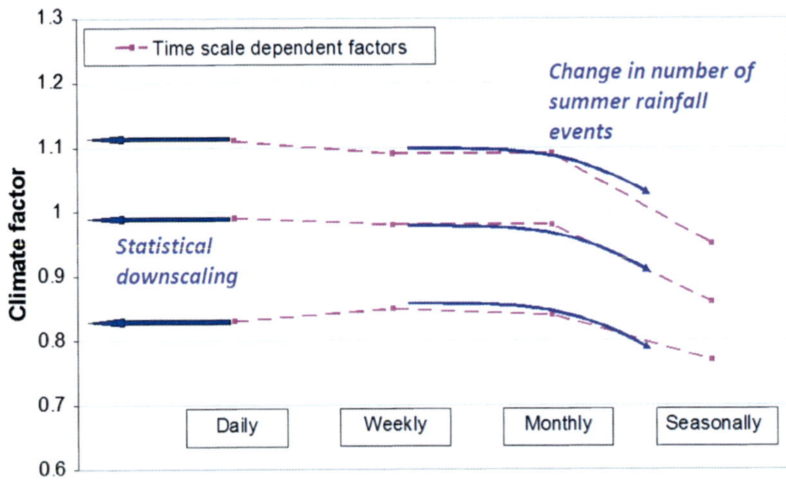

Figure 8.2 Climate factors for rainfall extremes (of return period higher than 0.1 year) at Uccle, Brussels, in the summer season, and their dependence on time scale (adapted from Ntegeka *et al.* 2008). The three curves are for the highest, mean and lowest impacts from a set of 31 PRUDENCE RCM runs.

Giorgi *et al.* (2011) has shown using a suite of GCM and RCM results, that such combination of increase in (extreme) rainfall intensity and decrease in the number of rain storms is physically consistent with global warming. It is observed for many regions of the world, in both the late 20th century observations (period 1976–2000) and in climate model projections for the 21st century. They looked into the mean precipitation intensity during wet days (defined as days with precipitation amount above 1 mm/day), the mean annual precipitation of all days and the mean annual dry spell length. They concluded that the (extreme) precipitation intensity increases because of increased atmospheric water holding capacity (see the C-C relationship in Section 7.4). The mean precipitation, however, decreases for many regions of the world. This is because of the difference in time scale between, on the one hand, the slow and continuous surface evaporation that increase the atmospheric moisture and, on the other hand, the quick intermittent removal of atmospheric moisture by precipitation (Trenberth *et al.* 2003). Both the surface evaporation and precipitation rates increase due to global warming, but the increase in surface evaporation (from oceans and land surfaces) is lower than the increase in atmospheric moisture. This is because land areas do not evaporate at their potential rate. Another reason is that the oceans warm at a slower rate than the atmosphere, which decreases the ocean-air temperature gradient that determines evaporation. Therefore, as explained by Giorgi *et al.* (2011), a relatively long "recharge time" (and thus longer dry periods) would be needed for this moisture source to replenish the atmosphere to the local saturation levels necessary for precipitation.

As discussed previously, the currently available outputs from climate models most often have a daily time resolution, thus climate factors can be obtained directly only for daily and longer time scales. Extrapolation to finer time scales then requires downscaling (Figure 8.2). A limited number of studies have so far investigated results from RCM runs with sub-daily time steps.

Larsen *et al.* (2009) and Arnbjerg-Nielsen (2012) studied climate factors directly obtained from the hourly results of a 12-km resolution RCM. They found that the climate factors at the hourly scale are significantly higher in comparison with the climate factors at the daily scale, and that the difference increases with increasing return period (Figure 8.3).

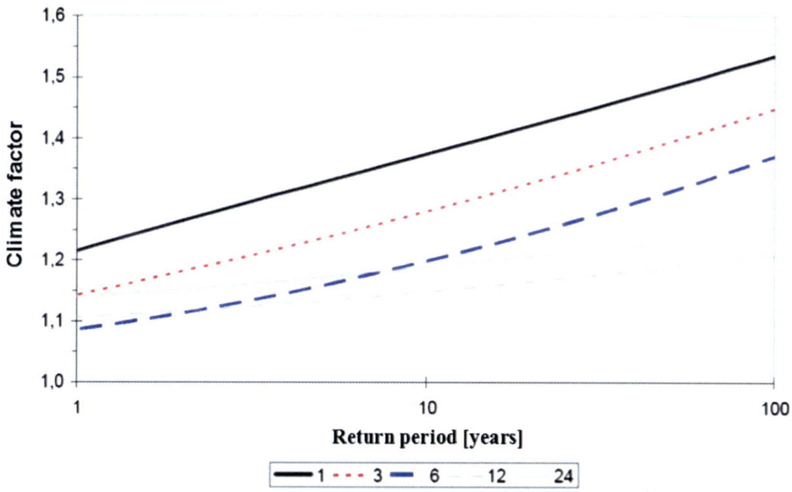

Figure 8.3 Estimated climate factors for Denmark for durations of 1, 3, 6, 12 and 24 hours (HIRHAM4, 2071–2100 vs. 1961–1990) (after Arnbjerg-Nielsen, 2012).

The process-based downscaling of Olsson *et al.* (2012a) (see Section 7.3) was tested and evaluated using published IDF relationships in Stockholm, Sweden. IDF curves obtained from point precipitation stochastically simulated by using the RCM cloud cover as an approximation of the grid "wet fraction" showed a clear improvement compared with IDF curves obtained from RCM rainfall results at grid scale. By further introducing a simple tuning procedure, the simulated IDF curves closely reproduced the observed ones. Olsson *et al.* (2012a) concluded based on this approach that the "local factors" associated with the 10-year intensity at sub-daily scales were generally 0.05–0.1 units larger than the "grid scale factors" (Figure 8.4). This suggests a 5–10% larger increase in local intensity than what can be inferred directly from RCM output. A practical approach may be to consider "grid-scale factors" as a conservative estimate of the increase at the local scale.

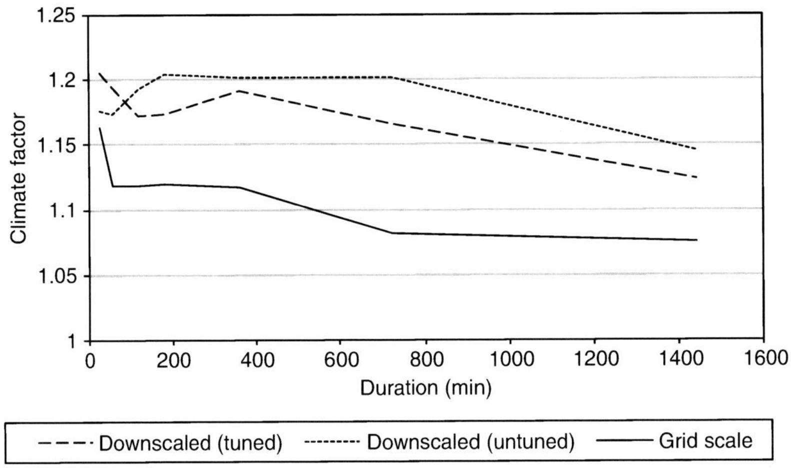

Figure 8.4 Climate factors for 10-year intensities in Stockholm, Sweden, for period 2071–2100 vs. 1961–1990 for durations between 30 min and 1 day (1440 min) estimated directly from grid-scale data (50 × 50 km) and after (untuned and tuned) process-based downscaling to local intensity. Global model: ECHAM4; regional model: RCA3. The curves represent averages over SRES emission scenarios A1B, A2 and B2 (after Olsson *et al.* 2012a).

Also for the city of Barcelona in Spain, Pouget *et al.* (2011) found that local climate factors differ from the GCM-based factors. They applied a statistical downscaling approach to the results of five GCMs and 4 SRES scenarios based on fractal theory, and found climate factors for hourly rainfall intensities and a 10-year return period in the range from 0.96 to 1.12 for 2033–2065, while the climate factors derived from the daily intensity GCM results and the same return period varied between 1.01 and 1.04.

Semadeni-Davies *et al.* (2008) derived climate factors for 6-hour rainfall intensities at Helsingborg in southern Sweden based on two RCM runs. The monthly climate factors were found to vary widely between a 50% decrease and a 500% increase for the period 2071–2100 (depending on season and intensity level), confirming that both season and intensity level need to be taken into account. Olsson *et al.* (2009) took this approach one step further by estimating a distribution of climate factors, being a continuous function of the rainfall intensity expressed as quantile. An application for future summer rainfall in Kalmar, southern Sweden, indicated for a selected RCM run climate factors up to about 1.2 for the highest intensities and factors smaller than 1 for low and medium intensities below the 80–90% quantile (Figure 8.5).

Willems and Vrac (2011) studied changes in IDF statistics for time scales between 10 minutes and 15 days, applying the method of Willems (2001) to 17 runs with the same GCM and SRES scenario, downscaled to 10-minute intensities of point rainfall for Uccle, Belgium. Results (Figure 8.6) showed that for the highest change the return period of the 2-year IDF curve may become approximately 1 year by 2071–2100; and the 100-year IDF curve may have a 10-year return period. This means that when an urban drainage system is currently designed for a 2 year return period to avoid urban flooding or sewer surcharge, these flood or surcharge events may occur approximately twice that often by the end of this century. It also would mean that to achieve identical surcharge or flood safety levels, either design storms have to be adjusted or design storms with longer return periods have to be used for the design. This assumes that no other changes such as land use or water management and planning are considered. Note that the wide range of impact results shown in Figure 8.6 (summarized in highest, mean and lowest impact) are due to differences in GCM initial condition only.

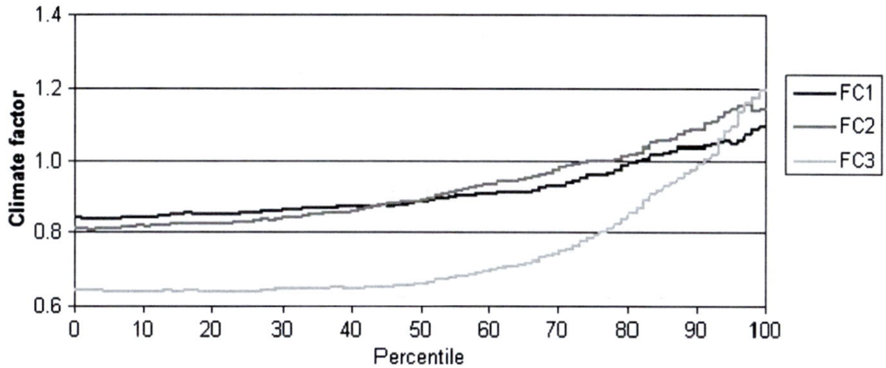

Figure 8.5 Continuous climate factors as a function of percentile (i.e. intensity level) for 30-min summer rainfall in Kalmar, southern Sweden, for periods 2011–2040 (FC1), 2041–2070 (FC2) and 2071–2100 (FC3) vs. 1961–1990. Global model: ECHAM4; regional model: RCA3; SRES emission scenario: A2 (after Olsson *et al.* 2009).

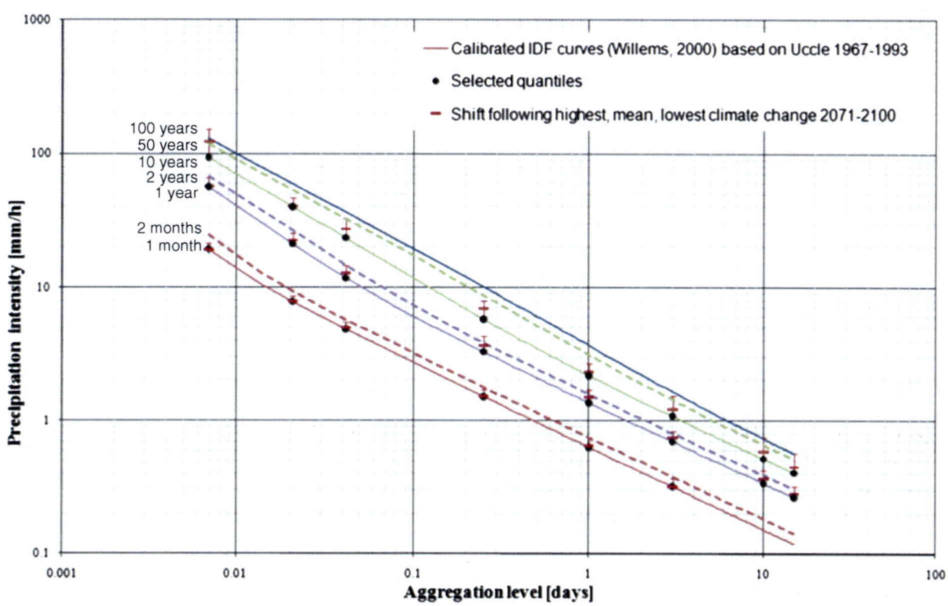

Figure 8.6 Change in IDF relationships for Uccle, Belgium, based on mean, highest and lowest change based on 17 ECHAM5 A1B runs for 2071–2100 and a quantile perturbation downscaling method (after Willems & Vrac, 2011). For the highest change, the 1-month IDF curve shifts upward to the 2-month IDF curve; the 1-year IDF curve shifts to the 2-year curve; the 10-year IDF curve shifts to the 50- or 100-year curve (or even above) depending on the aggregation level.

8.2 REGIONAL CHANGES IN RAINFALL EXTREMES

Results from Section 8.1 show that the changes in rainfall extremes strongly vary from study to study. They indeed vary depending on regional location and on local characteristics. In many parts of the world, the overall change is a combination of two processes: 1) a tendency towards fewer rainy days because the warmer air can contain more precipitation, and 2) when it does rain, there is a tendency towards higher rainfall intensities (Voss *et al.* 2002; IPCC, 2012). Therefore some regions may experience a higher risk of both flooding and droughts as contrasting extremes (Christensen & Christensen, 2003; Trenberth *et al.* 2005). Based on simulations for a set of climate models it is projected that rainfall extremes will become more frequent and severe in the tropical regions and higher latitude areas (Trenberth *et al.* 2003; Mailhot *et al.* 2007; Bates *et al.* 2008). For latitudes north of 45°N, climate models show a clear tendency for an increase in wintertime mean precipitation and a concomitant increase of high quantiles in daily rainfall. Obviously, there are strong regional differences in the net impacts, and indeed in some regions there may be a decrease in rainfall extremes. Meehl *et al.* (2005) examined the processes that explain the geographic differences. The increased rainfall intensity and frequency for the tropics is due to the general increase in water vapour associated with positive SST anomalies in the warmer climate. For the midlatitudes, the rainfall change is related to both the increased water vapor for areas with moisture convergence and to changes in atmospheric circulation. For northwestern and northeastern North America, northern Europe, northern Asia, the east coast of Asia, southeastern Australia and south-central South America, advective effects due to SLP changes are most important in explaining the increase in rainfall intensity.

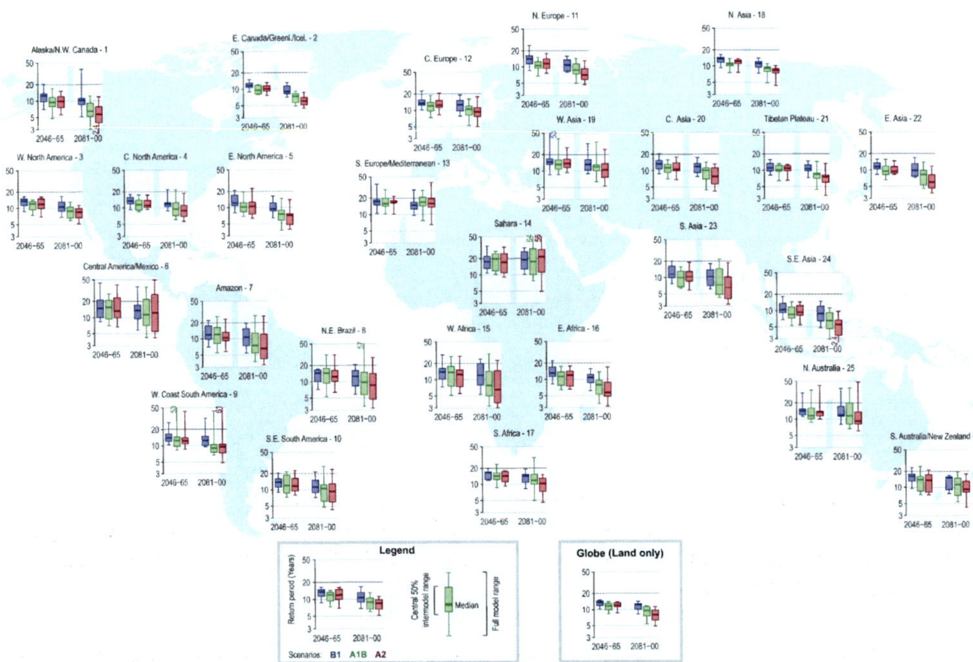

Figure 8.7 Projected return periods for daily precipitation intensity of 20-year return period (1981–2000) for two time horizons, 2046 to 2065 and 2081 to 2100, based on 14 GCMs and SRES emission scenarios B1, A1B and A2 (after IPCC, 2012; Retrieved 31 March 2012, from http://ipcc-wg2.gov/SREX/).

Figure 8.7 shows the continental results on the change in extreme daily precipitation quantiles. Rather than the change in the rainfall intensity, the figure plots the change in the return period for a rainfall intensity that has a 20-year return period for the late 20th century climate (1981–2000; based on annual maxima). The changes were derived from 14 GCMs contributing to the Coupled Model Intercomparison Project (CMIP3) of the World Climate Research Programme (IPCC, 2012) using the emission scenarios B1, A1B and A2. A 20-year return period rainfall intensity is "likely" to become a 5- to 15-year intensity by the end of the 21st century in many regions. The term "likely" refers to a 66–100% probability. In most regions the higher emissions scenarios (A1B and A2) lead to a stronger projected decrease in return period.

Other studies have made more detailed analyses of changes in smaller regions. Larsen *et al.* (2009) analysed the impact of climate change on hourly and daily rainfall quantiles for return periods between 2 and 100 years, across the whole of Europe. A sample result of that study is shown in Figure 8.8. Their analysis was based on one RCM simulation run with 12-km resolution using SRES scenario A2, conducted as part of the PRUDENCE project (see Section 5.3). The estimated climate factors are presented in Figure 8.9. The results show that there is a very high spatial heterogeneity that must be smoothed before the results are used to predict a change at any given location. Furthermore, the results indicate that climate change factors do depend on duration, return period and location (only dependence on location shown in Figure 8.9). Scandinavia seems to experience the highest increase and southern Europe a substantially lower increase. There is also a huge change noteable in the alpine/Mediterranean and Baltic areas. For the Baltic Sea, a climate factor as high as 2 is obtained, but this might be questionable given that the model has some problems simulating the temperature in the Baltic Sea (Christensen & Christensen, 2007). Based on the same RCM, May *et al.* (2005) have shown that changes are generally larger for heavy rainfall intensities than for average intensities.

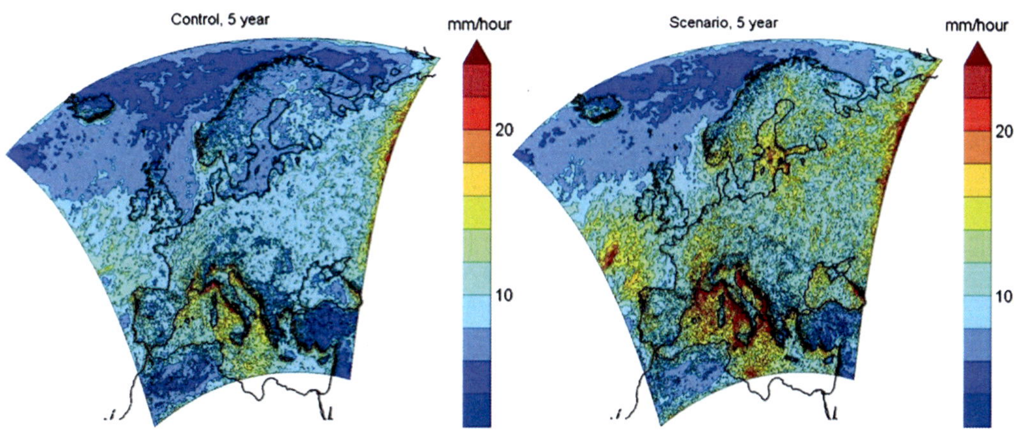

Figure 8.8 The spatial variability in rainfall intensity over Europe for a 1-hour rainfall intensity for a 5-year return period for the control period 1961–1990 (left) and future scenario period 2071–2100 (right), based on HIRHAM4 RCM and SRES scenario A2 (after Larsen *et al.* 2009). The increased precipitation can be seen as a general shift in colours.

The study by Larsen *et al.* (2009) is, as to the knowledge of the authors, so far the only study that looked at regional differences for rainfall extremes at sub-daily time scales. Despite the lack of regional studies for sub-daily rainfall extremes, several studies looked into the sub-daily local changes for a given station or city, or to the regional changes in daily rainfall for a given country or district (see e.g. Figure 8.10 for Belgium).

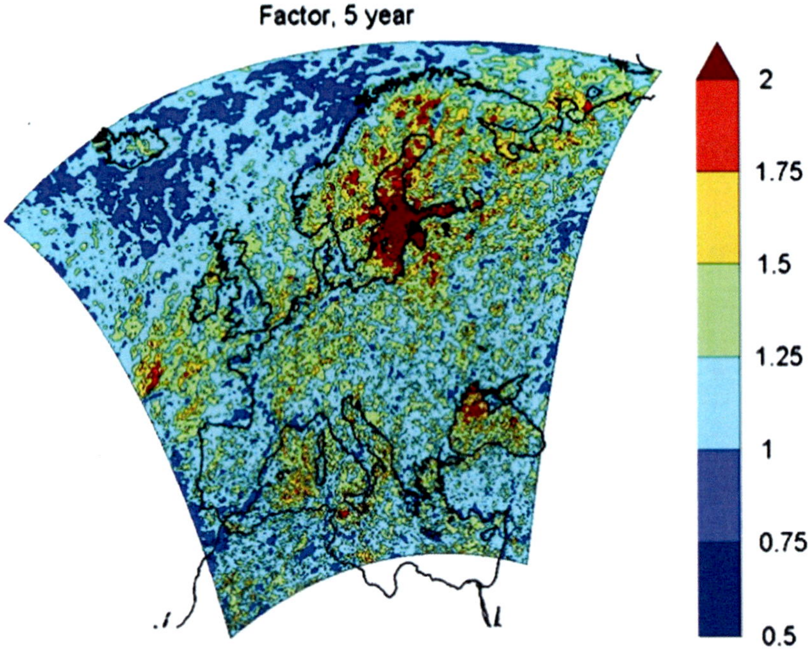

Figure 8.9 The spatial variability in the climate factor over Europe for a 1-hour rainfall intensity and a 5-year return period for 2071–2100 vs. 1961–1990, based on HIRHAM4 RCM and SRES scenario A2 (after Larsen *et al.* 2009).

Figure 8.10 shows that the rainfall changes for Belgium are rather uniform over the inland region. Only for the coastal polder and the High Fens plateau region, climate factors are about 10% higher. The climate factors for rainfall extremes in Belgium, derived by Ntegeka *et al.* (2008) therefore were assumed applicable for the entire country, except for the coastal and High Fens regions, where an enhancement factor of +10% is applied.

Construction of maps as Figure 8.10 where climate factors are presented after statistical analysis of a large number of RCM runs, face some difficulties. First difficulty is that different RCMs use different grids (as illustrated in Figure 8.11). Processing the model data individually using the intersection of the grid of each model with the study region proves inefficient in estimating the regional climate change signal when based on different RCMs. On the other hand, staying within the same RCM, it is not clear in which way the whole region is affected. Second difficulty is that the regional pattern of the change factor can be quite different for different RCMs. The approach adopted by Baguis *et al.* (2008) to solve these problems was to project all the RCM data on a common and high-resolution grid (7 km) and then calculate the perturbations based on the data collected over each cell of the new grid. These results were presented for the highest, mean and lowest climate factors for the 31 PRUDENCE RCM results in Figure 8.10. An interesting topic for future investigation would be to test whether the regional changes in daily, supra-daily and sub-daily rainfall are identical or similar.

In The Netherlands, the daily rainfall intensity for a 10 year return period is expected to increase by 10 to 50% (van Luijtelaar *et al.* 2008) by 2100. A similar range of rainfall changes is reported by Arnbjerg-Nielsen (2012) who found that within the next 100 years, rainfall intensities would increase for Denmark by 10 to

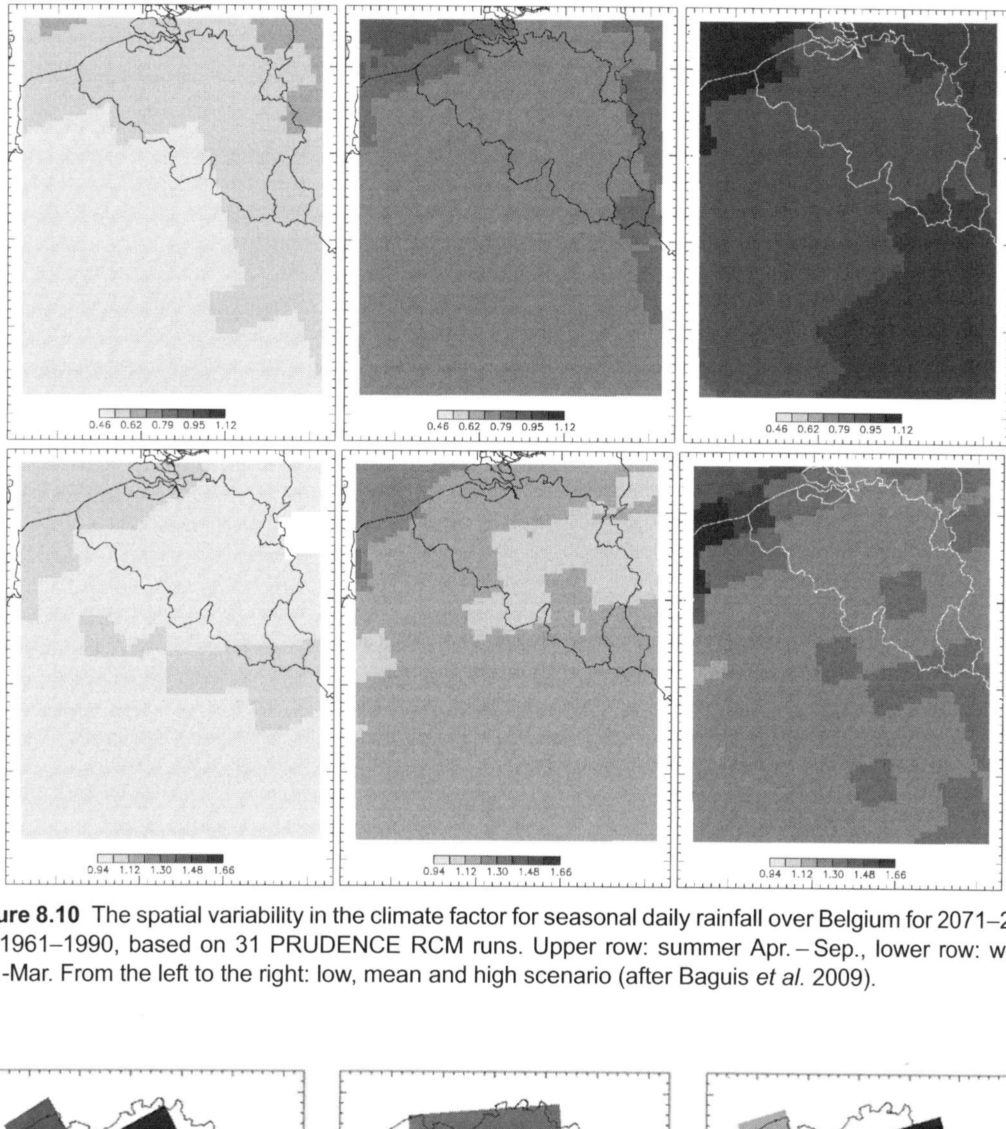

Figure 8.10 The spatial variability in the climate factor for seasonal daily rainfall over Belgium for 2071–2100 vs. 1961–1990, based on 31 PRUDENCE RCM runs. Upper row: summer Apr. – Sep., lower row: winter Oct.-Mar. From the left to the right: low, mean and high scenario (after Baguis *et al.* 2009).

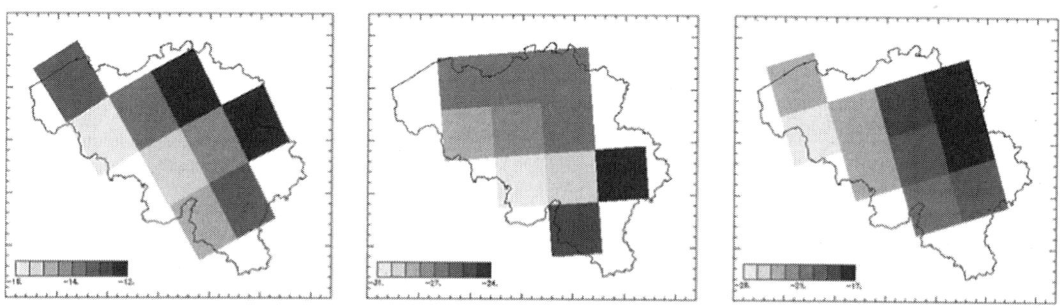

CNRM summer **HC summer** **KNMI summer**

Figure 8.11 Sample of regional climate factors for three RCMs with different grids for Belgium (after Baguis *et al.* 2009).

50% (for return periods between 2 and 100 years, and durations between 10 minutes and 24 hours). For a location north of Copenhagen, Denmark, Sunyer and Madsen (2009) concluded that besides the number of extreme rainfall events, the variance of the extreme rainfall intensities will also increase. They projected an increase of 62% in the number of extreme rainfall events and an increase of 101% in the daily variance by 2100. Based on daily rainfall outputs from 4 RCMs, Sunyer *et al.* (2012) found climate factors were in the range 0.74 to 1.61 and 0.63 to 2.06 by 2100 for return periods of 10 and 100 years, respectively (see Figure 7.5). For the city of Roskilde in Denmark, Arnbjerg-Nielsen *et al.* (2012) stated that a 40% increase in the design rainfall intensities would imply that a 100-year return period storm would represent an 18-year event within the next 100 years. Also for Denmark, Grum *et al.* (2006) showed that by 2100, extreme rainfall events would occur at least twice as frequently as in the recent past. This means that the return period of a given rainfall intensity would be halved. Similar results have been obtained recently for Belgium by Willems and Vrac (2011) who found that rainfall IDF relationships shifted to such an extent that storms commonly used for urban drainage designs could have their return periods halved by 2100.

For Kalmar City in Sweden, Olsson *et al.* (2009) found that high 30-minutes rainfall intensities in summer and autumn would increase by between 20 and 60% by the year 2100. In a recent national analysis (unpublished) using 6 RCM runs, compared with period 1981–2010 the 10-year 30-minute intensity was found to increase by ~5% until 2011–2040, by ~15% until 2041–2070 and by ~25% until 2071–2100, averaged over all scenarios and all of Sweden. Generally, the increase in western Sweden was higher than in the east.

For the UK, for rainfall volumes and intensities of relevance for urban drainage impact assessments, Ashley *et al.* (2008a) and Tait *et al.* (2008), report on increases of around 20% and higher spatial clustering by 2085.

In the Ruhr area in western Germany, Staufer *et al.* (2008) pointed out that 10% to 15% of seasonal precipitation volumes could shift from summer to winter by 2100, while snow days would almost cease to exist. The intensity of the extreme rainfall events with durations less than 1 hour would change such that a current 5-year event would become a 3-year event.

For the state of Washington in the USA, Rosenberg *et al.* (2010) analysed the WRF model based RCM nested in two GCMs for hourly precipitation with spatial resolutions of 20 and 36 km. Changes in annual maxima for different durations between 1 hour and 10 days between the 1970–2000 and 2020–2050 periods were studied. The annual maximum 1-hour intensity increased with 14.3% for one GCM and decreased with 6.3% for the other GCM.

In Canada, Waters *et al.* (2003) concluded that by the end of this century the future 5-year, 1-hour rainfall intensity for Kingston, southern Ontario might become equal to the current 10-year, 1-hour rainfall depth or a 17% increase over the current 5-year, 1-hour value.

For Japan, Nakamura *et al.* (2008) found based on a 5-km grid cloud-system-resolving RCM, increases in the frequency of 1-hour rainfall intensities exceeding 30 mm/h with about 90%. The changes were strongest for Kyushu island in Soutern Japan. The high changes in rainfall extremes for Kyushu island in summer were confirmed by Kanada *et al.* (2005). They explained that the extremely high changes for that area are due to the confluence of disturbances from the Chinese continent and the East China Sea, which will occur more frequently in the future.

Interestingly, Nakamura *et al.* (2008) found that 1-km resolution simulations, which explicitly resolve individual cloud cells, led to smaller increases in east Asia Monsoon rainfall extremes than 5-km resolution simulations. It was explained by many authors that due to the complex topography and the unique weather and climate system, the climate change signal over east Asia exhibits a complex spatial structure which can be simulated only with high resolution RCMs (e.g. Gao *et al.* 2008).

For developing regions, local urban drainage impact studies are rare, despite the fact that these regions my get the strongest impacts. However, some exist. For the city of Khulna in Bangladesh, ADB (2011) reported an increase of 6-hour rainfall intensities for 55 mm or more from 4.2 times a year to 4.25 (B1 scenario) or 5.9 (A2 scenario) times a year for 2050. These values were obtained after statistical downscaling of one GCM run by quantile mapping.

Several other studies exist that reported on changes in rainfall extremes, but without an explicit focus on urban drainage applications. They most often focus on daily or supra-daily time scales. Some studies with daily results are reported hereafter, because they give an idea of the changes in rainfall extremes in other regions of the world. Sub-daily changes might differ from these daily changes, but most of the aforementioned studies indicate same order of magnitude for the changes.

In southeastern Australia, Schreider *et al.* (2000) considered an increase in daily summer precipitation between 0% and 8% by 2030 and between 0% and 20% by 2070. This was based on five GCMs (CIG, 1996).

For California in the USA, Bell *et al.* (2004) found based on one RCM run strong regional differences. The changes range from 2.4 fewer heavy rainfall events per year to 2.5 additional heavy rainfall events per year when doubling the CO_2 concentrations in the model. Also in western USA, Leung *et al.* (2004) concluded that daily rainfall extremes during the cold season increase in the northern Rockies, the Cascades, the Sierra Nevada and British Columbia by up to 5 to 15 mm/day (15–20%) for 2040 to 2060, although mean rainfall was mostly reduced.

In South America, Marengo *et al.* (2009a, 2009b) and Nuñez *et al.* (2008) derived from RCMs an increase in the frequency of rainfall extremes in Southeastern South America (southern Brazil, Northern Argentina) and western-central Amazonia by the end of the 21st century.

In the Artic region, Saha *et al.* (2006) found an increase of the 90% quantile daily precipitation by 30–50% over Siberia and the north Atlantic by 2050. Also the number of extreme precipitation days increases. Their results were, however, based on only one RCM simulation.

In Africa, Nyeko-Ogiramoi *et al.* (2012) did find for Lake Victoria catchments in Uganda changes in daily rainfall extremes of 100% or more for the 2090s and a return period of 10 years. They analysed 24 different GCM runs. Changes in the number of wet days were as high as 300% or even more for some GCM runs by the 2090s. Differences in GCM results were, however, extremely high, probably because data limitation constrained the calibration of GCMs for that continent.

For east Asia, several studies reported that during the Monsoon period (May-Sept.) an increase of the frequency and intensity of daily summer rainfall extremes are expected till 2100: for example Boo *et al.* (2006), Im *et al.* (2008, 2011) for Korea; Kanada *et al.* (2005), Nakamura *et al.* (2008) for Japan; Zhang *et al.* (2006), Fu *et al.* (2008) for China. Most of these studies are based on a single RCM simulation. Zhang *et al.* (2006) simulated for the B2 SRES scenario an overall increasing trend in extreme precipitation by 2100 over most of China, including the southeast coastal zone, the middle and lower reaches of the Yangtze River and northern China.

8.3 UNCERTAINTY IN RAINFALL CHANGES

Chapter 6 has shown that climate change impact investigations based on only a few climate model simulations may not be reliable. An ensemble approach is recommended where rainfall changes are analysed for a large set of climate models, with different climate forcing scenarios, and maybe even a set of initial conditions and statistical downscaling models. In this way, the uncertainty (in the GCM/RCM results due to limited knowledge of the precipitation processes at small spatial scales, the limited resolution of the models, the future climate forcing, the uncertain initial conditions, the future relationship between the coarse and fine scale rainfall intensities, etc.) can be assessed. As described in previous sections, the uncertainty generally increases with decreasing duration and increasing return period. It also strongly depends on the location. For instance, in the tropics the change due to the anthropogenic factors is large relative to the model uncertainty. Also the natural variability is higher at higher latitudes, even though the magnitudes of the trends might be larger at higher latitudes (Hegerl *et al.* 2007).

Impact modelling based on results from the entire set of available models may, however, be difficult in practice for urban impact studies, because of the large computational resources often associated with hydraulic modelling.

One pragmatic approach would be to summarize the different impact results in a limited set of (tailored) scenarios. This was undertaken by Boukhris *et al.* (2008), who applied high, mean and low scenarios based on the highest, average and lowest climate factors for the entire set of potential scenarios considered. Another option is to use the limits of confidence intervals computed from the whole set of scenarios. This was done by Ntegeka *et al.* (2008) to simplify the climate scenarios by constructing 3 sets of scenarios to represent the range of model results: "high/wet", "mean/mild" and "low/dry" (Figure 8.12). The high scenario may be referred to as wet, and is thus adapted to studies of the risk of flooding, while the low scenario may be referred to as dry, and is thus critical for dry weather conditions. It is notable that their mean scenario represented mean conditions and not necessarily the best future guess. Their definition of high/mid/low was not unique, but rather it was tailored to the application: it depended on time scale, return period and season/month, and was based on the expected hydrological impacts. A similar approach was followed by the CSIRO in Australia by defining scenarios for "most wet" and "most dry" years (CIG, 1996).

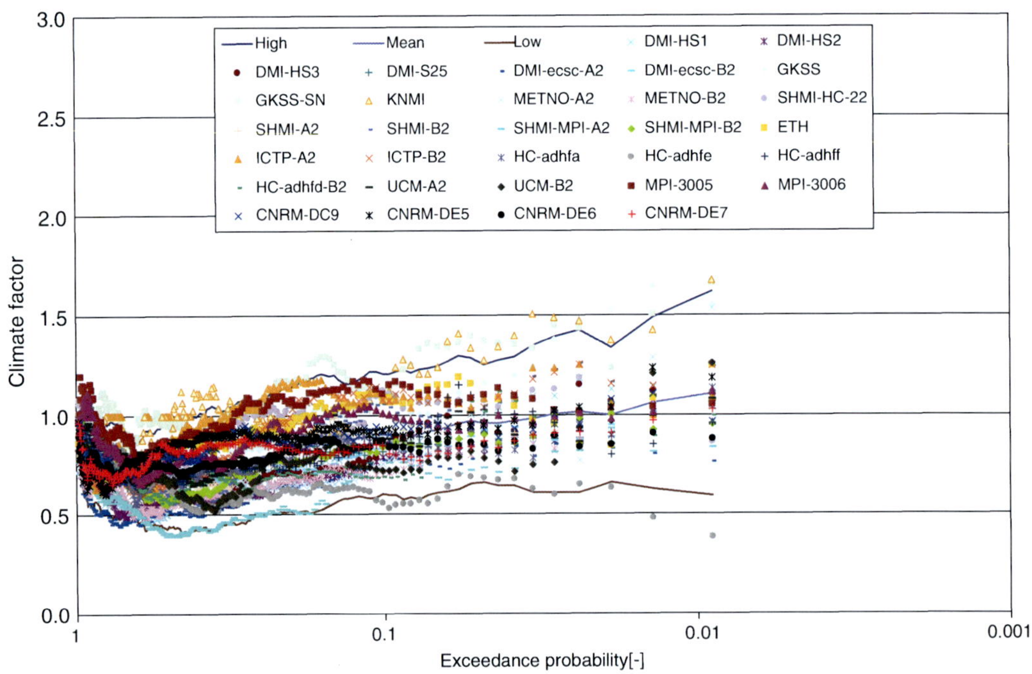

Figure 8.12 Climate factors versus return period for 31 PRUDENCE RCM wet day rainfall intensities as a function of the exceedance probability at Uccle, Brussels, for the month of July. The constructed low, mean and high scenarios are also shown (adapted from Ntegeka *et al.* 2008).

The tailored scenarios by Ntegeka *et al.* (2008) were derived for Belgium, based on 31 RCMs runs from the PRUDENCE project and 21 GCM simulations from the IPCC AR4 Archive (see Section 5.3) and four SRES emission scenarios A2, B2, A1 and B1. They found that the percentage increase in daily summer rainfall extremes by 2100 may vary from −17% to +12% (on average for return periods higher than 0.1 year) or within a much wider range for higher return periods (see Figure 6.10). Also Nguyen *et al.* (2008b) found that the use of different GCMs may switch the climate factor from a positive to a negative one.

While providing a simplified view of the range of model results with a limited set of derived climate scenarios is very useful for impact studies, it is also relatively difficult due to the need for adapting the selection of scenarios to the variables of interest and their application. For example, in hydrological applications where multiple hydrological variables and/or multiple sites are considered, scenario changes have to be provided for each of these variables and sites. For instance, in many hydrological applications, besides the changes in rainfall also changes in temperature or potential evapotranspiration (ET_0) affect the impact assessment. In these cases, correlations between the changes in precipitation, temperature and ET_0 have to be considered, or at least the changes have to be coherent. This means that one has to ensure that the internal physical consistency of the climate system is preserved. The definition of high/mean/low scenarios then needs to be based on the combined effect of changes in these variables. In other words, the variables need to be combined to generate an impact, which can be classified as high, mean and low. For urban drainage applications, the situation is fortunately rather clear-cut as rainfall is by far the most important driver of high drainage system flow related impacts. Changes in temperature and ET_0 are less important when studying these impacts, thus the issues of correlation/coherency between variables and preservation of internal physical consistency is less relevant in such cases.

In cases where more variables/sites have to be considered for the impact study, another approach to obtain a reduced set of scenarios is by selecting a subset of all available GCM/RCM runs that are expected to lead to high/mean/low impacts. Because the results for the multiple variables/sites are obtained by the same model, the correlation between the variables due to the internal physical consistency of the climate system is preserved, at least within the physical constraints of the model. Selection of the relevant subset is naturally difficult as impacts are hard to assess beforehand, but may be facilitated by using previous evaluation and experience from the full ensemble.

Next to the climate model projections, also the statistical downscaling step might introduce significant uncertainties. To investigate whether application of different downscaling techniques might yield further information about the actual uncertainty, Arnbjerg-Nielsen (2012) compared three different statistical downscaling methods for estimation of climate factors for Denmark: one based on the BLRP generator, one which made direct use of RCM precipitation results for precipitation extremes and one which was based on climate analogues for monthly averages of temperature and precipitation. He studied the climate factor for return periods between 2 and 100 years, and for durations between 10 minutes and 24 hours, and found that the different methods exhibit systematic differences. The rainfall changes obtained by the BLRP method led to underestimations in comparison with two other methods. The changes were roughly 50% of the changes estimated by the other two methods, that is a climate factor around 1.1–1.2 rather than 1.2–1.4. This was explained by the fact that changes in the BLRP model were estimated based on general rain storm properties (mean cell intensity, mean cell duration, rate of rain storm arrival, etc.), but with less emphasis on changes in the statistics of rainfall extremes. However, Burton *et al.* (2008) showed that the almost identical weather generator model NSRP was capable of capturing hourly and daily rainfall extremes very well, and so the performance may be based on how the estimation of the key parameters was carried out. This emphasizes the need for a range of impact results to be presented based on ensembles of climate models, climate forcing scenarios, downscaling techniques, and that a careful evaluation of all other scenario uncertainties within the urban drainage context is required, as mentioned previously.

Larsen *et al.* (2009) pointed out that in addition to the uncertainties in climate scenarios, there might be additional uncertainties when statistical (extreme value) analysis is performed to calculate the rainfall changes or in the estimations of IDF relationships. Parameter estimation of extreme value distributions (Section 2.4), for instance, may indeed be significant. Figure 8.13 illustrates how this latter uncertainty affects the average climate factor for Denmark. The figure shows that the probability of the climate factor being larger than 1 is 68% for return periods between 2 and 20 years, and lower for higher return periods. This shows that care is required before concluding that rainfall extremes will increase under changing climate conditions, even if the climate models show positive climate factors.

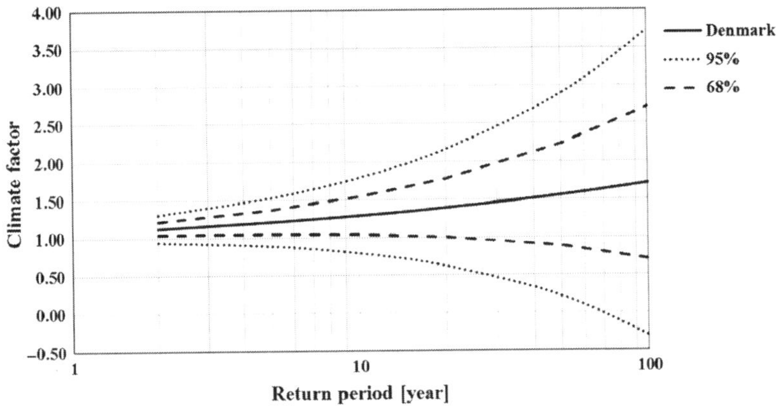

Figure 8.13 68% and 95% confidence intervals for the average climate factor for Denmark for 1-hour rainfall, due to the uncertainty in the calibration of extreme value distributions for the control and scenario periods (after Larsen *et al.* 2009).

When the full (or simplified/reduced) ensemble set of climate model runs, and possibly also several downscaling methods, are considered, question is whether they can be used to provide uncertainty estimates in the form of confidence assessments or likelihood statements, for instance on the urban rainfall extremes and related urban drainage impact results. When the climate projections made by different models show a higher congruence, the likelihood of accurately predicting the sign and value of change is higher (IPCC, 2007b). This is of course only the case when the confidence in the models is high. If the confidence is low, no likelihood assessment can be provided, even if the available climate projections display a high congruence (Risbey & Kandlikar, 2007).

Likelihood statements implicitly include confidence assessments of the models considered and the associated level of understanding. Confidence assessments can be made after comparison of model results with observations, similar to the methods used for testing the model reliability in Section 6.1. Wilby and Harris (2006), Tebaldi and Knutti (2007) and Tebaldi and Sanso (2009) have shown how weights can be given to different climate models based on such confidence assessments. In this way, probability distributions on climate change projections can be constructed. This is, however, not trivial given that different climate models share the same level of process understanding and sometimes even the same parameterization schemes and code, as explained before.

An interesting approach to deal with the necessary arbitrariness of assessing uncertainty on a randomly selected group of climate models was proposed by Leith and Chandler (2010). They adopted a hierarchical statistical approach in a Bayesian framework. Their approach started from the assumption that in broad terms all climate models represent essentially the same dynamical processes, and that climate model outputs are intended to provide plausible, rather than exact, scenarios that agree with actual climate statistically rather than in detail. Based on that assumption, it is expected that the time series outputs from different climate models have a similar structure, which can be described using the same form of statistical model, but with parameters that differ between climate models. They established ML estimators for the distribution of these parameters starting from a prior distribution and applying the theory of Bayes. GLMs were considered as statistical model. The method was, however, not tested for rainfall, neither for extremes, but for monthly averages of temperature, sea-level pressure and relative humidity. Leith and Chandler (2010) moreover reported a need for additional research regarding the influence of the choice of the prior distribution, on how to incorporate information on climate model performance based on historical observations, for example by downweighting under-performing climate models, and how to allow for the communal bias in all climate models that arises from knowledge being shared between modelling groups.

Due to these difficulties, most researchers prefer to work in the climate change context with scenario uncertainty rather than statistical uncertainty, where the scenarios are considered as hypothetical future conditions rather than predictions of the future.

8.4 DISCUSSION

This chapter has shown that there are strong regional differences in the impacts of climate change on small scale rainfall extremes. The results for Europe for example show that Northern Europe is expected to receive substantially higher changes in rainfall extremes than Southern Europe. The impact results also depend significantly on the climate change scenario considered. However, the review of the changes in urban rainfall extremes indicated that significant increases are expected for many regions of the world.

The impact estimates in this chapter were for all demonstrated cases provided in the form of impact ranges, rather than single deterministic values. The ranges were obtained by ensemble approaches, which were recommended in previous chapters, to account for the uncertainties involved in the climate forcing scenarios, the knowledge on climate physics, the numerical limitations of climate models and the statistical downscaling techniques.

However, detailed ensembles studies focusing on urban drainage have not yet been reported in the scientific literature, mainly because of the extensive efforts involved in data base compilation and statistical port-processing required to reach urban scales. Rather some arbitrary approaches based on limited ensembles have been applied, for example by using a limited set of tailored climate scenarios and one or few statistical downscaling methods. Whatever methods are adopted, the downscaled climate model results should be interpreted as indicative of the magnitude of rainfall intensification that could be expected over the next 20 to 100 years.

The overview provided in this chapter learns that typical increases in rainfall intensities at small urban hydrology scales range between 10% and 60% from historical control periods in the recent past (typically 1961–1990) up to 2100. Given that urban drainage systems receive runoff inflows mainly from paved urban areas, strong increases in sewer flows and other urban drainage related impacts are expected. But how large can we expect these impacts to become? A review is provided in the next chapter.

Chapter 9

Future impacts on urban drainage

The focus of this book so far has been on the technical aspects of quantifying climatic change in terms of its effects on precipitation. As described in the previous chapters, substantial changes in the water cycle can be anticipated due to climatic change occurring on a range of scales. This chapter focuses on describing the impacts in an urban drainage context. The performance of urban drainage systems is highly sensitive to changes in the water cycle, notably precipitation extremes. While Chapter 8 presented a literature overview of the regional differences in high rainfall extremes, this chapter describes the approaches for transferring the changes in rainfall to changes in the inputs for urban drainage models (Section 9.1) and analysing the impacts that these high rainfall extremes have on urban drainage flows, sewer floods, surcharges and overflows (Section 9.2). Section 9.3 discusses the other types of climate change related impacts on sewer and urban drainage systems, which have consequences for urban design, while Section 9.4 focuses on the uncertainties in impact analysis under changing conditions.

9.1 GENERATION OF RAINFALL INPUT FOR URBAN DRAINAGE IMPACT CALCULATION

Event-based versus continuous simulation based approaches

To estimate climate change impacts on urban drainage systems, the accepted and downscaled climate change scenarios need to be propagated through urban drainage models. This can be done directly using the rainfall time series derived from the climate model outputs as drainage model inputs. After simulation of both the series derived from the climate model control simulations (representing present climate) and the scenario simulations (representing future climate), changes in the urban drainage results can be investigated. Because local and fine scale variables are required as input to the urban drainage models, the statistical bias correction and downscaling step (or combined into one method; see Chapter 7) has to be applied to both the control and scenario simulation results. The bias correction should ensure that, for the control simulations, drainage results will be obtained that do not systematically differ from the results obtained after drainage model calibration and validation based on observations of the present precipitation regime. To avoid the latter problem, a different method can be applied where the estimated climate factors are applied to change or perturb the inputs of the urban drainage models. How such changes or perturbations can be done was presented in Section 7.2. These perturbed inputs are usually based on either historical observations or design storms. It depends on whether the impact study is undertaken in an event-based way through simulation of single historical events or synthetic design storms in the urban drainage model, or whether continuous simulations are used.

For sewer runoff and flow simulations, as is done in sewer design applications, event-based sewer impact calculations are most commonly applied (Butler & Davies, 2010). This is because of the direct relation between runoff flows from paved areas in an urban catchment or peak flows at given locations of the urban drainage system and the rainfall intensity at a given temporal scale (the catchment-system concentration time; see Section 2.6). For impact calculation of combined sewer overflow (CSO) frequencies and impacts on receiving surface waters, processes at a wider range of time scales interact which requires continuous model simulations (Harremoës, 1988; Rauch *et al.* 2002; Butler & Davies, 2010; Vaes *et al.* 2001). In this case, simulation results need to be post-processed for analysing flood, surcharge and/or overflow frequencies, or other types of impacts such as environmental and socio-economic impacts.

The urban drainage models can take the form of simulation models, which demand continuous rainfall time series as input, either for single events (short duration) or longer duration time series. Other models are (semi-)probabilistic where probability distributions of urban runoff discharge are calculated based on the rainfall event distribution. These models are per definition event-based. Examples of such models are the analytical probabilistic or semi-probabilistic models of Guo and Adams (1998a, 1998b, 1999), Adams and Papa (2000) and Bacchi *et al.* (2008), where rainfall events are characterized by probability distributions for the rainfall depth, storm duration and inter-arrival time. Based on these characteristics, event runoff volumes, peak flow discharges and CSO volumes can be simulated. Most climate change impact studies, however, make use of continuous simulation models. These typically take the form of full hydrodynamic models or more simplified reservoir type of models. The full hydrodynamic models involve computation of de Saint-Venant equations, which are a simplification of the general Navier-Stokes equation for one – dimensional flow (Chow, 1964; Chow *et al.* 1988). They are typically solved by means of a finite difference numerical scheme (Hall, 1984; Marsalek *et al.* 1998; Clemens, 2001; Butler & Davies, 2010). These basic hydrodynamic equations are based on conservation of mass and conservation of momentum, which are also considered as main processes in atmospheric models (see Section 4.1). As opposed to atmospheric models, water is modelled as an incompressible fluid.

Among the studies of which sewer impact results are presented hereafter, several studies were based on changes in design storms and event-based simulation of these storms in sewer system models. These include the studies by Niemczynowicz (1989), Waters *et al.* (2003), Nguyen *et al.* (2008b, 2010), Willems (2011), among others. Niemczynowicz (1989) applied changes up to 30% to two types of design storms: the Chicago storm and a block storm (constant intensity during the concentration time of the system). He simulated these changes in a model based on the SWMM software, for the sewer network of Lund, Sweden. The system was provided with combined sewers in the city center and separate sewers in the newer parts of the city and in the suburbs. For southern Ontario, Canada, Waters *et al.* (2003) changed 2-year 1-hour design storm intensities by 15% and simulated these for the the Malvern sewer network of a residential area in Burlington, also making use of the sewer simulation software SWMM. Nguyen *et al.* (2008b) derived for Quebec (Canada) IDF relationships for the current period as well as for future periods under various climate change scenarios given by two GCMs, after application of the spatial-temporal downscaling method by Nguyen *et al.* (2007) (see Section 7.3). On the basis of the derived IDF relationships, the design storms at a location of interest in the context of climate change and the resulting runoff characteristics from typical urban areas with various sizes, shapes and imperviousness levels were estimated (Nguyen *et al.* 2010). Also Willems (2011) applied changes to IDF relationships and design storms for impact studies on sewer floods. They updated composite design storms based on high-mean-low changes in IDF-relationships presented in Figure 8.6. The corresponding changes in rainfall intensities and composite design storms are shown in Figures 9.1 and 9.2.

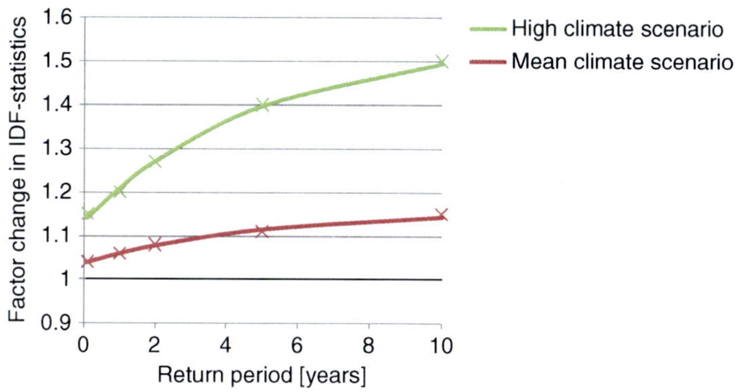

Figure 9.1 Factor change in rainfall intensity as a function of return period for high and mean climate scenarios (after Willems, 2011)

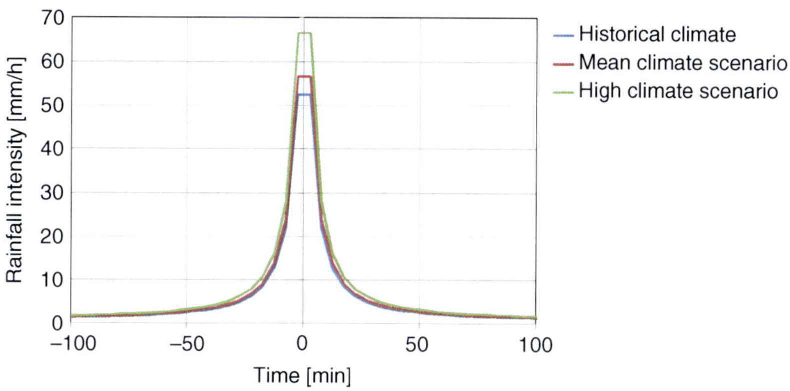

Figure 9.2 Change in the composite design storm for a 2-year return period for high and mean climate scenarios (after Willems, 2011)

Also in the class of event-based methods for propagation of rainfall changes, are the methods related to the use of probabilistic impact models. Two examples shown are the studies by Grossi and Bacchi (2008), who applied the semi-probabilistic model of Bacchi *et al.* (2008), and Cobbina *et al.* (2008) who applied the analytical probabilistic model of Adams and Papa (2000). They propagated changes in the distributions for the rainfall depth, storm duration and storm inter-arrival time to changes in the event runoff volumes, peak flow discharges and spill frequencies and volumes, and – based on these results – to changes in design values for storage facilities.

Other examples of sewer impact results shown are based on continuous simulation approaches, such as the results obtained by Semadeni-Davies *et al.* (2008), Olsson *et al.* (2009), Schreider *et al.* (2000) and Willems (2011). The approach by Semadeni-Davies *et al.* (2008) applied for Sweden their climate factors at a 6-hour scale to rescale a 10-year tipping-bucket series of observed 1-minute intensities. The series were simulated in the combined sewer system model for central Helsingborg, South Sweden, implemented in the MOUSE software. Later, Olsson *et al.* (2009) applied their continuous climate factor approach at the 30-minute scale to rescale a similar time series prior to climate impact simulations, also applying the sewer simulation software MOUSE, but for the city of Kalmar in southern Sweden. Using that method, contrasting future rainfall trends, such as decreases in total (seasonal) volumes but increases in the (short-term) extremes, were applied to the observed series. In southeastern Australia, Schreider *et al.* (2000) statistically downscaled changes in daily precipitation by means of two different approaches, the delta change approach with seasonal change factors and a stochastic Markov process based weather generator (Bates *et al.* 1993). Willems (2011) simulated for Belgium the impacts on sewer overflow frequencies and overflow frequencies of ancillary structures to these systems such as storage tanks using a continuous simulation reservoir-based model. This was done for a range of throughflow discharges that are commonly applied: from 1 to 50 l/(s.ha), assumed constant or linearly depending on the storage volume, and storage capacities which under current climate conditions correspond to overflow return periods between 0.5 and 20 years. The 10-minutes historical rainfall input series were perturbed by the quantile perturbation approach implemented by Willems and Vrac (2011).

All these studies have demonstrated the feasibility of linking GCM/RCM-based climate change scenarios with short duration rainfall extremes and runoff processes from small urban catchments, using different approaches.

Accounting for impact uncertainties

Whatever approach is implemented for the propagation of changes in rainfall to changes in urban runoff, it will become clear from the results presented in next section that drainage impacts depend significantly on the climate change scenario being considered. That is why the ensemble approach was recommended in Section 8.3. Instead of simulating the whole set of potential scenarios provided by the ensemble approach in the urban drainage model, simulations might be limited to selected scenarios, as discussed in Section 8.3. With such an approach, in case different climate forcing scenarios, GCM drivers, RCMs, downscaling methods, initial conditions, and so on, are considered, it also might be useful to test the significance of the differences in impact results for these different scenarios. In this way, one can gain a better insight in the main factors that control the differences in impact results between scenarios. In case a large ensemble set of scenarios is considered, they might be classified according to the type of scenario (climate forcing scenario, GCM driver, RCM, etc.) and the significance of the differences in impact results between classes tested. It might be useful to consider one representative scenario from each class rather than simulating the full range of scenarios. Climate scenarios moreover can be considered for various time horizons (e.g. 2020, 2050, 2100).

Next to the uncertainties in the climate models and climate forcing scenarios, uncertainties associated with the urban drainage impact model have to be taken into account. Some groups conducted research in this field and provided an assessment of the uncertainty in urban drainage impact models. Examples of such studies are: Schilling and Fuchs (1986) for a sewer stormwater modelling of a hypothetical network, Thorndahl and Willems (2008) and Thorndahl *et al.* (2008) for sewer surcharge, flooding and overflow results of an urban drainage system in Denmark, Dotto *et al.* (2012) for a separate stormwater system in Australia, Gaume *et al.* (1998) for an urban stormwater quality model of the combined sewer system of Quebec City, Willems (2006, 2008, 2012) for a sewer water quality model in Belgium, Arnbjerg-Nielsen and Harremoës (1996) for urban runoff and water quality modelling in Denmark, Freni *et al.* (2009) and Freni and Mannina (2010) for urban runoff and water quality modelling in Italy, Korving *et al.* (2005) for CSO volumes of a combined sewer system in The Netherlands, Willems and Berlamont (2002b) for the impact of combined sewer overflow results on the water quantity and quality state of a receiving river in Belgium, Mannina (2005) for combined sewer system modelling and impact analysis on a receiving river in Italy. See also Delectic *et al.* (2011) for a review study. However, it should be noted that these uncertainty assessments on the impact model are largely based on simulations for historical periods and after comparison with flow, water level and water quality observations. The uncertainty of the models in making extrapolations to more extreme conditions probably would be larger. Careful calibration of the impact models including extreme historical conditions might help (see Willems, 2009). However, as is also the case for a climate model or any type of model, their performance in making extrapolations cannot be assessed in a complete way. Van Steenbergen and Willems (2012) demonstrated that model performance for extreme conditions can be partly tested by analysing impact changes as a function of different classes of rainfall changes. They developed such a test by dividing the available historical rainfall and flow series in events, by defining pairs of events belonging to different classes of relative rainfall changes, and testing the performance of the runoff model in describing flow changes due to given rainfall changes.

Uncertainty assessment of the urban drainage impact model by Willems & Berlamont (2002b) learned for a case-study of a combined sewer system in Belgium that sewer flows and CSO volumes for individual historical storms can have errors up to 40%. Errors of same order of magnitude were found by Gaume *et al.* (1998) for the combined sewer system of Quebec City, by Mannina (2005) and Freni *et al.* (2009) for combined sewer systems in Italy, and by Dotto *et al.* (2012) for a separate stormwater system in Australia. A large part of this model uncertainty is due to rainfall input uncertainty (Willems, 2008). Hoppe (2008) warned the climate change impact modeller that the changes in rainfall intensity due to climate change are of the same order of magnitude as the errors in rainfall input due to rain gauge measurement errors, wind induced losses, wetting and evaporation losses or other rainfall input estimation uncertainties such as the uncertainty involved in areal rainfall estimation. This does, however, not mean that sensitivity analyses based on relative changes in the (uncertain) model input would not be useful. As is the case for climate models, impact analysis of relative changes might be more reliable than simulation of absolute model output results.

Greater care must be given to the water quality impact models. Willems (2008, 2012), Mannina (2005), Freni *et al.* (2009) and Dotto *et al.* (2012) all have shown that the total uncertainty in the outputs of sewer water quality models is an order of magnitude higher than for the sewer flow models. These authors have shown that water quality concentrations such as biochemical oxygen demand, total suspended solids, total dissolved solids, settleable solids and ammonia, can have errors up to 200% or more (for instantaneous time moments and extremes). However, when quantiles are studied, the results have reduced uncertainties in comparison with the model results for individual storms (Willems, 2006).

9.2 IMPACTS ON URBAN DRAINAGE FLOWS, SEWER FLOODS, SURCHARGES AND OVERFLOWS

Impacts of climate change only

An urban drainage system may dampen or amplify the changes in precipitation, depending on the system characteristics. Niemczynowicz (1989) found for the sewer network of Lund, Sweden, that the relative change in urban runoff volumes is higher than for the rainfall input. They found for a 1-year return period storm that 30% increase in the 40-minute rainfall intensity leads to 66% – 78% increase in the CSO volume (depending on the return period between 1 and 10 years and the type of design storm). The flow volumes downstream the combined system increased by 20% – 51% and downstream the separate system by 32% – 45%. They explained this increase by the increased contribution of runoff from permeable surfaces and the increased leakages to conduits.

In Sweden, Olsson *et al.* (2009) found based on impact simulations in the sewer simulation model MOUSE an increase in the number of urban drainage system surface floods of between 20 and 45% for

Kalmar City 2100. For the city of Odense in Denmark, Mark *et al.* (2008) found that the flood depths and the number of buildings currently affected once in every 50 years would correspond to a reduced return period of 10 years in the future (based on the impacts discussed by Larsen *et al.* 2009, and Arnbjerg-Nielsen, 2012). For the city of Roskilde, which is also in Denmark, Arnbjerg-Nielsen & Fleischer (2009) found that a 40% increase in design rainfall intensities would lead to a factor 10 increase in the current level of damage costs related to sewer flooding. The actual change of cost will depend on catchment characteristics. In a similar study for another location with the same amount of increase in intensity, Zhou *et al.* (2012) reported a factor 2.5 increase in annual costs. One common conclusion, however, is that the main impact of an increase in rainfall extremes is not primarily related to the additional cost of the most extreme events, but rather to the damages occurring much more frequently.

A higher factor increase in sewer impacts in comparison with the factor increase in rainfall was also reported by Nie *et al.* (2009) for Fredrikstad in Norway. They concluded that the total volume of water spilling from the flooding manholes increases 2–4 times the increase in rainfall, and the total CSO volume increases 1.5–3 times as much as the increase in rainfall. They also show that the number of flooding manholes and number of surcharging sewers may change dramatically and irregularly with a slight change of rainfall.

For the case of Bergisches Land in North Rhine-Westphalia (Germany), Hoppe (2008) found that a 5% precipitation increase may lead to a considerable increase in the number of flooded nodes (+32%) in the model. The sewer impacts though depended on the current utilization of stormwater capacities in various parts of the system and the topographical variations in the region (Hoppe, 2008). For areas in the system for which capacities still existed, precipitation increases may not lead to new areas that will be flooded. If the capacities in these areas of the system were exceeded, the number of flooded nodes did increase, usually rapidly, depending on the topography of the system. The same applied to the downstream stormwater volumes mainly as influent to wastewater treatment plants but also as combined sewer overflow volumes. Overall, an increase in precipitation was shown to have very different impacts on different sections of the system, depending on the network's current hydraulic capacity utilization and on the topography.

Willems (2011) found that for sewer systems in Flanders, Belgium, that were designed for design storms with return periods in the range 2–20 years, the present-day composite storms increased for the high climate scenario with a factor 1.15 (+15%) up to 1.5 (+50%) depending on the return period in the range from 1 month to 10 years (Figures 9.1 and 9.2). For the mean scenario, these changes were less: from +4% to +15%. For the high scenario, the return period of sewer flooding increases with approximately a factor 2.

Impact results for a residential area in southern Ontario, Canada, by Waters *et al.* (2003) were based on the 15% increase in rainfall intensity. This rainfall increase resulted in a 19% increase in runoff volume, and a 13% increase in peak discharge, which caused 24% of the pipes in the system to surcharge.

For an urban catchment in British Colombia in Canada, using the SWMM model, Denault *et al.* (2006) found that peak design discharges by 2050 might increase by more than 100%. This finding was not, however, based on projections from climate models, but by extrapolation of historical trends in short-duration rainfall extremes.

In Australia, Schreider *et al.* (2000) performed hydrological modelling in three different catchments upstream of Sydney and Canberra. They found that a 20% increase in daily summer precipitation by 2070 only has a minor effect on urban flood damages.

Impacts of climate change and/versus urbanization

In the studies listed above, climate changes were mainly assessed with respect to their effects on sewer surcharge, urban flooding, overflow frequencies and risks, but Chapter 1 highlighted already that other changes may occur in an urban area, which might affect or strengthen the urban drainage impacts. One of these are the changes in pavement, hence surface runoff, due to urbanization trends, which should not be seen as isolated but related to population growth and increase in welfare, hence somehow interrelated with anthropogenic climate change.

Semadeni-Davies *et al.* (2008) performed an analysis of the combined impact of climate change and increased urbanization in Helsingborg, Sweden, and found that the combined effect on overflow volumes could reach a 4-fold increase. Using a similar approach, Olsson *et al.* (2010) analysed future loads (until 2100) on the main combined sewer system in Stockholm, Sweden, resulting from both climate change and population increase. Annual total inflow to the treatment plant was estimated to increase by 15 to 20%, CSO volumes to increase by 5 to 10% and critically high water levels to increase by 10 to 20% in the first half of the century. As a combined effect of evaporation increase and the way population change was modelled, no further increase of the total inflow was found until the end of the century. CSO volumes and high water levels were however found to increase further reaching 20 to 40% and 30 to 40%, respectively, in the second half of the century. Both of these studies also highlight the importance of not looking at climate change impacts separately, but combining them with other key non-stationary drivers that are of equal importance. In fact, the study by Semadeni-Davies *et al.* (2008) clearly shows that climatic changes are not the most important driver of increases in pollution levels, and that the increases in damage may be effectively counterbalanced by measures not solely related to urban drainage. This will be discussed further in Chapter 10.

In another study by Tait *et al.* (2008), it was confirmed that additional increase in urbanization (related to population growth and increase in welfare) next to climate change is of great importance to changes in urban runoff. They assumed for a system in the UK that by 2025 next to climate change the paved areas will increase by about 25% of their current values and that roof areas will increase by about 10%. Urban drainage model simulations showed that the CSO volumes for a typical urban area in the UK would increase by about 15 to 20% when only the increase in paved areas is considered. These changes are comparable to the changes expected from future climate change (see Section 9.2).

Also Huong and Pathirana (2011) have investigated how the importance of climate related trends relates to the impacts of increased urbanization. They did that for Can Tho, which is the biggest city in the Mekong River Delta in Vietnam and is a typical example of a city in the developing world. This city is not only affected by change in short-duration rainfall extremes but also by changes in coastal levels and river levels. Huong and Pathirana (2011) used a set of model simulations to assess the future impact of climate change driven by sea-level rises and tidal effects, increases in upstream river runoff due to climate change and increased urban runoff through increases in impervious areas. They also studied the enhancement of extreme rainfall due to urban growth driven changes in the micro climate of the city (e.g. urban heat island effect; Dettwiller & Changnon, 1976). The urban growth of the city was projected up to the year 2100 based on historical growth patterns, using a land use simulation model. The urban heat island effect on extreme rainfall was simulated with the limited area atmospheric model WRF (see also Chapter 4 and Appendix C), coupled with a detailed land-surface model with vegetation parameterization. Hence, their impact assessment was highly "integrated" and multi-disciplinary. Final conclusion of their assessment was that the combined influence of the different effects leads to an increase of about 20% in inundation depths and areas. The impact of climate change was found to be more important than the impact of urbanization.

While the effects on sewer surcharge, urban flooding, overflow frequencies and risks are the primary effects in urban drainage, there are many other secondary or indirect effects that are more difficult to quantify. Examples of such related impacts are sewer back-ups in residential homes, wastewater on the streets during urban flooding of combined systems, non-insured losses and psychological impacts of experiencing floods. The main conclusion is, that to a large extent the sewer system is designed to cope with exceedances of the capacity, but that the surrounding urban environments are not. Therefore increases in the amount and extent of design exceedances greatly impact assets such as buildings and key infrastructure in urban areas.

9.3 OTHER TYPES OF SEWER IMPACTS

In addition to the impact of local short-duration high rainfall extremes on sewer surcharge, flooding, and overflows, other types of impacts might also occur, which are not related to the change in peak storm intensities but to other changes in climatic variables, such as change in cumulative rainfall volumes, dry spell lengths, temperature, evaporation, and so on. The climate changes moreover will not only have water quantity effects such as urban drainage flooding, but will also affect water quality and therefore receiving water environments.

The most direct effect on receiving rivers is the increase in wash off from impermeable and permeable surfaces due to increase in runoff and the higher frequency of CSO spills or stormwater outflows. These effects were shown by Niemczynowicz (1989) for the sewer system of Lund, Sweden. They simulated the impact of rainfall and runoff changes on pollution loads of total suspended solids, biochemical oxygen demand, total phosphorous and heavy metals to receiving rivers from CSOs and direct stormwater discharges from the separate system. They found for the 30% rainfall increase scenario, an increase in the annual pollution load by 32% – 71% depending on the type of pollutant.

One of the additional expected climate change effects is the change in dry spells. As pointed out before, due to climatic change rainfall variability might increase, such that increases in the flood frequency may for some regions be combined with increases in the low flow frequency. In the case of longer dry spells or low flow periods, water and wastewater will stay longer in the sewer pipes during those periods. Particularly in low and flat regions, this will lead to impacts such as higher sedimentation rates, more pipe corrosion, deteriorating water quality, and odour problems (Bates *et al.* 2008). According to Ashley *et al.* (2008a) the prolonged sewer residence times could further lead to septicity where dissolved oxygen levels are low and allow more in-sewer chemical transformations.

Higher sedimentation rates will cause elevated pollution concentrations in first flushes (during the first storms after a long dry spell period) due to remobilization. This will lead to higher pollution loads in combined sewer overflows and in the influent to wastewater treatment plants. The latter will lead to higher solids loads to clarifiers, changed treatment efficiencies and higher pollution loads. Downstream of the treatment plants, the receiving water bodies are also likely to be suffering from the long dry spells and will therefore have reduced capacity to assimilate the changed effluent discharges. Prolonged dilute loading of wastewater treatment plants due to low-intensity long-duration precipitation events can also affect wastewater treatment due to the lack of substrate or nutrients for biomass maintenance. This has potential major impacts on overall treatment (Plosz *et al.* 2009). It is clear that all these effects will require more resources for operation and maintenance and possibly retrofitting of improved wastewater infrastructure.

Similar conclusions were obtained by Parkinson *et al.* (2005) for the specific impact of reduced low flows and higher sediment deposition on downstream treatment plants. The reduction in sludge production was found to be insignificant. Impacts on CSO loads were, however, more significant. The study of Parkinson *et al.* (2005) was not undertaken in the climate change context, but rather in a study of the impact of re-using low WC flush and greywater on reduction in dry weather flows and related impacts.

In addition to the impacts due to altered rainfall and flow regimes, other climatic change effects such as changes in temperature and wind speed might lead to additional impacts. Temperature changes might affect sewer quality processes (e.g. change in process kinetics due to temperature change). Ashley *et al.* (2008a) studied the impact of temperature changes on downstream treatment plants (due to changes in temperature dependent transformation processes) and effluent loads, but found the impacts to be very small. COD levels even reduced due to enhanced transformation rates.

Besides the effect on transformation processes, increases in the temperature created by anticipated climate change can increase the risk of sulphide production in the sewer pipes to undesirable levels. This could lead to increased odour problems. Other climatic changes include the changes in wind patterns as air temperatures increase. These can result in different amounts and characteristics of sediments on catchment surfaces that are available for wash-off into urban drainage systems. Other long term changes may include the amount and character of wastewater and trade inflows, but these changes will be driven more by long-term societal changes than by climate change (Tait *et al.* 2008).

There is a strong non-linearity between rainfall change and change in flood and overflow frequencies. Similarly, water quality impacts also depend on changes in rainfall or CSO volumes in a highly non-linear way. In the study of UKWIR (2003), it was found that CSO spill volumes can increase greatly with climate change, but that this generally only has a minor effect on receiving water quality. This was due to the reducing duration of rainfall events under climate change. Of course, these changes in CSO volumes should be modelled in an integrated way together with the changes in river flows and concentrations (for the same period of the year) in order to obtain a more complete picture of the impact. Given that CSO overflows mainly occur in the summer season in Europe and that river flows might reduce in that season, CSO dilution effects might be less, thus increasing the impacts on river quality.

The interaction between ground water and the flows in an urban drainage system is also expected to change. Changing groundwater levels may indeed affect groundwater infiltration into urban drainage systems, particularly for pipes that are located below the groundwater table. For pipes that are located higher, leakage of stormwater and sewage may occur into the surrounding soil. It is well known that groundwater inflow may negatively affect the operation of a sewerage system and will usually increase the costs of wastewater treatment (Karpf & Krebs, 2004). The increasing inflow may also increase the frequency of combined sewer overflows, hence significantly influencing the impact to receiving waters (Niemczynowicz, 1989). The impacts are likely to be further exacerbated by more rapid biological degradation of the emitted pollutants.

In addition, flows in the urban drainage system are also likely to be influenced by changes in snowmelt patterns in mountainous regions and in high latitude areas. Sea level rise will impact the risk of flooding greatly in low-lying coastal areas. While the main driver is sea surges, there is also an increased risk of indirect flooding due to snowmelt runoff in low lying coastal areas. Snowmelt by itself might be too low to cause flooding, but it could do so in combination with storm surges if discharge into the sea is not possible. ADB (2011) has shown for Khulna in Bangladesh that sewer floods and waterlogging will occur more frequently not only due to increase in rainfall intensities but more importantly also due to sea level rise. Higher mean sea levels would delay the drainage from the low lying areas. For an increase in 6-h rainfall extremes after statistical downscaling of 1 GCM run and 40 cm sea level rise by 2050, they found that a 10-year sewer flood event will become a 5-year event in 2030 and 10-year flood event in 2030 will become a 5-year event in 2050. In the area of damaging floods, the mean flood depth for a 10-year sewer flood event in the current climate increases from 41 cm to 63 cm in 2050. The population exposed to this damaging flood increases from 24% to 58% in 2050.

An overview of such qualitative effects of climate change on in-sewer processes is given in Table 9.1 (Ashley *et al.* 2008a).

The higher temperatures and longer dry spells associated with climate change might also affect landscape based stormwater management systems. These systems and their importance for climate change adaptation

Table 9.1 Summary of the implications of climate change on sewer processes and modelling in a UK context (from Ashley *et al.* 2008a)

Element of model	Implications for quantity modeling	Implications for quality modelling	Need for more information to reduce uncertainties	Potential ranges of relevance to urban drainage system	Model/analysis – what to look at
Rainfall	More extremes; Local time and space resolution not sufficient; Essential to use long series events with known statistics.	Depends on wind patterns and temperature/effects on finest particles. Anthropogenic effects will also affect this.	Rainfall volumes and peaks will change. How significant are the quality aspects of rainfall?	Intensities increased by typically 20%. Spatial change – more clustered.	Range of examination to look at effects: +20% in intensity.
Runoff	Local changes in surface cover/soil moisture affecting the runoff processes. Potentially increasing contributing area (permeable). More potential evapotranspiration. Expect more direct rainwater use reducing runoff.	More/less soil erosion desiccation of soils – higher/lower water table interaction. Warmer water into sewer network.	Runoff changes important in both quantity and quality. Need better estimates of runoff rates and interactions. Investigate use of rainwater.	e.g. New UK runoff model (Ashley *et al.* 2007) so far untested. Temperature increases of up to 3°C on current maxima.	Hydrological model performance in relation to soil moisture. Temp increase on pollutants and generation, surface build up in inter-storm dry periods.
Sanitary and other inflows	Decrease DWF where there is water stress – but infiltration may increase where soil moisture is high. Industrial flows to reduce with more reuse.	More concentrated pollutants as pollutant load may remain same, but water use may fall. More rainwater use at source changing the inflow quality as well as quantity.	Investigate implications of large increases in concentrations in inflows.	100% increase in the concentration (but not loads) of DWF pollutants, such as BOD, COD and AmmN.	Double concentrations of these inputs and check consequences.
In-sewer hydraulics sediment/dissolved	More inflows – higher flood risk as runoff overwhelms capacity. Interactive effects with sediment deposits will change conveyance (more and less).	More and different sediment inputs in high runoff peaks. Less transport of sediment in lower flow periods. Higher temperatures will lead to more biofilms – reducing dissolved organics but enhancing generation of gases and bioprocesses. More likely to get a rapid onset of anaerobic conditions.	Balance between the variations in inputs and consequences to be investigated.	Bioprocesses influenced by both temperature and pollutants. Temp ranges increase by 3°C, pollutants likely to increase concentrations. More sediment related phenomena.	Temperature up by 3°, increase pollutant concentrations by 100%. Check effect of 30% flow reductions; in northern hemispheres look at sediment data for warmer climates.

Receiving water	Both higher and lower flow different times. Also higher temperatures.	Periods with low oxygen concentration likely to be more prolonged and frequent. Different flora and fauna populations changing the ecological status. Bacterial die-off rates both enhanced and reduced.	Implications of interactive effects with discharges. Revised standards to comply with Water Framework Directive and Clean Water Act.	Potential for lower DO background levels. Lower flows may coincide with CSO spills depending on relative times of concentration.	Try halving the minimum background DO. Lower flows in rivers – by at least 20%.
CSOs and associated storage	Longer and shorter storage times depending upon inflows.	Potential for onset of septicity and other problems specified in total emission studies (e.g. Durschlag *et al.* 1991)	Requires climate change scenario definition as many studies already done.	Rainfall, temperature and input changes all as outlined in the table above.	Vary parameters by ranges defined above.
WWTP	More peaky flow events. Flows – more insystem storage (at CSOs) transmits more flows to treatment.	Wider changes in rate and quality of flow arriving at the works. Temperature and load changes influence performance. Treatment processes may be more efficient at higher temperatures.	Many studies done on WWTP performance under wide range of conditions. Information should be available.	As above (e.g. Bixio *et al.* 2001).	All as above.
Models	Hydraulic relationships and parameters. Check how sediment transport is modeled.	Check model equations and parameterization. Consider using a stochastic approach instead of deterministic.	Too little published information on uncertainty in sewer flow quality modeling.	As above.	Using ranges above, check implications for the equation, and parameterization.

will be further discussed in Chapter 10. These vegetated *Water Sensitive Urban Design* (WSUD) systems involve engineered structures that are designed to capture and retain larger proportions of precipitation in urban areas. This is also done for the benefits of enhancing social amenity through greening of cities, reducing urban heat island effects and enhancing urban biodiversity. In well-designed systems, this is achieved without the need for watering with municipal potable supplies. However, these systems will themselves be impacted by longer dry spells. Indeed under projected climate change, plants in these systems are likely to suffer from the combined effects of increased water and heat stress. To investigate this effect, Chowdhury and Beecham (2012) analysed dry and wet spell characteristics for the cities of Adelaide and Melbourne in Australia by means of Markov chain models. Their study showed that:

- The design of vegetated WSUD systems, such as bioretention basins and swales, should allow for increased inter-event dry periods to ensure plant survival;
- The statistical behaviour of spells, their distribution and critical length are important parameters that need to be considered in the design of integrated urban water management systems;
- Appropriate design of storage capacity of WSUD components is also of great concern not only for ensuring supply reliability but also for sustaining vegetation growth in bioretention basins and other vegetated WSUD systems.

9.4 DISCUSSION

This chapter has shown that there are clear indications that the expected increased urban rainfall extremes in the future will affect urban drainage systems. In summary, the consequences can be of the following types (Berggren, 2008):

- *Technical*: damage to pipes, facilities, pump stations, infrastructure, land (erosion and landslides), and poverty, which affects for example the system capacity, other parts of the technical infrastructure in the urban environment and inhabitants in the city.
- *Environmental*: Spread of pollutants, nutrients, and hazardous substances in the water, soil and/or air, affecting the ecosystems and species especially in the receiving waters.
- *Economic*: cost of damage, cost of treatment of a polluted environment, and secondary costs, for example if people are hindered from doing their job due to infrastructure failure (roads, railways, internet, etc), versus financing and economic aspects of adaptation measures.
- *Socio-cultural*: In the city/municipality/country, some areas might be more affected by damage and pollution than others, and if these are areas where poor people settle, then a class or social distinction will develop within society. City centres are typically also carriers of the cultural heritage and such values may be lost without specific climate change adaptation initiatives.
- *Health*: people become sick or are injured or killed by the damage and the polluted environment, and also in connection to drinking water quality.

Although some studies looked at the impact on the flood cost, most studies limited their impact calculations to the most primary type of impact: the impact on runoff peak flows, flood or surcharge frequencies, and CSO frequencies and volumes. While the changes in short-duration rainfall extremes from the baseline period 1961–1990 till 2100 are for most regions up to 10–60% (see Chapter 8), the impacts on flood and CSO frequencies and volumes show a much wider variation. They can be as high as a factor 4 increase or as low as 5% increase, depending on the system characteristics. Floods and overflows are due to exceedance of runoff or sewer flow thresholds and react to rainfall and changes in rainfall in a highly non-linear way. Due to mathematical reasons, when the exceedance probabilities are lower or the threshold higher – this means for systems with a higher safety level – the relative change often becomes higher. There is no need to explain that the impact ranges can even by wider when studying environmental or socio-economic impacts.

A general comment that has to be made here regarding the impact analysis presented in this chapter is that the projected impacts are uncertain, not only due to the uncertainties in the climate projections, but also due to uncertainties in the impact models. Review in Section 9.1 of the studies that dealt with uncertainty assessment of urban drainage impact models learned that the impact model uncertainties in runoff flows are in general an order of magnitude lower than the climate change impact uncertainties reported in Sections 9.2 and 9.3, although care is required when predicting the impacts of more extreme conditions. Impact models have been calibrated and validated based on historical series of limited length, which may have included only a few high extreme events. Extension of the model simulation to future conditions should therefore be made with caution, especially when considering system performance under extreme events that might become more common under climate change (Ashley *et al.* 2008a; Van Steenbergen & Willems, 2012). Table 9.1 summarizes the need for more information to reduce uncertainties in the model-based impact analysis of climate change on in-sewer processes.

In addition to the scenarios of future precipitation extremes, the impact analysis should also consider scenarios of other key variables of change over time such as level of urban development, degree of imperviousness, local management practices, and other factors in order to improve urban stormwater management.

There may be other factors or trends that might be as or more important than climate change. For example, Koutsoyiannis *et al.* (2009) described the chaos and uncertainty (internal variability) inherent in rainfall patterns and compared this to the rather more certain projections of population growth and energy shortages. They suggested that planners should at least give increased consideration to environmental, social and economic changes other than to climate change alone. This view was supported by Ward *et al.* (2011) who identified energy shortages as being at least as important as climate change.

The next and final key question considered in this book is: how can we adapt to these changing climates and their associated conditions?

Chapter 10

Climate change adaptation and flexible design

Climate change adaptation is defined as initiatives and measures to reduce the vulnerability of natural and human systems against actual or expected climate change effects (IPCC, 2007). Section 10.1 discusses the scope and purpose of such adaptation in the context of urban drainage. This involves new design philosophies and adaptation options, as further discussed in Section 10.2. Given the high uncertainty in the future climate projections and corresponding impact results, there is a need for flexible designs (Section 10.3). Section 10.4 finally discusses that adaptive management also involves active learning.

10.1 SCOPE AND PURPOSE OF ADAPTATION

According to Refsgaard *et al.* (2012), climate change adaptation options can be described in four dimensions: intent (autonomous or planned), timing (reactive, concurrent, or anticipatory), spatial scope (localized or widespread) and temporal scope (short or long term). Choosing a set of options defines a strategy for climate change adaptation. The Stern Review (Stern, 2006) further highlights the principle of no-regret options, meaning adaptation that contributes in a positive way to society regardless of whether estimated climate change impacts actually materialize, by building resilience to society. In this context he distinguishes between the developed and developing countries, mainly due to differences in planning horizons.

Choice of adaptation strategy

Both mitigation and adaptation to climate change is an extensive task that will influence cities in many ways. Urban drainage is but one of many sectors that will be impacted by climate change. Further, cultural differences and available resources may lead to differences in choice of strategy. The development of sewer systems in the middle of the 19th century was based on the following set of options:

- Intent: planned;
- Timing: reactive, acting upon cholera epidemics throughout many cities;
- Spatial scope: widespread;
- Temporal scope: long term.

Sewer systems constructed based on this strategy proved very efficient in improving public health and protecting private and public assets. A propagation of this strategy will most likely lead to the same type of solution, however allowing for larger pipes to maintain the level of protection of assets. In this context the main question is which level adaptation should aim at. Should the current frequency of floods be maintained, should the climate change impacts only partly be mitigated by partial adaptation, or should adaptation lead to an even higher safety level than what is presently accepted?

However, other strategies could also be chosen, leading to completely different adaptation options (Arnbjerg-Nielsen, 2011; Chocat *et al.* 2007). Within the urban drainage community especially the spatial scope is discussed, that is to study what role local solutions can play in climate change adaptation. One of the first structured analyses of local solutions was presented by Stahre (2006), highlighting that these solutions may be optimal given that synergies with other sectors are taken into account. As such it is recognized, that considering only the urban drainage sector, local solutions are more expensive than widespread solutions. However, as shown in one of the case studies, the local solutions have had positive impacts in the neighbourhood and together with other initiatives decreased crime rates, improved employment rate, and led to better integration between different groups of citizens in the area.

It is beyond the scope of this book to give a comprehensive overview of adaptation in general, or even a complete overview of all the ways urban drainage can and should contribute to mitigation and adaptation of climate change. An example of such an overview is found in EU (2010). The choice of action in a specific situation will depend on the overall strategy and this strategy will in turn be dependent on the development of society as a whole. As such the scope of the present chapter is to highlight why urban drainage systems must adapt to climate change impacts and how specific adaptation options will impact both the urban drainage sector and other sectors. In general the focus will be on planned and anticipatory actions on longer time scales as that is a common feature for adaptation within public planning of the urban drainage sector. The main distinction then seems to be between centralized and local systems.

Need for adaptation within the urban drainage sector

Conventional practice involves long-term planning and design of urban drainage systems based on historical meteorological observations and the assumption that these observations adequately represent future meteorological conditions. It is clear that this approach is no longer feasible. Potential changes in climate and their consequences have to be considered and there are also other drivers in society that threatens business as usual in the urban drainage community, as discussed by for example Arnbjerg-Nielsen (2011) and Chocat *et al.* (2007).

Based on a review of downscaled climate model predictions for extreme rainfall and their impact on urban drainage systems (as described in Chapters 8 and 9), climate change may result in more surcharging and flooding problems. The impact indicators considered include runoff peak discharges and volumes, number of nodes surcharged or flooded, number of properties flooded, duration of flood, economic loss due to flood damage, inflow to wastewater treatment plants, and (for combined systems) CSO volumes and frequencies, and their quantitative and qualitative impacts on receiving surface waters. The rapid conveyance of water outside of the city also leads to warmer cities, because evaporation of water during dry weather helps cooling the city.

However, as shown in Chapters 8 and 9, the impacts have significant regional differences and also depend on the climate model simulation considered. An ensemble approach where several climate models, climate forcing scenarios, and statistical downscaling techniques are considered was therefore recommended. This allows the order of magnitude of the uncertainty levels caused by these approaches to be taken into account. As indicated before, the downscaled climate model results thus should in any case not be considered as predictions, but should rather be interpreted as indicative of the magnitude of rainfall change that could be expected for future decades. Urban planners and designers or rehabilitators of urban drainage infrastructure can use these results to begin to account for the effects of future climate change. This can be done by adjusting the design rainfall statistics or perturbing rainfall series, following one of the methods presented in Section 9.1. By means of urban drainage model simulations, changes in design values can be obtained in order to achieve given impact standards or constraint levels (i.e., allowed overflow frequency, required runoff control strategy). These constraint levels have an important influence on the update of the design values. When constraints become more stringent, small changes in rainfall statistics require large changes in design parameters (Cobbina *et al.* 2008). Obviously also the response behaviour of the urban drainage systems to changes in rainfall intensities strongly affects the required changes in design values. This highly depends on the system characteristics, such as the topography (flat/steep gradients) and the system configuration, which influences the hydraulic capacity utilization.

Some studies have investigated the change in design values. For existing sewer systems in the UK, the UKWIR (2003) study derived relationships between the proportion of storage volumes required by 2080 and the proportion of present day rain to reduce the flood risk to current standards. Regionally, the results varied significantly from a minimal amount to some 230% of current values. The annual CSO spill frequency and volume were found to increase by varying amounts across the UK, with smaller increases in spill frequency than in spill volumes (Ashley *et al.* 2008a). The models suggested that the storage volumes required to maintain performance at current levels of spill volume were 10 times greater than those required to maintain current levels of spill frequency. Willems (2011) considered changes in spill frequencies in Belgium and found that up to 30% additional storage would be required to limit the overflow frequency for a "high" climate scenario (see Section 8.3) by 2100 to the current level.

For Brescia, Italy, Grossi & Bacchi (2008) calculated the retention storage required for a fictitious drainage basin for each of their climate scenarios, which involve an increase in the rain storm intensity. They considered two options: increasing the average rainfall volume per event by 20%, or keeping the average rainfall volume per event the same but decreasing the average storm duration by 20%. They combined these changes with unchanged annual rainfall volumes, computed by multiplying the number of events per year and the average rainfall volume per event, or increased annual rainfall volumes by 10% or decreased by 20%. The required water retention storage increased from 22615 mm (under historical climate conditions) to values up to 27092 mm for a 10-year return period, and from 30018 mm to 35976 mm for a 50-year return period.

Waters *et al.* (2003) studied for southern Ontario, Canada, three retrofit options for the Malvern urban catchment that provide the required peak discharge reductions of 13% corresponding to a 15% rainfall increase scenario. The first option was disconnection of full/half roof areas. This provided peak discharge reductions of 39 and 18%, for full and half roof areas respectively. As second option, surface depression storage has to be increased by 45 m^3 per impervious hectare to provide a peak discharge reduction of 14%. The third option they studied is reduction of the rate of stormwater inputs to the sewer system by providing 40 m^3 of street detention storage per impervious hectare to reduce the peak discharge by 13%.

Whatever adaptation measures are considered, they have to be based on the expected level of performance (or acceptable level of risk) and the expected lifetime of the infrastructure/system. The resulting design criterion ensures that the service level remains above the selected "acceptable" level over a predefined portion of the infrastructure lifetime. It is argued that the definition of new design criteria should be part of a global adaptation strategy combining various measures to maintain an acceptable level of service in a long-term perspective (Mailhot & Duchesne, 2010). It also is clear that constraint levels (i.e., allowed overflow frequency, required runoff control strategy) have an important influence on the update of the design values. When constraints become more stringent, small changes in rainfall statistics require large changes in design parameters (Cobbina *et al.* 2008).

Apart from which scenario(s) to select for adjusting the design rainfall, and which design criteria to be taken, a further question of course is how extensive and how urgent are the required adaptations to cope with the projected climate change. The answer to this question is particularly difficult because of the high uncertainties associated with future climate change and its subsequent impacts. The solution depends on various factors including the severity of potential impacts, the socio-economic, environmental and public health consequences, the operational lifetime of the urban drainage system, and the options for flexibility in future adaptation.

Firm adaptation plans have already been established in Denmark and Sweden. Mark *et al.* (2008) reports on the costs of these adaptation plans. For Denmark, if all Danish sewers were to be upgraded today to comply with the rainfall foreseen in emission scenario A2, then the additional yearly investment in Danish sewers would be in the order of magnitude of 20% of the yearly maintenance budgets. Arnbjerg-Nielsen and Fleischer (2009) report for the city of Roskilde in Denmark that currently a total of 15 million DKK (2 million EUR) is spent annually on rehabilitation and maintenance of the city sewers. This will be increased by 1.75 million DKK (0.2 million EUR) each year to account for the expected gradual extension of the combined sewer systems. When design rainfall intensities are increased by 40%, the additional annual cost would be about 50 million DKK (6.7 million EUR). Figure 10.1 shows how the flood frequency and the sewer flood damage costs would increase for that city in relation to the return period.

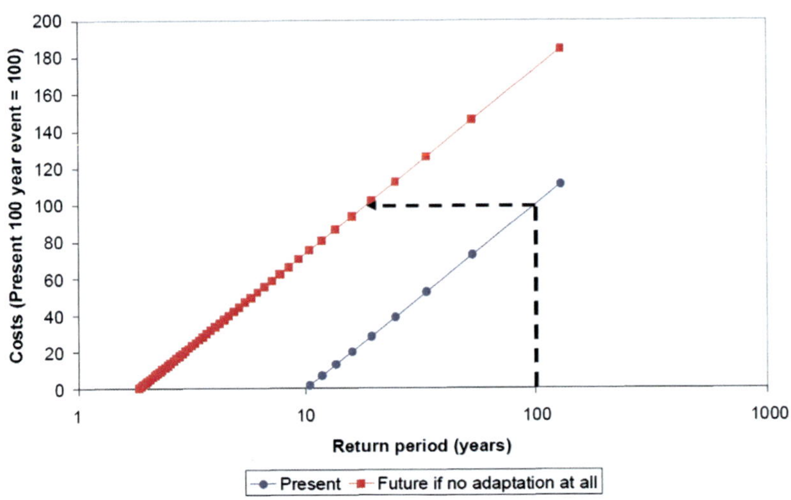

Figure 10.1 Change in sewer flood damage cost for the city of Roskilde in Denmark when the design rainfall intensities increase with 40% (after Arnbjerg-Nielsen & Fleischer, 2009).

For Sweden, the costs for adaptation of the wastewater systems have been estimated at 25 billion SEK (2.9 billion EUR) for the next 25 years, on top of the regular rehabilitation, which is estimated to be 50 billion SEK (5.7 billion EUR) for the same period (Svenskt Vatten, 2007). Recommended regional climate factors, describing the expected future increase of both annual precipitation volumes and short-term maximum intensities, are included in the revised guidelines for precipitation data used in sewer system design (Svenskt Vatten, 2011). The importance of taking climate change into account and aim at flexible solutions (Section 10.3) is emphasised in the guidelines for sewer system design.

An example from a developing region is provided by ADB (2011). They computed the adaptation cost for Khulna in Bangladesh, which is the third largest city in Bangladesh. The region is highly vulnerable and has a low resilience to climate change due to large population, high population density, inadequate infrastructure, low level of social development and lack of institutional capacity. Drainage is already a serious problem with frequent floods during the rainy season, and delayed drainage in low-lying areas during high coastal levels. Sewer flooding moreover causes problems of health risk by contaminated wastewater. The adaptation cost for improving the current drainage system of the city was estimated to be 25.7 million USD to make 80% of each ward free from damaging floods (defined as areas with flood depth higher than 30 cm) for a 5-year return period by 2050. About 7 million USD would be required to improve the current system, and an additional 18.7 million USD to cope with the climate change by 2050. The cost is higher, 26.7 million USD, for a 10-year event. They also made a rough cost-benefit analysis, of course based on several assumptions, and came to a benefit-to-cost ratio of 2.9 for a 5-year return period and about 5 for a 10-year return period. Socio-economic developments were taken into account in the analysis, based on scenarios for population growth, future water demand, GPD growth rate and change in impervious area due to urban development.

10.2 NEW DESIGN PHILOSOPHIES AND ADAPTATION OPTIONS IN URBAN DRAINAGE

Given that stormwater infrastructure generally has a long operational lifetime (up to 100 years) and that the socio-economic, environmental and public health impacts of potential climate change are severe, adoption of the precautionary principle is advisable. In other words it would be prudent to account for possible climate change in the design. When doing so, optimization based on a given future scenario is highly questionable given the high uncertainties in the future climate projections. Therefore, no-regret adaptation strategies or policies that build generic resilience to extreme weather conditions should be considered (Pathirana, 2008).

Parts of the urban drainage system with insufficient capacity to convey future design flows can be upgraded over the next few decades as part of a program of routine and scheduled replacement and renewal of aging infrastructure. Upgrading at the time or later replacement can be undertaken for relatively little additional cost and yet this will yield very high socio-economic benefits (Denault *et al.* 2002; Arnbjerg-Nielsen & Fleischer, 2009). This helps ameliorate the implications of climate change for drainage infrastructure, because these changes can be managed through long-term planning that accounts for future increases in flows.

Even less trivial is that long-term planning does not only require climate trends to be taken into account but also climate oscillations. Section 3.2 explained that (multi-decadal) climate oscillations make it difficult to detect climate change trends in historical time series. They also make it difficult to create awareness on the potential impacts of climate change. Long periods of relatively lower probabilities for extreme precipitation events and related floods also lower public and political awareness. It is moreover difficult to explain to the public the difference between natural variability and patterns, and the superimposed effects of climate change. Due to this complexity and difficulty, a risk exists that awareness of flood vulnerability only appears in irrational waves of short-term actions that do not include long-term perspectives. Decision makers need to be aware that climate oscillations may produce decades with relatively less extreme rainfall in their region, but the risks associated with extreme rainfall are expected to peak again, and that anthropogenic climate change might increase the magnitude of the subsequent peak.

Decentralized local storage to cope with the increased rainfall variability

The use of vegetated areas and other storage options where human exposure can be foreseen cannot be recommended for storage of combined sewage due to the hygienic risks to humans. Also stormwater drained from roads might be significantly polluted, which should be taken into account when considering adaptation solutions.

In many cities in the world, the rate of renewal of urban infrastructure is currently low, but this may need to change in the future, in order for communities to cope with population growth and climate change. Urban drainage systems need to be adapted in light of these societal and climate evolutions. At the same time these changes need to be consistent with the concepts of integrated water resources management. In many regions of the world, climate trends will lead to changes in the temporal variability of rainfall intensities: more extreme high-intensity rain storms, but lower cumulative volumes due to a decrease in the number of low-intensity rain storms. For southern and central Europe, climate impact modellers indeed anticipate a decrease in the total rainfall volume in the summer season due to a decrease in the number of rainy days, but if it rains the rain might be more intense, causing an increased variability of runoff discharges (Middelkoop *et al.* 2001; IPCC, 2007b; EEA, 2007; Boukhris *et al.* 2008; Kyselý & Beranová, 2009). This leads at the same time to more floods and more droughts or water availability problems. Both types of problem need to be dealt with in a combined way. For example, by implementing upstream storage of stormwater in infiltration ponds, stormwater will be kept out of urban drainage systems, hence potentially reducing the risk of urban flooding. Instead, the stored rainwater can be used for soil infiltration, thereby increasing groundwater tables and reducing low flow problems in rivers. In Australia, such water sensitive urban design is becoming a mainstream practice (Argue, 2004; Wong, 2005; Beecham, 2012).

The term Water Sensitive Urban Design (WSUD) was first referred to in various Australian publications exploring concepts and possible structural and non-structural practices in relation to urban water resource management during the early 1990s. Parallel design philosophies, such as Sustainable Drainage Systems (SuDS) were also developing in the UK and other EU countries. In the USA, China and Japan, WSUD is known as Low Impact Urban Design (LIUD), or just Low Impact Development (LID). WSUD, LID and SuDS embrace the concept of integrated land and water management and in particular integrated urban water cycle management. This includes the harvesting and/or treatment of stormwater and wastewater to supplement (normally non-potable) water supplies. More generally WSUD focuses on the interaction between the urban built form and the natural water cycle. It may be regarded as an alternative to the traditional "catch and convey" approach to urban drainage. WSUD incorporates the economic, social and environmental dimensions of managing the urban water cycle. It can use both structural and non-structural approaches tailored to each specific development to deliver the best outcomes against triple-bottom lined sustainability objectives (Diaper *et al.* 2007). In terms of typical structural WSUD components Beecham (2010) includes green roofs, bio-filtration swales, bioretention basins, rainwater tanks, infiltration trenches and basins, permeable pavements, wetlands and ponds.

However, non-structural measures are equally important. According to Pezzaniti *et al.* (2009), the guiding principles of WSUD are centred on:

- Reducing potable water demand through water efficient appliances, rainwater and grey water reuse;
- Minimising wastewater generation and treatment of wastewater to a standard suitable for effluent re-use opportunities and/or release to receiving waters;
- Treating urban stormwater to meet water quality objectives for reuse and/or discharge to surface waters;
- Using stormwater in the urban landscape to maximise the visual and recreational amenity of developments.

The legitimacy and acceptance of WSUD in the community is a critical issue to address in successful implementation and operation of WSUD systems. In many cases, WSUD can mean the local community has to play a more active role in managing water services that were previously managed centrally. This includes capital investment and maintenance of rainwater and greywater systems, behavioural changes to conserve water, and appropriate use of alternative water sources. There is the need for ongoing community education and engagement processes to ensure that the community understands the importance of WSUD and its role in integrated urban water management. This understanding, along with an appropriate suite of incentives, can increase the motivation of the local community to adopt WSUD features and behaviour. These community and other stakeholder issues are discussed further in Section 9.4.

Whether it is WSUD, LID, SuDS or another new design philosophy, a common feature of WSUD systems is often the disconnection of impervious surfaces combined with promotion of stormwater retention and infiltration through technologies such as green roofs, rainwater tanks, permeable pavements and bioretention systems (Kazemi *et al.* 2009, 2010 and 2011). Indeed, an interesting new development in permeable pavements is to incorporate retention storage within the pavement system itself, thereby creating a truly multi-functional land use. For example, in addition to its transport functionality, a car space in a permeable paved car park can serve as a flood attenuation structure, a treatment device for improving water quality and a storage reservoir for water reuse (Myers *et al.* 2011). This concept is shown in Figure 10.2. The potential for urban drainage systems, through WSUD, to be integrated into multi-functional urban land use is described by Beecham (2012).

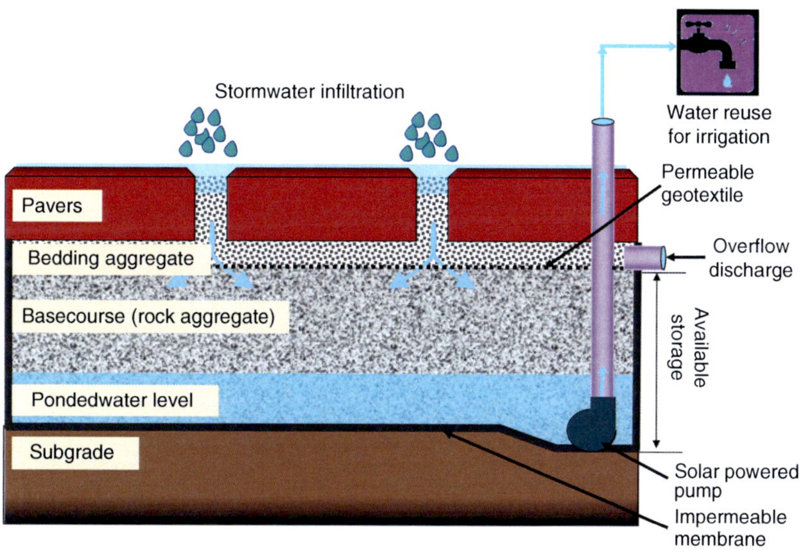

Figure 10.2 Permeable pavement system with water reuse (after Myers *et al.* 2011).

Sports fields, recreational parks, playgrounds, and other types of green spaces in the city can be used for stormwater storage during extreme storm events. Given that these storm events only occur a few times per year, these open spaces can be given multiple functions as described above. Each of these storage volumes has a limited size and thus limited effect on the downstream runoff, but the combined effect of a great number of such decentralized facilities may significantly reduce or attenuate the surface runoff (Sieker, 1998). Arisz and Burrell (2006) highlighted in this context that most storm drainage systems comprise minor and major drains. The minor drainage system is designed to convey stormwater runoff from more frequent storms (smaller, less severe events) thereby providing safe and convenient use of streets, parking lots, and other developed areas. For economic reasons, the hydraulic capacity of the minor drainage system is limited (generally the hydraulic capacity is designed to convey a flow with a return period of between 5 years and 10 years), and during extreme events there will be overflow in the street and roadway system. The major drainage system is designed to convey stormwater runoff from less frequent storms (larger, more severe events) when the capacity of the minor storm drainage system is exceeded. The major storm drainage system generally consists of open channels, rivers and streams, roadways, and detention/retention ponds. Arisz and Burrell (2006) stressed that the creation of a major drainage system and the potential upgrading of the hydraulic capacity of the major system components (drainage channels, retention/detention ponds, stormwater storage areas) requires space. It is easiest to designate space for the major drainage system during the planning and layout stages of a new development, as it is often very difficult to do so retroactively in existing developments.

In Figure 10.3, a few options are shown to create space for major drainage system or for decentralized urban stormwater storage. Since the major concern is avoiding loss of assets due to floods of very rare events the pollution is a minor issue compared to the situation where WSUDs are used to retain the bulk of the stormwater.

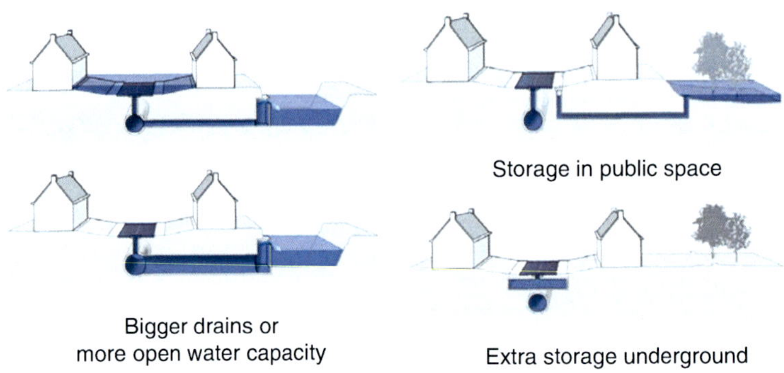

Figure 10.3 Examples of decentralized urban adaptation measures to cope with the future increase in extreme rainfall (after van Luijtelaar *et al.* 2008).

The option chosen depends on the allowed frequency of urban flooding, damage, cost, natural conditions, and the possibility of combining the measures with other land uses in the public space such as open public green spaces, drain replacements, road reconstructions and urban renewal. Among the natural conditions are the soil infiltration capacity and the topography. It is clear that in more flat urban areas, there is potentially a larger storage capacity on roads and in green areas.

Other benefits associated with multi-functional land use can include enhanced landscape amenity and biodiversity. Developments designed and implemented with a WSUD approach will often be perceived as having a higher landscape amenity, than similar developments designed without water sensitive features. Landscape amenity is improved through integrating design features, such as flood retardation, into the natural landscape in a way that both satisfies the need to mitigate the impacts of flooding events, while also providing green-space for public recreation. This green-space is important for social cohesion, and physical and mental wellbeing. WSUD also offers the chance to use alternative water sources for irrigation of gardens and public open space, which can improve the amenity of the local area, particularly when the use of drinking water is restricted for these non-potable uses. These factors whilst difficult to quantify, impact the overall well-being of the community and are also translated into a higher price premium for developments that incorporate WSUD (Diaper *et al.* 2007). Community and developer interest in WSUD can be driven by the indirect benefits associated with the implementation of such concepts in urban developments:

- Improved amenity;
- Improved well being, productivity and health;
- Potential for improved development or developer reputation, and
- Potential for increase in property market value.

One of the public health benefits of vegetated WSUD systems is amelioration of the urban heat island effect (UHI) by lowering surface and air temperatures (Lucke *et al.* 2011), largely through evapotranspiration. UHI exacerbates heat wave intensity in urban areas, and has caused more deaths in the USA than hurricanes, tornados, earthquakes, and floods combined (Golden *et al.* 2008). The functional characteristics of vegetated WSUD are important in current urban development for mitigating both climate change and UHI effects. Urban vegetation can also play a significant role in stormwater management, particularly by reducing runoff and enhancing water quality by absorbing and biofiltering stormwater flows. This provides essential protection for sensitive and fragile receiving waters. In addition, vegetated WSUD systems can provide habitat and bio-links for urban flora and fauna and can provide aesthetic and amenity values for urban dwellers, thereby improving the quality of life for urban residents. In a systematic series of studies, Kazemi *et al.* (2009, 2010, 2011) clearly demonstrated that bioretention systems improve urban biodiversity compared with even conventional urban green spaces from which they have often been created. This finding challenges the statement that urbanisation always results in biodiversity losses and biotic homogeneity and rather they demonstrated that urbanisation, through proper planning and management of new development proposals, can contribute to the conservation of biodiversity.

More efficient use of available storage capacities

In most cities, however, a major constraint is that the upstream capacity to capture and delay precipitation extremes are not sufficient to substantially mitigate flood risks due to climate extremes. Hence local retention measures can be implemented along with downstream control mechanisms, which can include both structural [such as storage reservoirs, levees, and implementation of real-time control (RTC) systems] and non-structural (such as contingency planning, education, and citizen awareness) measures.

Figure 10.4 gives an example (for the city of Bergisches Land in Germany) of how sewer overflow discharges change with changes in precipitation ($\pm$20% change to the actual situation) versus the effects of various types of system changes: change in throttle pipe dimensions ($\pm$20%), change in sewer infiltration volumes [+0.15 l/(s.ha)], the installation of a RTC system to reduce combined sewer overflow volumes and pollution loads, and disconnection of paved areas combined with decentralized percolation of stormwater (by 27.5% of paved areas). It becomes clear from the figure that the disconnection of paved areas has the strongest effect, and can compensate for the precipitation increases. Also RTC is useful for that system. For the city of Aalborg in Denmark, Arnbjerg-Nielsen *et al.* (2008) found that flooding a single large local area used for recreational purposes once in 10 years, would reduce the costs of flooding in a very efficient way (even though the capacity of the main sewer pipes remained unchanged; only some flood control actions and installation of outlet facilities from the storm sewer system to the green area are needed).

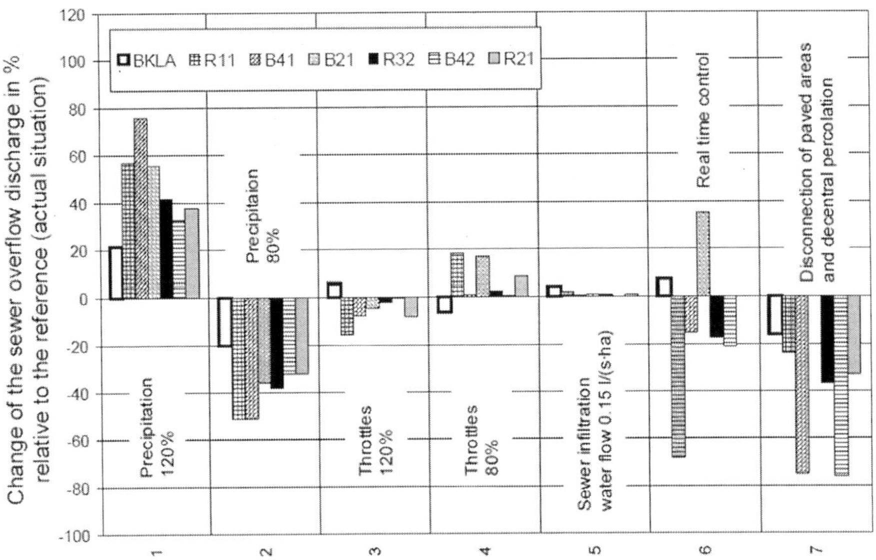

Figure 10.4 Percentage change in sewer overflow discharges for seven overflow structures in the 320 ha large sewer system of Bergisches Land in North Rhine-Westphalia (Germany) (after Hoppe, 2008).

Limitations of retro-fitting cities with decentralized storage and/or larger sewer systems

In some cases it can be foreseen that in a high climate scenario it would be impossible to retrofit urban infrastructure to limit the return period of flooding to acceptable levels (e.g. 2 to 10 years at the most critical location of the urban drainage system). In these cases, it has to be accepted that the frequency of sewer surcharge and urban flooding might be (too) high in the future. Adaptation would then involve reducing the vulnerability to compensate for the increase in exposure. This involves limitation of the damage the flood might cause. In some cases removal of buildings may be necessary, but often small changes in system, road and building designs – mainly close to the most critical locations of the system (which often only involve a limited number of locations) – can significantly reduce flood damages. Examples of such changes are: increase of door steps of buildings, increase of road borders or sloping kerbs, adequate height differences between the ground floor and ground level, small walls, no underground parking lots, drainage facilities in cellars, and so on. It is clear that this does require tackling mainly the genuinely vulnerable areas (the "hot spots"). Accurate high-resolution topographical information can rather easily be obtained and provides much insight in identification of these vulnerable areas and also helps to shape optimal solutions. In very flat areas, it is indeed a challenge to identify the vulnerable locations, where small errors in topographical elevations could be fatal, causing or preventing serious urban flood damage.

Another option is to plan for secondary drainage pathways to be used when the primary pathway, the sewer, has insufficient capacity. These secondary pathways will have to convey the runoff to areas that may be flooded without causing costly damages (e.g. Mark *et al.* 2008). All this requires municipalities to aim for a water-aware layout of their aboveground infrastructure, to cope safely with extreme quantities of rainfall (van Luijtelaar *et al.* 2008; Zhou *et al.* 2012). Examples of adaptation measures that meet this requirement are shown in Figure 10.5. The measures chosen depend on frequency and quantity of water in the street, damage, cost, present drainage and buildings, development plans, and the possibility of combining the measures with other operations in the public space (such as drain replacements, road reconstructions and urban redevelopment). Also the degree to which the hazard of flooded streets is accepted is of importance. In a questionnaire to 203 municipalities in the Netherlands, van Luijtelaar *et al.* (2008) found that only a minority of municipalities consider flooding of residential streets a problem (taking into account that the water remains on the streets maximum for a few hours after the rain storm). The same was concluded for (exceptional) flooding of parks, gardens or paths behind dwellings, loading docks of industrial premises, and so on.

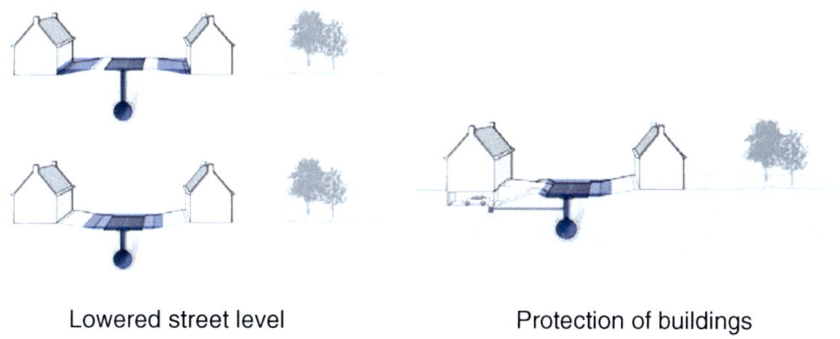

Figure 10.5 Two examples of options for coping with water in the streets and water-aware planning of the municipalities' aboveground infrastructure to prevent damage in case of increased frequencies of urban flooding (after van Luijtelaar *et al.* 2008).

Urban flood forecasting and warning

Another option to deal with the higher urban flood frequencies, in which sewer surcharge and flooding cannot be limited to acceptable levels, is to warn people about the floods by means of flood forecasting and warning systems. It is beyond the scope of this book to give an introduction to the large field of forecasting and warning systems, including how to link planning processes to operational management, including contingency planning. An overview and introduction to the relevant actions can be found at the United Nations programme on Disaster Risk Reduction (UN, 2012).

Water quantity-quality interaction

In addition to adaptation measures that focus on water quantity control, it is important to realize that adaptation for water quality control is also required. Increased stormwater pollution loads in the influent of treatment plants (e.g. during first flush) and in receiving water bodies often requires downstream pollution control such as storage-sedimentation basins, screening or chemical treatment at combined sewer overflows, and so on. Contaminated stormwater results in a range of negative social and environmental impacts such as eutrophication, loss of fish and wildlife habitat, aesthetic degradation, beach closings, and restrictions on fish and wildlife consumption.

Interaction with receiving waters

Urban drainage systems impact on receiving water systems, but the interaction may also be in the reverse direction. Rainfall intensities might increase at the same time as both urban runoff and river flow discharges. In these cases, floods along stormwater pipes will also increase due to backwater effects from the receiving river or due to limited drainage capacities. The same problems occur in systems that drain into the sea (due to sea level rise) (e.g. Domingo *et al.* 2010).

Another potential problem is the change in groundwater levels in the catchment, which can lead to increased infiltration into the urban drainage pipe network or leakage of sewer water into the soil (see Section 9.3). This underlines that climate change impact and adaptation studies should not consider urban drainage as an isolated system, but should take into account the interactions with groundwater and receiving waters.

Cross-disciplinary approach

These new concepts of urban drainage management put more emphasis on the cross-disciplinary skills of water engineers, designers, planners. For example, designers will in the future be challenged by social, cultural, technological and business developments, and there will be a constant need to find novel solutions. In their professional practice, good designers of urban drainage systems will be distinguished by their capacities for innovation. Required skills will include initiative and originality as well as critical ability. In terms of field of practice skills, designers will require greater understanding of biological systems and their interaction with water infrastructure (Beecham, 2010). Effective communication skills will also be required to raise public awareness and to engage with stakeholders. This higher level of multidisciplinary cooperation and stakeholder involvement will be further discussed in Section 10.4. From an engineering viewpoint, more creativity will be required to look for the best site-specific solutions rather than generic nationwide plans. Setting a general level of protection, or a strict application of standards, may lead to excessive or unnecessary measures being taken (van Luijtelaar *et al.* 2008). For example, a single design threshold across an entire city will indeed not adequately address the risks associated with higher frequencies of extreme rain storms and the high uncertainties of potential climate change impacts (Schimek *et al.* 2008).

As pointed out by van Luijtelaar *et al.* (2008), the new concepts also ask for a much broader approach than the usual design framework based on hydrologic and hydrodynamic model simulations. More types of information have to be collected and brought together. Information about the possible causes and effects of urban flooding could be presented as a multi-layer map that includes GIS data of the location of local ground level depressions, location of accessible basements under buildings, roads that function like open channels, observations of residents, and so on. Overlaying this information could lead to a better understanding of the mechanisms of flooding and how best to take precautionary measures. The information can also be used if further development is planned and measures can be taken to climate proof existing hot spots. The multi-layer GIS-based "urban flood risk landscape" that is created, can also be used in feeding back relevant information about possible flood threats to the public and to local stakeholders, thus helping to stimulate public willingness to support local measures (see Section 10.4 for more discussion on the role of local stakeholders).

Hydrological and hydrodynamic models play an important role in that process, including models that allow to provide two-dimensional flood maps. Whereas some years ago hydrodynamic models used for the design of sewer systems were not yet suitable for the analysis of flooding in urban areas, nowadays they can be combined with two-dimensional surface runoff modelling using high-resolution digital terrain model data (Maksimović *et al.* 2009; Leandro *et al.* 2009; Simões *et al.* 2011; Seyoum *et al.* 2012). In this way, the models do not only indicate where and when flooding will start, but also provide information on the flood depth and where the flooded water will flow. The latter is difficult to simulate, partly because of the level of detail and data required, including all small topographical variations, types of obstacles (including traffic/speed humps), potential flow paths (such as drive-in garages), and so on. Strong progress is, however, made in this field both in ongoing research and software development.

The above discussion shows that in the context of climate change, knowledge and expertise will play a more important role in the future design and operation of urban drainage systems. This means that next to good knowledge on urban drainage design methods and advanced skills in the use of simulation models, the role of good practical and local knowledge will increase in the decision making process (van Luijtelaar *et al.* 2008). This is illustrated by Figure 10.6, which is based on the learning stages of Dreyfus and Dreyfus (1985). The figure shows that the role of expertise and craftsmanship is becoming more important, and that this requires experts that use models in a flexible way to design change measures (developments) based on facts (contexts, e.g. climate change).

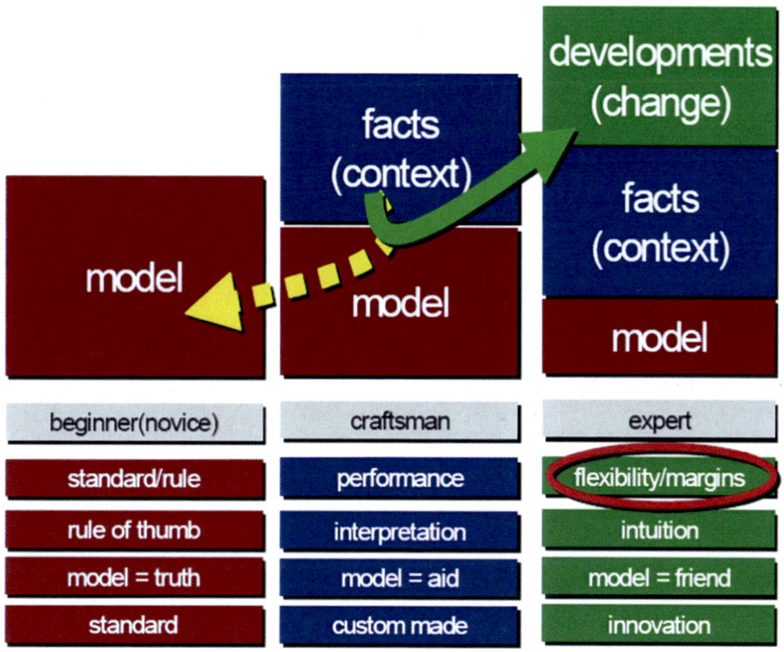

Figure 10.6 Learning stages of Dreyfus (after van Luijtelaar *et al.* 2005).

10.3 COPING WITH UNCERTAINTY: FLEXIBLE DESIGNS

Although the needs for future adaptation are in many cases clear, new designs or investments should account for the high uncertainties associated with future climate change projections (within the time horizon of the operational lifetime of the urban drainage systems). This requires a change in philosophy in terms of the future planning, design and rehabilitation of urban drainage systems or in fact of the entire built environment. It requires flexible/adaptive approaches that are resilient. New types of flexible designs, which allow for changes in response to new knowledge, have to be considered. Examples of design changes are increasing pumping capacity and incorporating additional storage capacity to cope with the impacts of a future climate. Depending on the design objectives, the vulnerability and envisaged life cycle of new facilities, the possibility of upgrading should indeed be considered in today's design practices (Verworn *et al.* 2008).

In order to be able to do this, designers have to take into account the ensemble range of future climate projections, based on current knowledge of the climate system and an understanding of societal trends.

This means that different plausible scenarios have to be simulated in urban drainage models, based on up-to-date climate models and climate forcing scenarios. Sustainable urban water management requires actions, measures and designs to be made that are climate-resilient, or in other words "which we will not regret in the future". The solutions or measures for adaptation should indeed be robust and able to cope with a variety of future changes, within the uncertainty band on future projections. They need to be "no regret" in that they should be potentially reversible, or abandonable, if they are found to be inappropriate, ineffective or inefficient in the light of the future knowledge (Ashley *et al.* 2008b; Brekke *et al.* 2009).

Adaptive approaches mainly originate from the ecological sector where high uncertainties in the knowledge of processes and responses are inherent, even beyond the climate change context. In ecology, traditional engineering design approaches are often not applicable because of scientific uncertainties. As described by Fotherby *et al.* (2008), this also involves challenging modifications to the roles of the scientist, the manager and the engineer. As explained below, this list should be expanded to include property owners and users, insurers, city and green area planners, socio-economists and politicians.

10.4 ADAPTIVE MANAGEMENT AND ACTIVE LEARNING

The concepts of integrated water management that are well established today require strong integration and cooperation between all stakeholders including water managers, city planners, social scientists and economical experts. This clearly becomes even more important in the context of climate change and adaptive management. Ashley *et al.* (2008b) quoted Williams *et al.* (2007) that "without active stakeholder involvement, an adaptive management process is unlikely to be effective". While traditional approaches in the urban drainage field include large engineering flood defence infrastructure, designed and implemented by engineers and funded through specific governance and institutional arrangements, climate change impacts and its associated high uncertainties require new ways of working between stakeholders. Schimek *et al.* (2008) point out that this requires a shift from a capital-focused approach to a portfolio approach that adopts the principles of asset management.

According to Ashley *et al.* (2008b), this requires a commitment to ongoing *active learning* and partnerships and collaborations via learning alliances. Active learning (also called *social learning* by Pahl-Wostl *et al.* 2008) leads to a shared understanding of the water system and the challenges facing this system. It can develop the capacity of various stakeholder groups to accept different interests and points of view on risk and performance of water systems. It also enables utilization of various types of response and if necessary at different times of implementation. Consequently it can lead to collective management of water resources in a sustainable way. The usefulness and importance of adaptive learning is shown by the fact that flood receptors (people, nature) have historically been able to adapt. The history of human society indeed shows that communities have been aware of the need to live with flooding and have adapted through experience.

Ashley *et al.* (2008b) point out that stakeholder participation and active learning processes require that professional and institutional stakeholders behave differently in regard to the public and community stakeholders by being more inclusive and willing to share knowledge at the appropriate levels. Many urban communities have to change their current held view that flood risks are unacceptable and that these risks have to be dealt with by governmental agencies. Also the role of the insurance industry has to be considered. In many European countries, citizens are routinely insured against flood risks. The same citizens, however, have to be aware that insurers are now re-appraising their position in areas where local flooding is becoming more common and where flood insurance may become too expensive in the future (Ashley *et al.* 2008b).

An example of stakeholder cooperation is the increasing demand for upstream stormwater storage and infiltration, which requires close cooperation with city planners. Given that it involves a decentralized approach of rainwater source control (see Section 10.2), which often involves locating storages on private land, the role of individual citizens is again shown to be of high importance (Staufer *et al.* 2008). This often introduces the challenge of dealing with a far greater number of participants, who are often supported by political representatives.

Following van Luijtelaar *et al.* (2008), 40% of the municipalities in The Netherlands say that public acceptance of water in the streets has clearly declined. This shows that it is necessary that adaptation strategies are combined with communication of adequate information on the controlled storage and discharge of water in the streets with the aim to prevent property damage. Through such stimulation of public awareness, individual property owners could also be convinced to contribute by restricting the impervious areas and by allowing rainwater infiltration on their own land. They also could be provided with relevant information about possible risks, thereby allowing the public to protect their own properties. Citizens and property owners could also help in keeping drainage inlets clear of leaves and debris during the flood season, to clear their own gutters, and to generally maintain the upstream property drainage systems.

Social scientists are able to quantify the risk awareness of the population and its willingness to take part in the change (e.g. German Klimanet project; Staufer *et al.* 2008). In cooperation with city planners, ecologists and social scientists, engineers can determine how the installation of additional stormwater storage and infiltration ponds can be integrated into the new urban form, thereby contributing to community wellbeing. They can indeed be seen in connection with planning of new recreational areas or other shared public spaces.

Given that these local scale measures will also affect more downstream water systems, cooperation needs to be organized between the local "municipality level" stakeholders (including city planners and potentially local property owners) and river basin or catchment authorities. As outlined above, the same applies to cooperation with insurers, spatial planners and other stakeholders. Adaptive management can also help to overcome conflicts which currently exist between these stakeholders, or between groups of stakeholders. This can include conflicts between upstream and downstream water managers or between water managers and insurers.

These adaptation strategies have to go hand-in hand with "institutional adaptation" as explained by Crabbé and Robin (2006). These authors focused on the bureaucracies of Canada and the financial and physical responsibility that local municipalities will need to bear in adapting infrastructure to climate change. Attention was drawn on the institutional costs and the potential increases in both revenues and expenditures for local and regional governments when adapting to climate change. They also pointed out that there might be institutional barriers, such as a lack of skilled scientists and engineers and reliance on management-by-crisis rather than long-term management and planning. Crabbé and Robin (2006) proposed potential approaches that include easily understandable climate-change reporting, increased citizen participation and financial assistance from regional governments.

10.5 DISCUSSION

This chapter explained how we can cope with the climate trends towards more extreme short-duration rainfall extremes and related impacts in urban hydrology. Strong emphasis was put on the adaptive approach, which is commonly used in the ecological sphere, particularly in ecosystem rehabilitation and restoration planning, where uncertainties are very high and are scientifically acknowledged. Due to similarities with the high uncertainties in future climate projections, application of this essentially ecologically focused approach to urban water and flood management would provide more timely achievement of objectives and more cost effective and feasible solutions. As shown throughout the chapter, adaptation involves the following three aspects.

- The need to change current water management strategies due to climatic evolutions. These include operational and infrastructural changes, but also changes in public awareness and expectation.
- The need to have flexible, adaptable, resilient and abandonable approaches. The approach in the past of using large infrastructure cannot be sustained as this will generally be unaffordable on the scale needed. It will also "lock in" the use of the large infrastructure for many decades and perhaps even centuries. The performance of any solution to the flood risk problem will degrade over time; hence, there is a need to respond over time as the external climate drivers evolve and as assets deteriorate with time (Ashley *et al.* 2008b).
- The ability to try different approaches or solutions, abandoning those that are found to be ineffective or inefficient.

Given that climate change occurs gradually (most of the projections summarized in the Chapters 8 and 9 were for the next 50 to 100 years), there is no need to invest heavily today in upgrading all infrastructure as soon as possible. Instead it is recommended that the potential consequences of current climate projections are assessed and that this knowledge is incorporated into current and future maintenance plans.

Points of interest for adaptation are favourably located at the head of the urban drainage system. This may involve source control through disconnection of impermeable surfaces, increased retention and infiltration, as well as upstream pollution control. There will be a need for more natural urban drainage approaches and installation of "green" stormwater infrastructure, all of which requires a change in design philosophy. There will be an increasing need to incorporate roads and parks into the active urban drainage system. This may be a new situation within parts of Europe, but as Grum *et al.* (2006) and Parkinson (2003) pointed out that this is common practice in tropical regions. It also fast becoming mainstream practice in Australia (Beecham, 2012). In addition, the design and management of urban drainage systems needs to change. There is a current trend of reducing the role of experts. There needs to be more focus on controlling local damage, for which specific knowledge and expertise is required which is in contrast to the general perception.

Important to note in this respect is that, in contrast to many water management sectors, adaptation of urban drainage systems is mainly undertaken at the municipal level. Municipalities and authorities responsible for local drainage as well as wastewater collection and treatment are also responsible for addressing climate change effects in their master plans for drainage, sanitation, development and area planning. These authorities have the means to install alternative decentralized stormwater retention, local infiltration and controlled surface flood ways and to regulate these measures through planning and building laws. Through these means, municipalities can prevent undesirable building developments. Through long-term area planning, a more controlled approach to planning can be considered. Hence, climate adaptation measures can also help achieve other municipal sustainability goals and should not be considered only as stand-alone actions to address climate change (Richardson, 2010).

The majority of the book is devoted to understanding the complex hydrological cycle with focus on extreme precipitation and notably how to predict the frequency and size of future extreme precipitation. However, this chapter clearly shows that there are other equally important drivers within the context of design and performance evaluation of urban drainage infrastructure, namely intense city development and change in people's perceptions and preferences over time. The last 160 years there has been one dominating concept of urban drainage: an underground piped network where the water is moved by means of gravity. The service levels have been set by engineering communities without much public debate.

This concept is now being challenged by both deteriorating infrastructure in the developed world and difficulties in generating the capital costs needed for establishing the infrastructure in the developing world. In combination with a higher demand for quick return of investment and higher awareness of future changes there is a recognised need to rethink the concepts of urban drainage. Climate change impacts on precipitation clearly show that a business-as-usual approach is not feasible in many regions of the world. The issues raised in the present chapter point out, that it is in many situations necessary to re-evaluate the entire concept of urban drainage rather than "just" upgrading the technical solutions we have implemented over the last 150 years. It is necessary to establish and maintain hygienic barriers and to build cities that interact with water in a healthy, environmentally friendly, and cost-efficient way. This may include the use of sewer systems, but perhaps other solutions should be investigated as well.

Chapter 11

Concluding remarks

11.1 KEY MESSAGES FROM THIS BOOK

This book provides an overview of the impacts of climate change on short-duration rainfall extremes and the related consequences at the scales relevant to urban hydrology. Also the state-of-the-art practices to study these impacts are presented. Key messages provided by the different chapters are:

- **Chapter 1:**
  - Current discussions of climate change impacts on hydrology generally focus on large scale hydrological impacts, for instance on floods, low flows, groundwater recharge, and droughts along major rivers.
  - Future scenarios offer little information on the fine temporal and spatial scale hydrological impacts of relevance for urban drainage applications.
  - This is despite the fact that higher risk of flooding caused by direct runoff from rainfall is a key impact of climate change.
  - Investigation of such impact poses particular difficulties due to the small scales involved.
- **Chapter 2:**
  - Statistical modelling of the total short-term rainfall process is mainly performed using either point process or multifractal theory. Extreme value theory is used to describe extremes, either annual maxima or peaks-over-threshold values.
  - For design purposes, the statistical properties of rainfall extremes are compactly described in the form of Intensity-Duration-Frequency (IDF) curves which are often used together with design storms.
  - A key objective of rainfall analysis is to develop tools for scale shifts. Areal Reduction Factors (ARFs) are used to convert between point extremes and spatial averages. Multifractals, expressed in the form of a random cascade process, are generically suited for rainfall disaggregation.
  - Rainfall statistics and stochastic models that are typically used in support of urban drainage studies, such as IDF curves, design storms and rainfall generators, are typically derived from historical series, under the intrinsic assumption that these series are stationary. This assumption is likely to be violated in a changing climate.

- **Chapter 3:**
  - Difficulties exist in identifying climate change trends in historical series of rainfall extremes because of limited data sets as well as short- and long-term persistence and instrumental or environmental changes.
  - Extrapolations of historical trends to future decades can be made but the future trends will have a higher degree of uncertainty.
  - For these reasons, as is the case for any modelling application, it is recommended to combine the empirical data with the results from physically-based modelling by means of climate models.
- **Chapter 4:**
  - Climate models make projections on the response of the atmosphere to external forcing including GHG concentration. Such projections are inherently probabilistic and it is important to treat them as such in further analyses.
  - Global Climate Models (GCMs) produce results at spatial and temporal scales that are far too coarse for urban drainage applications.
  - GCMs typically have poor accuracy in simulating precipitation extremes, because of the simplified representation of clouds, convection and land-surface processes at their coarse spatial scale, and because of the complexity of the feedbacks or mesoscale circulations that are not resolved in the models.
  - GHG emission scenarios used to date are based on a broad range of socio-economic scenarios. New scenarios are based on sets of future pathways of GHG concentrations (Representative Concentration Pathways; RCPs) in the atmosphere and levels of radiative forcing.
- **Chapter 5:**
  - Increasing the resolution of a climate model for a particular area of interest can be achieved with dynamical downscaling using a high-resolution Regional Climate Model (RCM) or Limited Area Model (LAM) nested in a GCM.
  - This type of models may be run on personal computers, not only at the synoptic scale but also at the smaller meso-scale area or even at the scale of a single city. Fine-scale LAM simulations may provide useful insight in for example how local geographical features affect fine-scale rainfall intensities.
  - Currently, urban drainage impact analysis generally have to rely on the synoptic-scale RCM simulations from public databases, given that only these provide easy access to sufficiently long simulation runs (30 years or more) for an ensemble of scenario simulations.
- **Chapter 6:**
  - Existing assessments of RCM rainfall output generally conclude that biases exist and that the grid-scale representation of short-duration rainfall extremes differs widely from local observations. This results in the necessity for bias correction and downscaling using empirical-statistical methods, before using the data in impact modelling.
  - Both biases and future changes differ between projections, depending on global model, regional model and emission scenario. An ensemble approach therefore is recommended where a set of scenarios is considered, leading to a range of plausible impacts.
  - Good modelling practise indeed includes consideration of uncertainties, but this becomes even more important in climate change impact analysis, certainly in urban hydrology.
  - At the same time, it must be recognised that the total uncertainty of climate projections is likely to be larger than that exhibited by an ensemble of models, because the models share the same level of process understanding and sometimes even the same parameterization schemes and code.
- **Chapter 7:**
  - Several methods for statistical downscaling exist, each with their own assumptions, all based on the combined use of historical data and climate model results.

  - Methods exist that do not make direct use of the rainfall results of climate models, but that project rainfall changes from changes in other more reliable climate model outputs such as atmospheric circulation and temperature.
  - There is generally limited possibility to validate the statistical downscaling assumptions. Good practice therefore involves assessment of the uncertainties associated with the downscaling step.
  - Application of different downscaling methods should be recommended, but limited to the methods for which assumptions – that can only be partly tested – are found valid for the specific study area.
  - Statistical bias correction, temporal and spatial downscaling can be performed separately, or combined. Due to lack of accurate long-term spatial rainfall statistics, spatial downscaling and bias correction of rainfall intensities at given temporal scales are difficult to separate, hence are commonly combined.
- **Chapter 8:**
  - Due to the difficulties and uncertainties in climate change impact modelling and analysis on the urban scales, caution must be exercised when interpreting climate change scenarios.
  - Typical increases in rainfall intensities at small urban hydrology scales range between 10% and 60% from historical control periods in the recent past (typically 1961–1990) up to 2100.
  - Consideration of an ensemble approach where several climate forcing scenarios, climate models, initial states and statistical downscaling techniques are considered, allows the order of magnitude of the uncertainty associated with each aspect to be assessed.
  - Whatever methods are adopted, the resulting change should not be interpreted as an exact number but only as indicative of the expected magnitude of change within the next 20 to 100 years.
- **Chapter 9:**
  - Climate change impacts on extreme short-duration rainfall events may have significant impacts in terms of surcharge of urban drainage systems and pluvial flooding. Results so far indicate more problems with sewer surcharging, sewer flooding and more frequent CSO spills.
  - There are also many other types of severe consequences at the scales relevant to urban hydrology, such as sedimentation, environmental/water quality, damage to infrastructure, and even socio-economic and cultural effects.
  - Next to regional differences in extreme rainfall and other meteorological changes, the precise impacts will also depend on local topography and on urban planning practices.
  - Extreme rainfall changes in the range 10–60% may lead to changes in flood and CSO frequencies and volumes in the range 0–400% depending on the system characteristics. This is because floods and overflows are due to exceedance of runoff or sewer flow thresholds and react to rainfall (changes) in a highly non-linear way.
- **Chapter 10:**
  - Urban planners and designers of urban drainage infrastructure can use the projected changes in precipitation and other key input to start accounting for the effects of future climate change. The sections of the urban drainage system with insufficient capacity to convey future design flows can be upgraded over the next few decades as part of a program of routine and scheduled replacement and renewal of aging infrastructure.
  - The large uncertainties that currently exist should not be an argument for delaying climate change impact investigations or adaptation actions. Instead, uncertainties should be accounted for and flexible and sustainable solutions aimed at. An adaptive approach has to be established that both provides inherent flexibility and reversibility and also avoids closing off options. This is different from the traditional engineering approach, which is rather static and is often based on design rules set by engineering communities without much public debate.

- This adaptive approach involves active learning, hence recognizing that flexibility is required as understanding increases.

11.2 FUTURE DEVELOPMENTS

As highlighted from the start of the book, climate science evolves very rapidly in connection with major international initiatives and collaborations. The 5th Assessment report of IPCC (AR5) is scheduled to be published in September 2013 and for example the CMIP5 and CORDEX projects are currently providing the scenario simulations for AR5. This means that a new generation of climate scenarios, based on a new climate forcing concept (RCP) and generally of a higher spatial resolution than previous scenarios, will be available for analysis and impact modelling. So far, new scenarios have generally overall agreed well with earlier ones with respect to future changes of key climate variables (including rainfall), but in some cases a better reproduction of historical climate is obtained, increasing the overall credibility of the simulation.

Besides analysing new scenarios it is important to follow the development of climate models closely and to critically evaluate their outputs with respect to reproduction of local rainfall extremes for urban hydrological climate change impact assessment. Significant research is taking place in these areas. High-resolution models with improved descriptions of fine-scale processes are being developed and within a few years simulations on single km^2 grid-scales will be available. Expected advances on the global scale include climate simulations that are "in phase" with low-frequent climate oscillations and so-called Earth System Models with a more complete description of the biogeosphere are under development. It should be noted, however, that due to the current limited physical understanding of the rainfall generating processes at small (e.g. convective) scales, both dynamic downscaling and statistical post-processing will remain needed for high-quality climate-related impact assessment studies.

This book shows that the topic of climate change and urban drainage involves several aspects, each with their own limitations and uncertainties. Future developments therefore have to focus on all these aspects in a balanced way, ranging from climate models to statistical methods and urban drainage practises. These developments may, however, evolve much faster if cooperations would be enhanced. Several chapters in this book highlighted that large gaps exist between what urban hydrologists need and what climate modellers currently can offer. Better mutual understanding of demands and offers would help to develop demand-driven scenario simulations by climate modellers, targeted to the needs of the urban water decision makers. At the same time it would help urban drainage experts to use simulation results from climate models in a proper way in their impact investigations and to develop statistical tools that accurately address the scale gaps. This requires good understanding of the climate model features and the atmospheric-oceanic-land surface processes that control the gaps. Also intensive cooperation between researchers and practitioners is important. Climate model impacts in urban hydrology are highly uncertain and decision makers have to cope with that uncertainty. They also have to deal with the difference between short-term climate variability and longer-term climate change trends. This requires good understanding and scientific support, for example on application of the risk concept. They have to move from traditional approaches of static design and upgrading of technical solutions, to flexible, adaptive and resilient designs. Chapter 10 shows that also enhanced cooperation between urban water managers, spatial planners, urban designers and other land and water related sectors becomes mandatory as part of efficient climate adaptation strategies. Also active involvement of local and political communities, public awareness and understanding of service levels and limitations becomes more important. *Enhanced cooperation* thus is the key message here, next to research, education and training. We hope that this book contributes to that process.

References

Adamowski K. and Bougadis J. (2003). Detection of trends in annual extreme rainfall. *Hydrol. Process.*, **17**, 3547–3560.

Adams B. J. and Papa F. (2000). Urban Stormwater Management Planning with Analytical Probabilistic Models. John Wiley & Sons, New York, p. 358.

ADB. (2011). Adapting to Climate Change: Strenghtening the Climate Resilience of the Water Sector Infrastructure in Khulna, Bangladesh. Asian Development Bank, Mandaluyong City, Philippines, p. 32.

Aguilar E., Aziz Barry A., Brunet M., Ekang L., Fernandes A., Massoukina M., Mbah J., Mhanda A., do Nascimento D. J., Peterson T. C., Thamba Umba O., Tomou M. and Zhang X. (2009). Changes in temperature and precipitation extremes in western central Africa, Guinea Conakry, and Zimbabwe, 1955–2006. *J. Geophys. Res.*, **114**, D02115.

Alexander L. V., Zhang X., Peterson T. C., Caesar J., Gleason B., Klein Tank A. M. G., Haylock M., Collins D., Trewin B., Rahimzadeh F., Tagipour A., Rupa Kumar K., Revadekar J., Griffiths G., Vincent L., Stephenson D. B., Burn J., Aguilar E., Brunet M., Taylor M., New M., Zhai P., Rusticucci M. and Vazquez-Aguirre J. L. (2006). Global observed changes in daily climate extremes of temperature and precipitation. *J. Geophys. Res.*, **111**, D05109.

Alexander L. V. and Arblaster J. M. (2009). Assessing trends in observed and modelled climate extremes over Australia in relation to future projections. *Int. J. Climatol.*, **29**(3), 417–435.

Alila Y. (1999). A hierarchical approach for the regionalization of precipitation annual maxima in Canada. *J. Geophys. Res.*, **104**(D24), 31,645–31,655.

Allan R. P. and Soden B. J. (2008). Atmospheric warming and the amplification of precipitation extremes. *Science*, **321** (5895), 1481–1484.

Anderson C. J. *et al.* (2003). Hydrological processes in regional climate model simulations of the central United States flood of June-July 1993. *J. Hydrometeor.*, **4**(3), 584–598.

Apipattanavis S., Podestá G., Rajagopalan B. and Katz R. W. (2007). A semiparametric multivariate and multisite weather generator. *Water Resour. Res.*, **43**, W11401.

Argue J. R. (ed.). (2004). 1st edn, WSUD: Basic Procedures for Source Control of Stormwater – A Handbook for Australian Practice. Urban Water Resources Centre, University of South Australia, pp. 201–226.

Arisz H. and Burrell B. C. (2006). Urban drainage infrastructure planning and design considering climate change. EIC Climate Change Change Technology Conference (2006 IEEE), 10–12 May 2006, Ottawa, Canada.

Arnbjerg-Nielsen K. (2006). Significant climate change of extreme rainfall in Denmark. *Water Sci. Technol.*, **54**(6–7), 1–8.

Arnbjerg-Nielsen K. (2011). Past, present, and future design of urban drainage systems with focus on Danish experiences. *Water Sci. Technol.*, **63**(3), 527–535.

Arnbjerg-Nielsen K. (2012). Quantification of climate change effects on extreme precipitation used for high resolution hydrologic design. *Urban Water J.*, **9**(2), 57–65.

Arnbjerg-Nielsen K. and Harremoës P. (1996). The importance of inherent uncertainties in state-of-the-art urban storm drainage modelling for ungauged small catchments. *J. Hydrol.*, **179**(1–4), 305–319.

Arnbjerg-Nielsen K. and Fleischer H. S. (2009). Feasible adaptation strategies for increased risk of flooding in cities due to climate change. *Water Sci. Technol.*, **60**(2), 273–281.

Arnbjerg-Nielsen K., Fleischer H. S., Gudiksen S. and Laden B. (2008). Identifying economically and technically feasible adaptation strategies for mitigation of flood risks in cities. Two case studies from Denmark. Proceedings of the 11th International Conference on Urban Drainage, 31 August – 5 September 2008, Edinburgh, Scotland, p. 10.

Arnell N. W. (2003). Effects of IPCC SRES emissions scenarios on river runoff: a global perspective. *Hydrol. Earth Syst. Sci.*, **7**, 619–641.

Arnell V., Harremoës P., Jensen M., Johansen N. B. and Niemczynowicz J. (1984). Review of rainfall data application for design and analysis. *Water Sci. Technol.*, **16**(8–9), 1–45.

Aryal S. K., Bates B. C., Campbell E. P., Li Y., Palmer M. J. and Viney N. R. (2009). Characterizing and modeling temporal and spatial trends in rainfall extremes. *J. Hydrometeor.*, **10**, 241–253.

ASCE. (1983). Annotated Bibliography on Urban Design Storms. American Society of Civil Engineers, Urban Water Resources Research Council, New York, p. 41.

Ashley R. M., Blanksby J. R., Cashman A., Jack L., Wright G., Packman J., Fewtrell L. and Poole A. (2007). Adaptable urban drainage – addressing change in intensity, occurrence and uncertainty of stormwater (AUDACIOUS). *J. Built Environ.*, **33**(1), 70–84.

Ashley R. M., Clemens F. H. L. R., Tait S. J. and Schellart A. (2008a). Climate change and the implications for modeling the quality of flow in combined sewers. Proceedings of the 11th International Conference on Urban Drainage, 31 August – 5 September 2008, Edinburgh, Scotland, p. 10.

Ashley R. M., Newman R., Molyneux-Hodgson S. and Blanksby J. (2008b). Active learning: building the capacity to adapt urban drainage to climate change. Proceedings of the 11th International Conference on Urban Drainage, 31 August – 5 September 2008, Edinburgh, Scotland, p. 10.

Austin P. M. and Houze R. A. (1972). Analysis of structure of precipitation patterns in New England. *J. Appl. Meteorol.*, **11**, 926–935.

Bacchi B., Balistrocchi M. and Grossi G. (2008). Proposal of a semi-probabilistic approach for storage design facility design. *Urban Water J.*, **5**(3), 195–208.

Back A. J., Uggioni E., Oliveira J. L. R. and Henn A. (2011). Application of mathematical modeling for simulation of hourly rainfall. Proceedings of the 12th International Conference on Urban Drainage, 10–16 September 2011, Porto Alegre, Brazil, p. 8.

Baguis P., Ntegeka V., Willems P. and Roulin E. (2009). Extension of CCI-HYDR Climate Change Scenarios for INBO, Technical Report by KU Leuven and Royal Meteorological Institute of Belgium, for the Flemish Institute for Nature and Forest Research (INBO) & Belgian Science Policy Office, Brussels, Belgium, January 2009, p. 31; available 31 March 2012 on http://www.kuleuven.be/hydr/CCI-HYDR.htm.

Baguis P., Roulin E., Willems P. and Ntegeka V. (2010). Climate change scenarios for precipitation and potential evapotranspiration over central Belgium. *Theoret. Appl. Climatol.*, **99**(3–4), 273–286.

Baker M. B. and Peter T. (2008). Small-scale cloud processes and climate. *Nature*, **451**, 299–300.

Bastin G., Lorent B., Duque C. and Gevers M. (1984). Optimal estimation of the average areal rainfall and optimal selection of raingauge locations. *Water Resour. Res.*, **20**(4), 463–470.

Bates B. C., Charles S. P. and Fleming P. M. (1993). Simulation of daily climatic series for the assessment of climate change impacts on water resources. In: Engineering Hydrology, C. Y. Kuo (ed.), Amer. Soc. Civ. Eng., New York, USA, pp. 67–72.

Bates B. C., Kundzewicz Z. W., Wu S. and Palutikof J. P. (eds.). (2008). Climate Change and Water, Technical Paper of the Intergovernmental Panel on Climate Change. IPCC Secretariat, Geneva, p. 210.

Bárdossy A. and Caspary H. J. (1990). Detection of climate change in Europe by analyzing European atmospheric circulation patterns from 1881 to 1989. *Theoret. Appl. Climatol.*, **42**, 155–167.

Barring L., Holt T., Linderson M. L., Radziejewski M., Moriondo M. and Palutikof J. P. (2006). Defining dry/wet spells for point observations, observed area averages, and regional climate model gridboxes in Europe. *Clim. Res.*, **31**(1), 35–49.

Beaulant A.-L., Joly B., Nuissier O., Somot S., Ducrocq V., Joly A., Sevault F., Deque M. and Ricard D. (2011). Statistico-dynamical downscaling for mediterranean heavy precipitation. *Quart. J. Roy. Stat. Soc.*, **137**(656), 736–748.

Beecham S. (2012). Development of multifunctional landuse using water sensitive urban design. Chapter 18. In: Designing for Zero Waste: Consumption, Technologies and the Built Environment', S. Lehmann and R. Crocker (eds.), 'Routledge Press, ISBN: 978-1-84971-434–1, pp. 374–384.

Beecham S. (2010). Water sensitive urban design. Chapter 23. In: 'Adelaide: Water of a City', C. B. Daniels (ed.), Wakefield Press, ISBN: 978-1-86254-861–9, pp. 451–470.

Beecham S. and Chowdhury R. (2012). Effects of changing rainfall patterns on WSUD in Australia. *J. Water Manage.*, **165**(5), 285–298.

Beguería S. and Vicente-Serrano S. M. (2006). Mapping the hazard of extreme rainfall by peaks over threshold extreme value analysis and spatial regression techniques. *J. Appl. Meteor. Climatol.*, **45**, 108–124.

Bell J. L., Sloan L. C. and Snyder M. A. (2004). Regional changes in extreme climatic events: a future climate scenario. *J. Clim.*, **17**, 81–87.

Bendjoudi H., Hubert P., Schertzer D. and Lovejoy S. (1997). Interprétation multifractale des courbes intensité-durée-fréquance des précipitations. *Géosciences De Surface (Hydrologie)/Surface Geosciences (Hydrology)*, **325**, 323–326.

Benestad R. E., Hanssen-Bauer I. and Chen D. (2008). Empirical-Statistical Downscaling. World Scientific Publishing Co. Pte. Ltd., Hasensack, New Jersey. ISBN 981-281-912-6.

Bengtsson L. (2008). Extreme daily rainfall and trends in Skåne and the West Coast (in Swedish). *Vatten*, **64**, 31–39.

Bengtsson L. and Milotti S. (2008). Intensive storms in Malmö, Sweden. *Hydrol. Process.*, **24**(24), 3462–3475.

Benjamin J. and Cornell C. A. (1970). Probability, Statistics and Decision for Civil Engineers. McGraw–Hill, New York, p. 684.

Berg P., Haerter J. O., Thejll P., Piani C., Hagemann S. and Christensen J. H. (2009). Seasonal characteristics of the relationships between daily precipitation intensity and surface temperature. *J. Geophys. Res.*, **114**, D18102.

Berggren K. (2008). Indicators for urban drainage system: assessment of climate change impacts. Proceedings of the 11th International Conference on Urban Drainage, 31 August – 5 September, Edinburgh, Scotland, p. 7.

Berggren K., Olofsson M., Viklander M., Svensson G. and Gustafsson A. (2012). Hydraulic impacts on urban drainage systems due to changes in rainfall, caused by climatic change. *Journal of Hydrologic Engineering*, **17**(1), 92–98.

Bermudez P. and Kotz S. (2010a). Parameter estimation of the generalized pareto distribution – Part I. *Journal of Statistical Planning and Inference*, **140**(6), 1353–1373.

Bermudez P. and Kotz S. (2010b). Parameter estimation of the generalized pareto distribution – Part II. *Journal of Statistical Planning and Inference*, **140**(6), 1374–1397.

Bilham E. G. (1962). The Classification of Heavy Falls of Rain in Short Periods. Great Britain Meteorological Office, H.M. Stationery Office, London, UK, p. 19.

Bixio D., van Hauwermeiren P., Thoeye C. and Ockler P. (2001). Impact of cold and dilute sewage on prefermentation – a case study. *Water Sci. Technol.*, **43**(11), 109–117.

Blanckaert J. and Willems P. (2006). Statistical analysis of trends and cycles in long-term historical rainfall series at Uccle. Proceedings of the 7th International Workshop on Precipitation in Urban Areas, St.Moritz, 7–10 December 2006, pp. 124–128.

Bo Z., Islam S. and Eltahir E. A. B. (1994). Aggregation-disaggregation properties of a stochastic rainfall model. *Water Resour. Res.*, **30**, 3423–3435.

Boo K.-O., Kwon W.-T. and Baek H.-J. (2006). Change of extreme events of temperature and precipitation over Korea using regional projection of future climate change. *Geophys. Res. Lett.*, **33**, L01701.

Booij M. J. (2002). Extreme daily precipitation in western Europe with climate change at appropriate spatial scales. *Int. J. Climatol.*, **22**, 69–85.

Boukhris O., Willems P. and Vanneuville W. 2008. The impact of climate change on the hydrology in highly urbanized Belgian areas. In: Proceedings International Conference on 'Water & Urban Development Paradigms: Towards an Integration of Engineering, Design and Management Approaches', Leuven, 15–17 September 2008 J. Feyen, K. Shannon and M. Neville (eds.), CRC Press, Taylor & Francis Group, 271–276.

Bouttier F. and Courtier P. (1999). Data Assimilation Concepts and Methods. ECMWF Meteorological Training Course Lecture Series, ECMWF, Reading, UK, p. 59.

Brekke L. D., Kiang J. E., Olsen J. R., Pulwarty R. S., Raff D. A., Turnipseed D. P., Webb R. S. and White K. D. (2009). Climate Change and Water Resources Management – A Federal Perspective. U.S. Geological Survey Circular 1331, U.S. Army Corps of Engineers, U.S. Dept. of Interior and NOAA, USGS, Reston, VA, p. 65.

Brissette F. P., Khalili M. and Leconte R. (2007). Efficient stochastic generation of multi-site synthetic precipitation data. *J. Hydrol.*, **345**, 121–133.

Bruce J. P. and Clark P. H. (1966). Introduction to Hydrometeorology. Pergamon Press, Oxford, UK, p. 319.

Brunetti M., Maugeri M. and Nanni T. (2001). Changes in total precipitation, rainy days and extreme events in northeastern Italy. *Int. J. Climatol.*, **21**, 861–871.

Buishand T. A. and Brandsma T. (2001). Multisite simulation of daily precipitation and temperature in the rhine basin by nearest-neighbor resampling. *Water Resour. Res.*, **37**(11), 2761–2776.

Bulygina O. N., Razuvaev V. N., Korshunova N. N. and Groisman P. Y. (2007). Climate variations and changes in extreme climate events in Russia. *Environ. Res. Lett.*, **2**, 045020.

Burian S. J., Nix S. J., Durrans S. R., Pitt R. E., Fan C.-Y. and Field R. (1999). Historical development of wet-weather flow management. *J. Water Resour. Plann. Manage.*, **125**(1), 3–13.

Burian S. J., Durrans S. R., Tomic S., Pimmel R. L. and Wai C. N. (2000). Rainfall disaggregation using artificial neural networks. *J. Hydrol. Eng.*, **5**, 299–307.

Burlando P. and Rosso R. (1991). Extreme storm rainfall and climatic change. *Atmos. Res.*, **27**, 169–189.

Burlando P. and Rosso R. (1996). Scaling and multiscaling models of depth-duration-frequency curves for storm precipitation. *J. Hydrol.*, **187**, 45–64.

Burton A., Kilsby C. G., Fowler H. J., Cowpertwait P. S. P. and O'Connell P. E. (2008). RainSim: a spatial-temporal stochastic rainfall modelling system. *Environ. Modell.Softw.*, **23**, 1356–1369.

Butler D. and Davies J. (2010). Urban Drainage. 3rd edn, Spon Press, Taylor & Francis, 656 p.

Buzzi A., Tartaglione N. and Malguzzi P. (1998). Numerical simulations of the 1994 piedmont flood: role of orography and moist processes. *Mon. Weather Rev.*, **126**(9), 2369–2383.

Cârsteanu A., Foufoula-Georgiou E. and Venugopal V. (1999). Event-specific multiplicative cascades for temporal rainfall. *J. Geophys. Res.*, **104**(D24), 31611–31622.

Cavazos T., Turrent C. and Lettenmaier D. P. (2008). Extreme precipitation trends associated with tropical cyclones in the core of the North American monsoon. *Geophys. Res. Lett.*, **35**, L21703.

Cayan D. R., Redmond K. T. and Riddle L. G. (1999). ENSO and hydrologic extremes in the western United States. *J. Clim.*, **12**, 2881–2893.

Chandler R. E. (2011). GLIMCLIM: Generalised Linear Modelling of Daily Climate Sequences. University College of London, London, UK; Retrieved 31 March 2012, from http://www.ucl.ac.uk/~ucakarc/work/glimclim.html.

Chandler R. E. and Wheater H. S. (2002). Analysis of rainfall variability using generalized linear models: a case study from the west of Ireland. *Water Resour. Res.*, **38**(10), 1192.

Chandler R. E. and Bate S. (2007). Inference for clustered data using the independence log-likelihood. *Biometrika*, **94**, 167–183.

Chandler M. A., Richards S. J. and Shopsin M. F. (2005). EdGCM: Enhancing climate science education through climate modeling research projects. Proceedings of the 85th Annual Meeting of the American Meteorological Society, 14th Symposium on Education, 8–14 January 2005, San Diego, USA.

Charles S. P., Bates B. C., Whetton P. H. and Hughes J. P. (1999). Validation of downscaling models for changed climate conditions: case study of southern Australia. *Clim. Res.*, **12**, 1–14.

Chocat B., Ashley R., Marsalek J., Matos M. R., Rauch W., Schilling W. and Urbonas B. (2007). Toward the sustainable management of urban storm-water. *Indoor Built Environ.*, **16**(3), 273–285.

Chow V. T. (1951). A general formula for hydrologic frequency analysis. *Trans. Amer. Geophys. Un.*, **32**, 231–237.

Chow V. T. (1964). Handbook of Applied Hydrology; A Compendium of Water-Resources Technology. McGraw–Hill, New York, p. 1494.

Chow V. T., Maidment D. R. and Mays L. W. (1988). Applied Hydrology. McGraw–Hill, New York, USA, p. 570.

Chowdhury R. and Beecham S. (2010). Australian rainfall trends and their relation to the Southern Oscillation Index. *Hydrol. Process.*, **24**(4), 504–514.

Chowdhury R. and Beecham S. (2012). Characterisation of rainfall spells for urban water management. *Int. J. Climatol.*, doi: 10.1002/joc.3482.

Christensen J. H. and Christensen O. B. (2003). Climate modeling: severe summertime flooding in Europe. *Nature*, **421** (45), 805–806.

Christensen J. and Christensen O. B. (2007). A summary of the PRUDENCE model projections of changes in European climate by the end of this century. *Clim. Change*, **81**(1), 7–30.

Christensen J. H., Carter T. R., Rummukainen M. and Amanatidis G. (2007). Evaluating the performance and utility of regional climate models: the PRUDENCE project. *Clim. Change*, **81**, 1–6.

Christensen J. H., Kjellström E., Giorgi F., Lenderink G. and Rummukainen M. (2010). Weight assignment in regional climate models. *Clim. Res.*, **44**, 179–194.

Chun K., Onof C. and Segond M. L. (2009). Disaggregation of climate model outputs. CD-ROM Proceedings of the 8th International Workshop on Precipitation in Urban Areas, 10–13 December 2009, St.Moritz, Switzerland, pp. 144–148.

CIG. (1996). Climate Change Scenarios for the Australian Region. Climate Impact Group, CSIRO, Division of Atmospheric Research, Australia.

CISL. (2011). CISL Research Data Archive: ds083.2 / NCEP FNL Operational Model Global Tropospheric Analyses, Continuing from July 1999. Research Data Archive, Data Support Section, Computational and Information Systems Laboratory, National Center for Atmospheric Research, Boulder, Colorado, USA; Retrieved 31 March 2012, from http://dss.ucar.edu/datasets/ds083.2/].

Clemens F. H. L. R. (2001). Hydrodynamic Models in Urban Drainage: Application and Calibration. Doctoral dissertation, TUDelft, The Netherlands, p. 372.

Cobbina A., Adams B. J. and Guo Y. (2008). The impact of climate change on Toronto precipitation characteristics and stormwater infrastructure. Proceedings of the 11th International Conference on Urban Drainage, 31 August – 5 September, Edinburgh, Scotland.

Coles S. (2001). An Introduction to Statistical Modeling of Extreme Values. Springer, London, UK, p. 208.

Colle B. A. (2004). Sensitivity of orographic precipitation to changing ambient conditions and terrain geometries: an idealized modeling perspective. *J. Atmos. Sci.*, **61**(5), 588–606.

Collier Ch.C. (1996). Applications of Weather Radar Systems. Praxis Publishing Ltd., p. 390.

Connolly R. D., Schirmer J. and Dunn P. K. (1998). A daily rainfall disaggregation model. *Agric. For. Meteorol.*, **92**, 105–117.

Coulibaly P., Dibike Y. B. and Anctil F. (2005). Downscaling precipitation and temperature with temporal neural networks. *J. Hydrometeorol.*, **6**(4), 483–496.

Covey C., Achuta Rao K. M., Cubasch U., Jones P., Lambert S. J., Mann M. E., Phillips T. J. and Taylor K. E. (2003). An overview of results from the coupled model intercomparison project (CMIP). *Global and Planet. Change*, **37**, 103–133.

Cowpertwait P. S. P. (1995). A generalized spatial-temporal model of rainfall based on a clustered point process. *Proc. R. Soc. Lond. A Math. Phys. Sci.*, **450**(1938), 163–175.

Cowpertwait P. S. P. (2004). Mixed rectangular pulses models of rainfall. *Hydrol. Earth Syst. Sci.*, **8**, 993–1000.

Cowpertwait P. S. P. and O'Connell P. E. (1997). A regionalised neyman-scott model of rainfall with convective and stratiform cells. *Hydrol. Earth Sys. Sci.*, **1**, 71–80.

Cowpertwait P. S. P., O'Connell P. E., Metcalfe A. V. and Mawdsley J. A. (1996). Stochastic point process modelling of rainfall. I. Single-site fitting and validation. *J. Hydrol.*, **175**, 17–46.

Cowpertwait P. S. P., Kilsby C. G. and O'Connell P. E. (2002). A space-time Neyman-Scott model of rainfall: empirical analysis of extremes. *Water Resour. Res.*, **38**, 1131.

Cowpertwait P. S. P., Isham V. and Onof C. (2007). Point process of rainfall: developments for fine-scale structure. *Proc. R. Soc. Lond. A*, **463**, 2569–2587.

Crabbé P. and Robin M. (2006). Institutional adaptation of water resource infrastructures to climate change in Eastern Ontario. *Clim. Change*, **78**(1), 103–133.

Daley R. (1991). Atmospheric Data Analysis. Cambridge Atmospheric and Space Science Series. Cambridge University Press, Cambridge, UK, and New York, USA, p. 457.

Deb S. K., Kishtawal C. M., Pal P. K. and Joshi P. C. (2008). Impact of TMI SST on the simulation of a heavy rainfall episode over Mumbai on 26 July 2005. *Mon. Weather Rev.*, **136**(10), 3714–3741.

de Elía R., Caya D., Côté H., Frigon A., Biner S., Giguère M., Paquin D., Harvey R. and Plummer D. (2008). Evaluation of uncertainties in the CRCM-simulated North American climate. *Clim. Dyn.*, **30**(2–3), 113–132.

Deidda R. (1999). Multifractal analysis and simulation of rainfall fields in space. *Phys. Chem. Earth (B)*, **24**, 73–78.

Deidda R. (2000). Rainfall downscaling in a space-time multifractal framework. *Water Resour. Res.*, **36**(7), 1779–1794.

DeGaetano A. T. (2009). Time-dependent changes in extreme-precipitation return-period amounts in the continental United States. *J. Appl. Meteor. Climatol.*, **48**, 2086–2099.

De Jongh I. L. M., Verhoest N. E. C. and De Troch F. P. (2006). Analysis of a 105-year time series of precipitation observed at Uccle Belgium. *Int. J. Climatol.*, **26**(14), 2023–2039.

Deletic A., Dotto C. B. S., McCarthy D. T., Kleidorfer M., Freni G., Mannina G., Uhl M., Henrichs M., Fletcher T. D., Rauch W., Bertrand-Krajewski J.-L. and Tait S. (2011). Assessing uncertainties in urban drainage models. *Phys. Chem. Earth*, in press; doi: 10.1016/j.pce.2011.04.007.

Demuzere M., Werner M., van Lipzig N. P. M. and Roeckner E. (2009). An analysis of present and future ECHAM5 pressure fields using a classification of circulation patterns. *Int. J. Climatol.*, **29**(12), 1796–1810.

Denault C., Millar R. G. and Lence B. J. (2002). Climate change and drainage infrastructure capacity in an urban catchment. Annual Conference of the Canadian Society for Civil Engineering, Montréal, Québec, Canada, 5–8 June 2002, p. 10.

Denault C., Millar R. G. and Lence B. J. (2006). Assessment of possible impacts of climate change in an urban catchment. *J. Am. Water Resour. Assoc.*, **42**, 685–697.

den Elzen M. G. J., Hof A. F. and Roelfsema M. (2011). The emissions gap between the Copenhagen pledges and the 2° C climate goal: options for closing and risks that could widen the gap. *Glob. Environ. Change*, **21**, 733–743.

Desbordes M. (1978). Urban runoff and design storm modelling. Proc. First International Conference on Urban Storm Drainage, UK, Pentech Press, London, UK, pp. 353–361.

Dettwiller J. and Changnon S. A. (1976). Possible urban effects on maximum daily rainfall at Paris, St. Louis and Chicago. *J. Appl. Meteorol.*, **15**, 517–519.

Déqué M. *et al.* (2007). An intercomparison of regional climate simulations for Europe: assessing uncertainties in model projections. *Climatic Change*, **81**, 53–70.

Di Baldassarre G., Castellarin A. and Brath A. (2006). Relationships between statistics of rainfall extremes and mean annual precipitation: an application for design-storm estimation in northern central Italy. *Hydrol. Earth Syst. Sci.*, **10**, 589–601.

Diaper C., Tjandraatmadja G. and Kenway S. (2007). Sustainable Subdivisions: Review of Technologies for Integrated Water Services. CRC Constructed Innovation, Brisbane, ISBN 978-0-9803503-9-5.

Dibike Y. B., Gachon P., St-Hilaire A., Ouarda T. B. M. J. and Nguyen V.-T.-V. (2008). Uncertainty analysis of statistically downscaled temperature and precipitation regimes in Northern Canada. *Theoret. Appl. Climatol.*, **91**, 149–170.

D'onofrio A., Boulanger J.-P. and Segura E. C. (2010). CHAC: a weather pattern classification system for regional climate downscaling of daily precipitation. *Clim. Change*, **98**(3–4), 405–427.

Domingo N. D. F., Sunyer M. A., Hansen F., Madsen H., Mark O. and Paludan B. (2010). Modelling of sea level rise and subsequent urban flooding due to climate changes. SimHydro 2010: Hydraulic modeling and uncertainty, 2–4 June, 2010, Nice, France.

Dotto C. B. S., Mannina G., Kleidorfer M., Vezzaro L., Henrichs M., McCarthy D. T., Freni G., Rauch W. and Deletic A. (2012). Comparison of different uncertainty techniques in urban stormwater quantity and quality modelling. *Water Res.*, **46**(8), 2545–2558.

Douglas E. M., Vogel R. M. and Kroll C. N. (2000). Trends in floods and low flows in the United States: impact of spatial correlation. *J. Hydrol.*, **240**, 90–105.

Doyle J. D. and Durran D. R. (2001). The dynamics of mountain-wave- induced rotors. *J. Atmos. Sci.*, **59**, 186–201.

Dreyfus H. L. and Dreyfus S. E. (1985). From Socrates to expert systems: the limits of calculative rationality. In: Philosophy and Technology II: Information Technology and Computers in Theory and Practice, C. Mitcham and A. Huning (eds.), Boston Studies in the Phylosophy of Sciences Series.

Durchschlag A., Hartel L., Hartwig P., Kaselow M., Kollatsch D., Otterpohl R. and Schwentner G. (1991). Total emissions from combined sewer overflow and wastewater treatment plants. *Eur. Water Pollut. Control*, **1**(6), 13–23.

Durran D. R. and Klemp J. B. (1982). On the effects of moisture on the Brunt-Väisälä a frequency. *J. Atmos. Sci.*, **39**, 2152–2158.

Durran D. R. and Klemp J. (1983). A compressible model for the simulation of moist mountain waves. *Mon. Weather Rev.*, **111**, 2341–2361.

Easterling D. R., Karl T. R., Lawrimore J. H. and Del Greco S. A. (1999). United States Historical Climatology Network Daily Temperature, Precipitation, and Snow Data for 1871–1997. Report ORNL/CDIAC-118, NDP-070, Carbon Dioxide Information Analysis Center, Oak Ridge National Laboratory, U.S. Department of Energy, Oak Ridge, USA, p. 84.

Econopouly T. W., Davis D. R. and Woolhiser D. A. (1990). Parameter transferability for a daily rainfall disaggregation model. *J. Hydrol.*, **118**, 209–228.

EEA. (2007). Climate change and water adaptation issues. Technical Report no. 2/2007, European Environment Agency, Copenhagen.

Enfield D. B., Mestas-Nunez A. M. and Trimble P. J. (2001). The Atlantic multidecadal oscillation and its relation to rainfall and river flows in the continental U.S. *Geophys. Res. Lett.*, **28**, 2077–2081.

Einfalt T., Arnbjerg-Nielsen K., Golz C., Jensen N. E., Quirmbach M., Vaes G. and Vieux B. (2004). Towards a roadmap for use of radar rainfall data use in urban drainage. *J. Hydrol.*, **299**, 186–202.

Einfalt Th., Quirmbach M., Langstädtler G. and Mehlig B. (2011). Climate change tendencies observable in the rainfall measurements since 1950 in the federal land of North Rhine-Westphalia and their consequences for urban hydrology. *Water Sci. Technol.*, **63**(11), 2633–2640.

Endo N., Matsumoto J. and Lwin T. (2009). Trends in precipitation extremes over Southeast Asia. *SOLA*, **5**, 168–171.

EU. (2010). European Commission White Paper: Adapting to Climate Change: Towards a European Framework for Action, European Commission, Brussels; Retrieved 31 March 2012, from http://eur-lex.europa.eu/LexUriServ/LexUriServ.do?uri=OJ:C:2011:081E:0115:0128:EN:PDF.

Fahrmeir L. and Tutz G. (1994). Multivariate Statistical Modeling Based on Generalized Linear Models. Springer Verlag, New York.

Fee E., Johansson D. J. A., Lowe J., Marbaix P., Matthews B. and Meinshausen M. (2010). Scientific Perspectives after Copenhagen. Information Reference Document, EU Presidency, European Union, Brussels, Belgium.

Fernandez Mills G. (1995). Principal component analysis of precipitation and rainfall regionalization in Spain. *Theoret. Appl. Climatol.*, **50**, 169–183.

Fitzgerald D. L. (1989). Single station and regional analysis of daily rainfall extremes. *Stoch. Hydrol. Hydraul.*, **3**, 281–292.

Flatau P. J., Walko R. L. and Cotton W. R. 1992. Polynomial fits to saturation vapor pressure. *J. Appl. Meteorol.*, **31**, 1507–1513.

Fontaine T. A. (1991). Predicting measurement error of areal mean precipitation during extreme events. *Water Resour. Bull.*, **27**(3), 509–520.

Fotherby L. M., McBain S. M. and Aumen M. G. (2008). Adaptive management as a framework for ecosystem restoration. ASCE proceedings of the World Environmental and Water Resources Congress, Honolulu.

Foufoula-Georgiou E. and Guttorp P. (1987). Assessment of a class of Neyman-Scott models for temporal rainfall. *J. Geophys. Res.*, **92**(D8), 9679–9682.

Fowler H. J. and Wilby R. L. (2010). Detecting changes in seasonal precipitation extremes using regional climate model projections: implications for managing fluvial flood risk. *Water Resour. Res.*, **46**(W0525).

Fowler H. J., Kilsby C. G. and O'Connell P. E. (2000). A stochastic rainfall model for the assessment of regional water resource systems under changed climatic conditions. *Hydrol. Earth Sys. Sci.*, **4**, 261–280.

Fowler H. J., Kilsby C. G., O'Connell P. E. and Burton A. (2005). A weather-type conditioned multi-site stochastic rainfall model for the generation of scenarios of climatic variability and change. *J. Hydrol.*, **308**(1–4), 50–66.

Fowler H. J., Ekstrom M., Blenkinsop S. and Smith A. P. (2007). Estimating change in extreme European precipitation using a multimodel ensemble. *J. Geophys. Res.-Atmos.*, **112**(D18104).

Fraedrich K. and Larnder C. 1993. Scaling regimes of rainfall time-series. *Tellus*, **45A**, 289–298.

Frei C. and Schär C. (2001). Detection probability of trends in rare events: theory and application of heavy precipitation in the Alpine region. *J. Clim.*, **14**, 1568–1584.

Freni G. and Mannina G. (2010). Uncertainty in water quality modeling: the applicability of variance decomposition approach. *J. Hydrol.* **394**(3–4), 324–333.

Freni G., Mannina G. and Viviani G. (2009). Urban runoff modelling uncertainty: comparison among Bayesian and pseudo-Bayesian methods. *Environ. Modell. Softw.*, **24**(9), 1100–1111.

Frei C., Christensen J. H., Deque M., Jacob D., Jones R. G. and Vidale P. L. (2003). Daily precipitation statistics in regional climate models: evaluation and intercomparison for the European Alps. *J. Geophys. Res.-Atmos.*, **108** (D3), 1–19.

Frei C., Schöll R., Fukutome S., Schmidli J. and Vidale P. L. (2006). Future change of precipitation extremes in Europe: intercomparison of scenarios from regional climate models. *J. Geophys. Res.- Atmos.*, **3**, 111(D06105).

Frich P., Alexander L. V., Della-Marta P., Gleason B., Haylock M., Klein Tank A. and Peterson T. (2002). Observed coherent changes in climatic extremes during the second half of the twentieth century. *Clim. Res.*, **19**, 193–212.

Foufoula-Georgiou E. and Guttorp P. (1987). Assessment of a class of Neyman-Scott models for temporal rainfall. *J. Geophys. Res.*, **92**(D8), 9679–9682.

Fowler H. J. and Kilsby C. G. (2003). A regional frequency analysis of United Kingdom extreme rainfall from 1961 to 2000. *Int. J. Climatol.*, **23**(11), 1313–1334.

Fowler H. J., Blenkinsop S. and Tebaldi C. (2007). Linking climate change modelling to impacts studies: recent advances in downscaling techniques for hydrological modeling. *Int. J. Climatol.*, **27**, 1547–1578.

Fu C., Jiang Z., Guan Z., He J. and Xu Z. (eds.). (2008). Regional Climate Change Studies of China. Springer–Verlag, Berlin Heidelberg, p. 476.

Fujibe F. (2008). Long-term changes in precipitation in Japan. *J. Disaster Res.*, **3**(1), 51–60.

Furrer E. M. and Katz R. W. (2008). Improving the simulation of extreme precipitation events by stochastic weather generators. *Water Resour. Res.*, **44**, W12439.

Galiatsatou P. and Prinos P. (2007). Outliers and trend detection tests in rainfall extremes. Proc. of 32nd IAHR Congress, SS10-15-O, Venice, Italy.

Gao X., Shi Y., Song R., Giorgi F., Wang Y. and Zhang D. (2008). Reduction of future monsoon precipitation over China: comparison between a high resolution RCM simulation and the driving GCM. *Meteorol. Atmos. Phys.*, **100**, 73–86.

Gaume E., Villeneuve J. and Desbordes M. (1998). Uncertainty assessment and analysis of the calibrated parameter values of an urban storm water quality model. *J. Hydrol.*, **210**, 38–50.

Gaume E., Sivakumar B., Kolasinski M. and Hazoumé L. (2006). Identification of chaos in rainfall temporal disaggregation: Application of the correlation dimension method to 5-minute point rainfall series measured with a tipping bucket and an optical raingage. *J. Hydrol.*, **328**(1–2), 56–64.

Gelati E., Christensen O. B., Rasmussen P. F. and Rosbjerg D. (2010). Downscaling atmospheric patterns to multi-site precipitation amounts in southern Scandinavia. *Hydrol. Res.*, **41**, 193–210.

Gershunov A. (1998). ENSO influence on intraseasonal extreme rainfall and temperature frequencies in the contiguous United States: implications for long-range predictability. *J. Clim.*, **11**, 3192–3203.

Gershunov A. and Cayan D. R. (2003). Heavy daily precipitation frequency over the contiguous United States: sources of climatic variability and seasonal predictability. *J. Clim.*, **16**(16), 2752–2765.

Giorgi F. (2002). Dependence of surface climate interannual variability on spatial scale. *Geophys. Res. Lett.*, **29**, 2101.

Giorgi F. (2005). Climate change prediction. *Climatic Change*, **73**(3), 239–265.

Giorgi F., Im E.-S., Coppola E., Diffenbaugh N. S., Gao X. J., Mariotti L. and Shi Y. (2011). Higher hydroclimatic intensity with global warming. *J. Clim.*, **24**, 5309–5324.

Ghil M., Allen M. R., Dettinger M. D., Ide K., Kondrashov D., Mann M. E., Robertson A. W., Saunders A., Tian Y., Varadi F. and Yiou P. (2002). Advanced spectral methods for climatic time series. *Rev. Geoph.*, **40**(1), 1003.

Glasbey C. A., Cooper G. and McGechan M. B. (1995). Disaggregation of daily rainfall by conditional simulation from a point-process model. *J. Hydrol.*, **165**, 1–9.

Gleason K. L., Lawrimore J. H., Levinson D. H. and Karl T. R. (2008). A revised U.S. climate extremes index. *J. Clim.*, **21**, 2124–2137.

Golden J. S., Hartz D., Brazel A., Luber G. and Phelan P. (2008). A biometeorology study of climate and heat-related morbidity in Phoenix from 2001 to 2006. *Int. J. Biometeorol.*, **52**, 471–480.

Gong D. and Wang S. (1999). Definition of antarctic oscillation index. *Geophys. Res. Lett.*, **26**, 459–462.

Goswami B., Venugopal V., Sengupta D., Madhusoodanan M. and Xavier P. K. (2006). Increasing trends of extreme rain events over India in a warming environment. *Science*, **314**, 1442–1444.

Gregersen I. B., Arnbjerg-Nielsen K. and Madsen H. (2010). Parametric analysis of regional trends in observed extreme rainfall in Denmark. Proceedings of the International Workshop on Advances in Statistical Hydrology, 23–25 May 2010, Taormina, Italy.

Gregersen I. B., Madsen H. and Arnbjerg-Nielsen K. (2011). Estimation of climate factors for future extreme rainfall: comparing observations and RCM simulations. Proceedings of the 12th International Conference on Urban Drainage, 10–16 September 2011, Porto Alegre, Brazil, p. 8.

Gregersen I. B., Madsen H., Rosbjerg D. and Arnbjerg-Nielsen K. (2012). A spatial and non-stationary model for the frequency of extreme rainfall events. *Water Resour. Res.*, submitted.

Grell G. A., Dudhia J. and Stauffer D. R. (1994). A Description of the Fifth-Generation Penn State/NCAR Mesoscale Model (MM5). NCAR Technical Note NCAR/TN–398 + STR, National Center for Atmospheric Research, Mesoscale and Microscale Meteorology Division, Boulder, Colorado, USA, p. 128.

Grell G. A., Schade L., Knoche R., Pfeiffer A. and Egger J. (2000). Nonhydrostatic climate simulations of precipitation over complex terrain. *J. Geophys. Res.- Atmos.*, **105**(D24), 29595–29608.

Grimes D. I. F., Pardo-Igúzquiza E. and Bonifacio R. (1999). Optimal areal rainfall estimation using raingauges and satellite data. *J. Hydrol.*, **222**, 93–108.

Grimm A. M. and Tedeschi R. G. (2009). ENSO and extreme rainfall events in South America. *J. Clim.*, **22**, 1589–1609.

Groisman P. Y. and Legates D. R. (1994). The accuracy of United States precipitation data. *Bull. Am. Meteorol. Soc.*, **75** (2), 215–227.

Groisman P. Y. A., Karl T. R., Easterling D. R., Knight R. W., Jamason P. F., Hennessy K. J., Suppiah R., Page C. M., Wibig J., Fortuniak K., Razuvaev V. N., Douglas A., Forland E. and Zhai P. M. (1999). Changes in the probability of heavy precipitation: important indicators of climatic change. *Clim. Change*, **42**, 243–283.

Groisman P. Y., Knight P. Y. A. G. R. W. and Karl T. R. (2001). Heavy precipitation and high streamflow in the contiguous United States: trends in the twentieth century. *Bull. Am. Meteorol. Soc.*, **82**, 219–246.

Grossi G. and Bacchi B. (2008). A tool to investigate potential climate change effects on the efficiency of urban drainage systems. Proceedings of the 11th International Conference on Urban Drainage, 31 August – 5 September 2008, Edinburgh, Scotland, p. 9.

Grum M., Jorgensen A. T., Johansen R. M. and Linde J. J. (2006). The effect of climate change on urban drainage: an evaluation based on regional climate model simulations. *Water Sci. Technol.*, **54**, 9–16.

Guhathakurta P., Sreejith O. P. and Menon P. A. (2011). Impact of climate change on extreme rainfall events and flood risk in India. *J. Earth Syst. Sci.*, **120**(3), 359–373.

Güntner A., Olsson J., Calver A. and Gannon B. (2001). Cascade-based disaggregation of continuous rainfall time series: the influence of climate. *Hydrol. Earth Syst. Sci.*, **5**, 145–164.

Guo Y. and Adams B. J. (1998a). Hydrologic analysis of urban catchments with event-based probabilistic models. 1. Runoff volume. *Water Resour. Res.*, **34**(12), 3421–3431.

Guo Y. and Adams B. J. (1998b). Hydrologic analysis of urban catchments with event-based probabilistic models. 2. Peak discharge rate. *Water Resour. Res.*, **34**(12), 3433–3443.

Guo Y. and Adams B. J. (1999). An analytical probabilistic approach to sizing flood control detention facilities. *Water Resour. Res.*, **35**(8), 2457–2468.

Gupta V. K. and Waymire E. (1990). Multiscaling properties of spatial rainfall and river flow distributions. *J. Geophys. Res.*, **95**(D3), 1999–2009.

Gupta V. K. and Waymire E. C. 1993 A statistical analysis of mesoscale rainfall as a random cascade. *J. Appl. Meteorol.* **32**, 251–267.

Gutowski W. J., Hegerl G. C., Holland G. J., Knutson T. R., Mearns L. O., Stouffer R. J., Webster P. J., Wehner M. F. and Zwiers F. W. (2008). Causes of observed changes in extremes and projections of future changes. In: Weather and Climate Extremes in a Changing Climate. Regions of Focus: North America, Hawaii, Caribbean, and U.S. Pacific Islands, T. R. Karl *et al.* (eds.), A Report by the U.S. Climate Change Science Program and the Subcommittee on Global Change Research, Washington, DC.

Haddad K., Rahman A., Green J. and Kuczera G. (2011). Design rainfall estimation for short storm durations using L-Moments and generalised least squares regression-application to Australian data. *Int. J. Water Resour. Arid Environ.*, **1**, 210–218.

Haerter J. O. and Berg P. (2009). Unexpected rise in extreme precipitation caused by a shift in rain type? *Nat. Geosci.*, **2**, 372–373.

Hall M. J. (1984). Urban Hydrology. Elsevier Applied Science, London and New York, p. 310.

Hamed K. H. and Rao A. R. (1998). A modified Mann – Kendall trend test for autocorrelated data. *J. Hydrol.*, **204**, 182–196.

Hanel M. and Buishand T. A. (2010). On the value of hourly precipitation extremes in regional climate model simulations. *J. Hydrol.*, **393**, 265–273.

Harremoës P. (1988). Stochastic models for estimation of extreme pollution from urban runoff. *Water Res.*, **22**(8), 1017–1026.

Harris D., Menabde M., Seed A. and Austin G. (1996). Multifractal characterization of rain fields with a strong orographic influence. *J. Geophys. Res.*, **101**, 26405–26414.

Harrison D. L., Driscoll S. R. and Kitchen M. (2000). Improving precipitation estimates from weather radar using quality control and correction methods. *Meteorol. Appl.*, **6**, 135–144.

Hawkins E. and Sutton R. (2009). The potential to narrow uncertainty in regional climate predictions. *Bull. Am. Meteorol. Soc.*, **90**, 1095–1107.

Hawkins E. and Sutton R. (2011). The potential to narrow uncertainty in regional precipitation change. *Clim. Dyn.*, **37**, 407–418.

Hay L. E., Clark M. P., Pagowski M., Leavesley G. H. and Gutowski W. J. (2006). One-way coupling of an atmospheric and a hydrologic model in Colorado. *J. Hydrometeor.*, **7**(4), 569–589.

Haylock M. and Nicholls N. (2000). Trends in extreme rainfall indices for an updated high quality data set for Australia, 1910–1998. *Int. J. Climatol.*, **20**, 1533–1541.

Haylock M. R. and Goodess C. M. (2004). Interannual variability of European extreme winter rainfall and links with mean large-scale circulation. *Int. J. Climatol.*, **24**, 759–776.

Haylock M. R., Peterson T. C., Alves L. M., Ambrizzi T., Anunciação Y. M. T., Baez J., Barroz V. R., Berlato M. A., Bidegain M., Coronel G., Corradi V., Garcia V. J., Grimm A. M., Karoly D., Marengo J. A., Marino M. B., Moncunill D. F., Nechet D., Quintana J., Rebello E., Rusticucci M., Santos J. L., Trebejo I. and Vincent L. A. (2006). Trends in total and extreme South American rainfall in 1960–2000 and links with sea surface temperature. *J. Clim.*, **19**, 1490–1512.

Haylock M. R., Hofstra N., Klein Tank A. M. G., Klok E. J., Jones P. D. and New M. (2008). A European daily high-resolution gridded dataset of surface temperature and precipitation. *J. Geophys. Res. (Atmos.)*, **113**, D20119.

Hegerl G. C., Zwiers F. W., Braconnot P., Gillett N. P., Luo Y., Marengo Orsini J. A., Nicholls N., Penner J. E. and Stott P. A. 2007. Understanding and attributing climate change. In: The Physical Science Basis. Contribution of Working Group I to the Fourth Assessment Report of the Intergovernmental Panel on Climate Change, S. Solomon *et al.* (eds.), Cambridge University Press, Cambridge, UK, and New York, USA, pp. 663–745.

Helsel D. R. and Hirsch R. M. (1992). Statistical Methods in Water Resources. Elsevier Publishers, Amsterdam.

Hershenhorn J. and Woolhiser D. A. (1987). Disaggregation of daily rainfall. *J. Hydrol.*, **95**, 299–322.

Hess P. and Brezowsky H. (1969). Katalog der Grosswetterlagen Europas, 2. neu bearbeitete und ergänzte Aufl. *Berichte des Deutschen Wetterdienstes*, **113**(15), Selbstverslag des Deutschen Wetterdienstes, Offenbach a. Main.

Hewitson B. C. and Crane R. G. (1996). Climate downscaling: techniques and application. *Clim. Res.*, **7**, 85–95.

Hingray B. and Ben Haha M. (2005). Statistical performances of various deterministic and stochastic models for rainfall disaggregation. *Atmos. Res.*, **77**, 152–175.

Hirsch R. M., Helsel D. R., Cohn T. A. and Gilroy E. J. (1992). Statistical analysis of hydrologic data. Chapter 17. In: Handbook of Hydrology, D. R. Maidment (ed.), McGraw–Hill, New York, pp. 17.1–17.55.

Hohenegger C., Brockhaus P. and Schär C. (2008). Towards climate simulations at cloud-resolving scales. *Meteorol. Z.*, **17**(4), 383–394.

Hoppe H. (2008). Impact of input uncertainties on urban drainage models: climate change – a critical issue? Proceedings of the 11th International Conference on Urban Drainage, 31 August – 5 September 2008, Edinburgh, Scotland.

Hosking J. R. M. (1990). L-moments: analysis and estimation of distributions using linear combinations of order statistics. *J. Roy. Stat. Soc. B*, **52**(1), 105–124.

Hosking J. R. M. and Wallis J. R. (1987). Parameter and quantile estimation for the generalized pareto distribution. *Technometrics*, **29**(3), 339–349.

Hosking J. R. M. and Wallis J. R. (1993). Some statistics useful in regional frequency-analysis. *Water Resour. Res.*, **29** (2), 271–281.

Hosking J. R. M. and Wallis J. R. (1997). Regional Frequency Analysis: An Approach Based on L-Moments. Cambridge University Press, New York, USA, p. 244.

Houze R. A. (1994). Cloud Dynamics. **53**, Academic Press, 573 p.

Huesler J., Li D. and Raschke M. (2011). Estimation for the generalized pareto distribution using maximum likelihood and goodness of fit. *Commun. Stat.-Theory Meth.*, **40**(14), 2500–2510.

Hundecha Y. and Bárdossy A. (2005). Trends in daily precipitation and temperature extremes across western Germany in the second half of the 20th century. *Int. J. Climatol.*, **25**, 1189–1202.

Hundecha Y. and Bárdossy A. (2008). Statistical downscaling of extreme of daily precipitation and temperature and construction of their future scenarios. *Int. J. Climatol.*, **28**, 589–610.

Hurrell J. W. (1995). Decadal trends in the North Atlantic Oscillation: regional temperatures and precipitation. *Science*, **269**, 676–679.

Hyndman R. J. and Grunwald G. K. (2000). Generalized additive modelling of mixed distribution markov models with application to Melbourne's rainfall. *Aust. N.Z.J. Statist.*, **42**, 145–158.

Im E.-S., Kwon W.-T., Ahn J.-B. and Giorgi F. (2007). Multi-decadal scenario simulation over Korea using a one-way double-nested regional climate model system. Part 2: future climate projection (2021–2050). *Clim. Dyn.*, **30**(2–3), 239–254.

Im E.-S., Jung I. W. and Bae D. H. (2011). The temporal and spatial structures of recent and future trends in extreme indices over Korea from a regional climate projection. *Int. J. Climatol.*, **31**(1), 72–86.

IPCC. (2007a). Climate change 2007: The physical science basis, Summary for policymakers. Contribution of Working Group I to the Fourth Assessment Report of the Intergovernmental Panel on Climate Change, IPCC Secretariat, Geneva, Switzerland, p. 42.

IPCC. (2007b). Climate change 2007: synthesis report. Contribution of Working Groups I, II and III to the Fourth Assessment Report of the Intergovernmental Panel on Climate Change R. K. Pachauri, A. Reisinger (eds.), IPCC Secretariat, Geneva, Switzerland, p. 104.

IPCC. (2012). Managing the risks of extreme events and disasters to advance climate change adaptation. A Special Report of Working Groups I and II of the Intergovernmental Panel on Climate Change C. B. Field, V. Barros, T. F. Stocker, D. Qin, D. J. Dokken, K. L. Ebi, M. D. Mastrandrea, K. J. Mach, G.-K. Plattner, S. K. Allen, M. Tignor, P. M. Midgley (eds.), Cambridge University Press, Cambridge, UK, and New York, USA, p. 582.

Ishizaki N. *et al.* (2012). Inter-comparison of the skill in the Japan region of five RCMs nested in JRA-25. *J. Meteorol. Soc. Jpn.*, in press.

Iwasaki H. (2012). Recent positive trend in heavy rainfall in eastern Japan and its relation with variations in atmospheric moisture. *Int. J. Climatol.*, **32**(3), 364–374.

Jenkinson A. F. and Collison F. P. (1977). An Initial Climatology of Gales Over the North Sea. Synoptic Climatology Branch Memorandum. No. 62, Meteorological Office, Bracknell, UK.

Jinno K., Kawamura A., Berndtsson R., Larson M. and Niemczynowicz J. (1993). Real-time rainfall prediction at small space-time scales using a two-dimensional stochastic advection-diffusion model. *Water Resour. Res.*, **29**(5), 1489–1504.

Johnson S. (1967). Hierarchical clustering schemes. *Psychometrika*, **32**, 141–154.

Jones P. D., Kelly P. M., Goodess C. M. and Karl T. (1989). The effects of urban warming on the Northern Hemisphere temperature average. *Am. Meteorol. Soc.*, **2**, 285–290.

Jones P. D., Hulme M. and Briffa K. R. (1993). A comparison of Lamb circulation types with an objective classification scheme. *Int. J. Climatol.*, **13**, 655–663.

Jothiyangkoon C., Sivapalan M. and Viney N. R. (2000). Tests of a spacetime model of daily rainfall in southwestern Australia based on nonhomogeneous random cascades. *Water Resour. Res.*, **36**(1), 267–284.

Kamruzzaman M., Beecham S. and Metcalfe A. V. (2012). Climatic influences on rainfall and runoff variability in the south-east region of the murray darling basin. *Int. J. Climatol.*, in press; doi: 10.1002/joc.3422.

Kamruzzaman M., Beecham S. and Metcalfe A. V. (2011). Non-stationarity in rainfall and temperature in the murray darling basin. *Hydrol. Process.*, **25**(10), 1659–1675.

Kanada S., Nakano M., Hayashi S., Kato T., Nakamura M., Kurihara K. and Kitoh A. (2008). Reproducibility of maximum daily precipitation amount over Japan by a high-resolution non-hydrostatic model. *SOLA*, **4**, 105–108.

Kang B. and Ramirez J. A. (2009). Use of artificial neural network for regionalizing numerical weather prediction and global climate model. CD-ROM Proceedings of the 8th International Workshop on Precipitation in Urban Areas, 10–13 December 2009, St.Moritz, Switzerland, pp. 139–143.

Kang B. and Ramírez J. A. (2010). A coupled stochastic space-time intermittent random cascade model for rainfall downscaling. *Water Resour. Res.*, **46**, W10534.

Karpf C. and Krebs P. (2004). Sewers as drainage systems – quantification of groundwater infiltration. NOVATECH 2004: 5th International Conference on Sustainable Techniques and Strategies in Urban Water Management, Lyon, France, 6–10 June 2004, pp. 969–975.

Karl T. R. and Knight R. W. (1998). Secular trends of precipitation amount, frequency, and intensity in the United States. *Bull. Am. Meteorol. Soc.*, **79**, 231–241.

Karl T. R., Diaz H. F. and Kukla G. (1988). Urbanization: its detection and effects in the United States climate record. *J. Clim.*, **1**, 1099–1123.

Katz R. W. and Parlange M. B. (1998). Conditioning on indices of atmospheric circulation. *Meteorol. Appl.*, **5**, 75–87.

Katz R. W., Parlange M. B. and Tebaldi C. (2003). Stochastic modeling of the effects of large-scale circulation on daily weather in the Southeastern U.S. *Clim. Change*, **60**, 189–216.

Kavvas M. L. and Delleur J. W. (1981). A stochastic cluster model of daily rainfall sequences. *Water Resour. Res.*, **17**(4), 1151–1160.

Kazemi F., Beecham S., Gibbs J. and Clay R. (2009). Factors affecting terrestrial invertebrate diversity in bioretention basins in an Australian urban environment. *Landscape Urban Plann.*, **92**, 304–313.

Kazemi F., Beecham S. and Gibbs J. (2010). Bioretention swales as multifunctional landscapes and their influence on Australian urban biodiversity: hymenoptera as biodiversity indicators. *Acta Hort.*, **881**, 221–228.

Kazemi F., Beecham S. and Gibbs J. (2011). Streetscape biodiversity and the role of bioretention swales in an Australian urban environment. *Landscape Urban Plann.*, **101**, 139–148.

Keifer C. J. and Chu H. H. (1957). Synthetic storm pattern for drainage design. *J. Hydraul. Div., Proc. ASCE*, **83**(HY4).

Kendall M. G. (1975). Rank Correlation Methods. Griffin, London, UK, p. 196.

Kerr R. A. (2000). A North Atlantic climate pacemaker for the centuries. *Science*, **288**(5473), 1984–1986.

Kyselý J. (2009). Trends in heavy precipitation in the Czech Republic over 1961–2005. *Int. J. Climatol.*, **29**(12), 1745–1758.

Kyselý J. and Beranová R. (2009). Climate-change effects on extreme precipitation in central Europe: uncertainties of scenarios based on regional climate models. *Theor. Appl. Climatol.*, **95**(3–4), 361–374.

Kew S. F., Selten F. M., Lenderink G. and Hazeleger W. (2011). Robust assessment of future changes in extreme precipitation over the rhine basin using a GCM. *Hydrol. Earth Syst. Sci.*, **15**, 1157–1166.

Khaled H. H. (2008). Trend detection in hydrologic data: the Mann-Kendall trend test under the scaling hypothesis. *J. Hydrol.*, **349**, 350–363.

Khan S., Kuhn G., Ganguly A. R., Erickson III D. J. and Ostrouchov G. (2007). Spatio-temporal variability of daily and weekly precipitation extremes in South America. *Water Resour. Res.*, **43**, W11424.

Kiktev D., Sexton D. M. H., Alexander L. and Folland C. K. (2003). Comparison of modeled and observed trends in indices of daily climate extremes. *J. Clim.*, **16**, 3560–3571.

Kilsby C. G., Jones P. D., Burton A., Ford A. C., Fowler H. J., Harpham C., James P., Smith A. and Wilby R. L. (2007). A daily weather generator for use in climate change studies. *Environ. Modell. Softw.*, **22**, 1705–1719.

Kishtawal C. M., Niyogi D., Tewari M., Pielke R. A. Sr. and Sheppard J. M. (2010). Urbanization signature in the observed heavy rainfall climatology over India. *Int. J. Climatol.*, **30**, 1908–1916.

Kistler R., Kalnay E., Collins W., Saha S., White G., Woollen J., Chelliah M., Ebisuzaki W., Kanamitsu M., Kousky V., Dool H. V. D., Jenne R. and Fiorino M. (2001). The NCEP/NCAR 50-year reanalysis. *Bull. Am. Meteorol. Soc.*, **82** (2), 247–267.

Kjellström E., Nikulin G., Hansson U., Strandberg G. and Ullerstig A. (2011). 21st century changes in the European climate: uncertainties derived from an ensemble of regional climate model simulations. *Tellus*, **63A**(1), 24–40.

Klein Tank A. M. G. and Können G. P. (2003). Trends in indices of daily temperature and precipitation extremes in Europe. *J. Clim.*, **16**, 3665–3680.

Klein Tank A., Wijngaard J. and van Engelen A. (2002). Climate of Europe: Assessment of Observed Daily Temperature and Precipitation Extremes. KNMI, De Bilt, The Netherlands. ISBN 90-396-2208-9.

Klemp J. B., Gill D. O., Barker D. M., Duda M. G., Wang W. and Powers J. G. (2008). A Description of the Advanced Research WRF Version 3. NCAR Technical Note NCAR/TN–475 + STR. Mesoscale and Microscale Meteorology Division, National Center for Atmospheric Research, Boulder, Colorado, USA, p. 125.

Knight J. R. (2005). A signature of persistent natural thermohaline circulation cycles in observed climate. *Geophys. Res. Lett.*, **32**, L20708.

Knote C., Heinemann G. and Rockel B. (2010). Changes in weather extremes: assessment of return values using high resolution climate simulations at convection-resolving scale. *Meteorol. Zeit.*, **19**(1), 23.

Knutti R., Furrer R., Tebaldi C., Cermak J. and Meehl G. A. (2010). Challenges in combining projections from multiple climate models. *J. Clim.*, **23**(10), 2739–2758.

Korving H. and Clemens F. (2005). Impact of dimension uncertainty and model calibration on sewer system assessment. *Water Sci. Technol.*, **52**(5), 35–42.

Kottegoda N. T., Natale L. and Raiteri E. (2003). A parsimonious approach to stochastic multisite modelling and disaggregation of daily rainfall. *J. Hydrol.*, **274**(1–4), 47–61.

Koutsoyiannis D. and Onof C. (2001). Rainfall disaggregation using adjusting procedures on a poisson-cluster model. *J. Hydrol.*, **246**, 109–122.

Koutsoyiannis D., Onof C. and Wheater H. S. (2003). Multivariate rainfall disaggregation at a fine timescale. *Water Resour. Res.*, **39**(7), **1173**, 1–18.

Koutsoyiannis D., Makropoulos C., Langousis A., Baki S., Efstratiadis A., Christofides A., Karavokiros G. and Mamassis N. (2009). Climate, hydrology, energy, water: recognizing uncertainty and seeking sustainability. *Hydrol. Earth Syst. Sci.*, **13**, 247–257.

Krahe P., Buiteveld H., Pfister L., Ritz C. and Sprokkereef E. (2005). Climate change and the effect on hydrology and water management in the rhine basin. *Water Sci. Technol.*, **51**(5), 23–25.

Krishnamurthy C. K .B., Lall U. and Kwon H.-N. (2009). Changing frequency and intensity of rainfall extremes over India from 1951 to 2003. *J. Clim.*, **22**, 4737–4746.

Kulkarni A. and von Storch H. (1995). Monte carlo experiments on the effect of serial correlation on the mann–kendall test of trend. *Meteorologische Zeitschrift*, **4**(2), 82–85.

Kundzewicz Z. W. and Robson A. (eds.). (2000). Detecting Trend and Other Changes in Hydrological Data. World Climate Programme Data and Monitoring (WCDMP-45, WMO/TD 1013). World Meteorological Organization, Geneva, Switzerland, p. 158.

Kunkel K. E. (2003). North American trends in extreme precipitation. *Nat. Hazards*, **29**, 291–305.

Kunkel K. E., Andsager K. and Easterling D. R. (1999). Long-term trends in extreme precipitation events over the conterminous United States and Canada. *J. Clim.*, **12**, 2515–2527.

Kunkel K. E., Easterling D. R., Kristovitch D. A. R., Gleason B., Stoecker L. and Smith R. (2010). Recent increases in US heavy precipitation associated with tropical cyclones. *Geophys. Res. Lett.*, **37**, L24706.

Kuzuha Y., Komatsu Y., Tomosuga K. and Kishii T. (2005). Regional flood frequency analysis scaling and PUB. *J. Japan Soc. Hydrol. Water Resour.*, **18**(4), 441–458.

Kyselý J. and Huth R. (2006). Changes in atmospheric circulation over Europe detected by objective and subjective methods. *Theoret. Appl. Climatol.*, **85**, 19–36.

Kysely J., Picek J. and Beranova R. (2010). Estimating extremes in climate change simulations using the peaks-over-threshold method with a non-stationary threshold. *Glob. Planet. Change*, **72**(1–2), 55–68.

Lall U., Rajagopalan B. and Torboton D. G. (1996). A nonparametric wet/dry spell model for resampling daily precipitation. *Water Resour. Res.*, **32**(9), 2803–2823.

Lamb H. (1972). British isles weather types and a register of the daily sequence of circulation patterns 1861–1971. *Geophys. Memoir.*, **16**(116), 85.

Langousis A. and Veneziano D. (2007). Intensity-duration-frequency curves from scaling representations of rainfall. *Water Resour. Res.*, **43**, W02422.

Larsen A. N., Gregersen I. B., Christensen O. B., Linde J. J. and Mikkelsen P. S. (2009). Potential future increase in extreme one-hour precipitation events over Europe due to climate change. *Water Sci. Technol.*, **60**(9), 2205–2216.

Lau J. and Sharpe W. (2003). Comparison between distributed point rainfall measurements and spatially correct rainfall data in urban drainage runoff and quantity predictions. Proceedings of the 6th International Workshop on Precipitation in Urban Areas, Pontresina, 4–7 December 2003.

Leander R. and Buishand T. A. (2007). Resampling of regional climate model output for the simulation of extreme river flows. *J. Hydrol.*, **332**, 487–496.

Leandro J., Chen A. S., Djordjević S. and Savić D. A. (2009). Comparison of 1D/1D and 1D/2D coupled (sewer/surface) hydraulic models for urban flood simulation. *J. Hydraul. Eng.*, **135**(6), 495–504.

Lebel T. and Laborde J. P. (1988). A geostatistical approach for areal rainfall statistics assessment. *Stoch. Hydrol. Hydraul.*, **2**, 245–261.

Lebel T., Bastin G., Obled C. and Creutin J. D. (1987). On the accuracy of areal rainfall estimation: a case study. *Water Resour. Res.*, **23**(11), 2123–2134.

Leijnse H., Uijlenhoet R. and Berne A. (2010). Errors and uncertainties in microwave link rainfall estimation explored using drop size measurements and high-resolution radar data. *J. Hydrometeor.*, **11**, 1330–1344.

Lenderink G. and Van Meijgaard E. (2008). Increase in hourly precipitation extremes beyond expectations from temperature changes. *Nat. Geosci.*, **1**, 511–514.

Lenderink G., Mok H. Y., Lee T. C. and van Oldenborgh G. J. (2011). Scaling and trends of hourly precipitation extremes in two different climate zones – Hong Kong and the Netherland. *Hydrol. Earth Syst. Sci.*, **15**, 3033–3041.

Lei M., Niyogi D., Kishtawal C., Pielke Sr., R. A., Beltrán-Przekurat A., Nobis T. E. and Vaidya S. S. (2008). Effect of explicit urban land surface representation on the simulation of the 26 July 2005 heavy rain event over Mumbai, India. *Atmos. Chem. Phys.*, **8**, 5975–5995.

Leith N. A. and Chandler R. E. (2010). A framework for interpreting climate model outputs. *J. Roy. Stat. Soc. C Appl. Stat.*, **59**, 279–296.

Lettenmaier D. P., Wood A. W., Palmer R. N., Wood E. F. and Stakhiv E. Z. (1999). Water resources implications of global warming: a US regional perspective. *Clim. Change*, **43**, 537–579.

Leung L. R., Qian Y., Bian X., Washington W. M., Han J. and Roads J. O. (2004). Mid-century ensemble regional climate change scenarios for the Western United States. *Clim. Change*, **62**(1–3), 75–113.

Licznar P., Łomotowski J. and Rupp D. E. (2011). Random cascade driven rainfall disaggregation for urban hydrology: an evaluation of six models and a new generator. *Atmos. Res.*, **99**(3–4), 563–578.

Livezey R. E. and Chen W. (1983). Statistical field significance and its determination by Monte Carlo techniques. *Mon. Weather Rev.*, **111**, 46–59.

Lo J. C. F., Yang Z. L. and Pielke R. A. (2008). Assessment of three dynamical climate downscaling methods using the Weather Research and Forecasting (WRF) model. *J. Geophys. Res.*, **113**(D9), D09112.

Lombard F. (1988). Detecting change-points by Fourier analysis. *Technometrics*, **30**(3), 305–310.

Lorenz E. N. (1975). Climate Predictability: The Physical Basis of Climate Modeling. WMO, GARP Publ. Ser. 16, Secretariat of the World Meteorological Organization, Geneva, Switzerland, pp. 132–136.

Lovejoy S. and Schertzer D. (1990). Multifractals, universality classes, satellite and radar measurements of clouds and rain. *J. Geophys. Res.*, **95**, 2021–2034.

Lu M. and Yamamoto T. (2008). Application of a random cascade model to estimation of design flood from rainfall data. *J. Hydrologic Eng.*, **13**(5), 385–391.

Łupikasza E. B., Hänsel S. and Matschullat J. (2011). Regional and seasonal variability of extreme precipitation trends in southern Poland and central-eastern Germany 1951–2006. *Int. J. Climatol.*, **31**(15), 2249–2271.

Mackay E. B., Challenor P. G. and Bahaj A. S. (2011). A comparison of estimators for the generalised pareto distribution. *Ocean Eng.*, **38**(11–12), 1338–1346.

MacQueen J. (1967). Some methods for classification and analysis of multivariate observations. In: Proceedings of 5th Berkeley Symposium on Mathematical Statistics and Probability Berkeley, University of California, 21 June – 18 July 1965, University of California Press, USA, pp. 281–297.

Madsen T. and Figdor E. (2007). When it Rains, it Pours: Global Warming and the Rising Frequency of Extreme Precipitation in the United States. Environment California Research and Policy Center, California, USA, p. 48.

Madsen H., Rasmussen P. F. and Rosbjerg D. (1997). Comparison of annual maximum series and partial duration series methods for modeling extreme hydrologic events. 1. At-site modeling. *Water Resour. Res.*, **33**(4), 747–757.

Madsen H., Mikkelsen P. S., Rosbjerg D. and Harremoës P. (2002). Regional estimation of rainfall-intensity-duration-frequency curves using generalised least squares regression of partial duration series statistics. *Water Resour. Res.*, **38**(11), 1239.

Madsen H., Arnbjerg-Nielsen K. and Mikkelsen P. S. (2008). Update of regional intensity-duration-frequency curves in Denmark: tendency towards increased storm intensities. *Atmos. Res.*, **92**(3), 343–349.

Mailhot A. and Duchesne S. (2010). Design criteria of urban drainage infrastructures under climate change. *J. Water Resour. Plann. Manage.*, **136**, 201–208.

Mailhot A., Duchesne S., Caya D. and Talbot G. (2007). Assessment of future change in intensity-duration-frequency (IDF) curves for Southern Quebec using the Canadian Regional Climate Model (CRCM). *J. Hydrol.*, **347**, 197–210.

Maksimović Č., Prodanović D., Boonya-aroonnet S., Leitão J. P., Djordjević S. and Allitt R. (2009). Overland flow and pathway analysis for modelling of urban pluvial flooding. *J. Hydraul. Res.* **47**(4), 512–523.

Mandelbrot B. (1982). The Fractal Geometry of Nature. W.H. Freeman & Co, San Francisco.

Mann H. B. (1945). Nonparametric tests against trend. *Econometrica*, **13**, 245–259.

Mannina G. (2005). Integrated urban drainage modelling with uncertainty for stormwater pollution management. Doctoral dissertation, Università di Catania, Italy.

Manton M. J., Della-Marta P. M., Haylock M. R., Hennessy K. J., Nicholls N., Chambers L. E., Collins D. A., Daw G., Finet A., Gunawan D., Inape K., Isobe H., Kestin T. S., Lefale P., Leyu C. H., Lwin T., Maitrepierre L., Ouprasitwong N., Page C. M., Pahalad J., Plummer N., Salinger M. J., Suppiah R., Tran V. L., Trewin B., Tibig I. and Yee D. (2001). Trends in extreme daily rainfall and temperature in Southeast Asia and the South Pacific: 1961–1998. *Int. J. Climatol.*, **21**(3), 269–284.

Mantua N. J., Hare S. R., Zhang Y., Wallace J. M. and Francis R. C. (1997). A Pacific interdecadal climate oscillation with impacts on salmon production. *Bull. Am. Meteor. Soc.*, **78**, 1069–1079.

Mason S. J., Waylen P. R., Mimmack G. M., Rajaratnam B. and Harrison J. M. (1999). Changes in extreme rainfall events in South Africa. *Clim. Change*, **41**, 249–257.

Mariën J. L. and Vandewiele G. L. (1986). A point rainfall generator with internal storm structure. *Water Resour. Res.*, **22**(4), 475–482.

Mark O., Svensson G., König A. and Linde J. J. (2008). Analyses and adaptation of climate change impacts on urban drainage systems. Proceedings of the 11th International Conference on Urban Drainage, 31 August – 5 September 2008, Edinburgh, Scotland.

Martins E. S. and Stedinger J. R. (2001). Generalized maximum likelihood pareto-poisson estimators for partial duration series. *Water Resour. Res.*, **37**(10), 2551–2557.

Maraun D., Rust H. W. and Osborn T. J. (2010). Synoptic airflow and UK daily precipitation extremes development and validation of a vector generalised linear model. *Extremes*, **13**(2), 133–153.

Maraun D., Wetterhall F., Ireson A. M., Chandler R. E., Kendon E. J., Widmann M., Brienen S., Rust H. W., Sauter T., Themeßl M., Venema V. K. C., Chun K. P., Goodess C. M., Jones R. G., Onof C., Vrac M. and Thiele-Eich I. (2010). Precipitation downscaling under climate change. Recent developments to bridge the gap between dynamical models and the end user. *Rev. Geoph.*, **48**, RG3003.

Marengo J. A., Jones R., Alves L. M. and Valverde M. C. (2009a). Future change of temperature and precipitation extremes in South America as derived from the PRECIS regional climate modeling system. *Int. J. Climatol.*, **15**, 2241–2255.

Marengo J. A., Ambrizzi T., da Rocha R. P., Alves L. M., Cuadra S. V., Valverde M. C., Torres R. R., Santos D. C. and Ferraz S. E. T. (2009b). Future change of climate in South America in the late twenty-first century: intercomparison of scenarios from three regional climate models. *Clim. Dyn.*, **35**(6), 1073–1097.

Marengo J. A., Rusticucci M., Penalba O. and Renom M. (2010). An intercomparison of observed and simulated extreme rainfall and temperature events during the last half of the twentieth century: Part 2: historical trends. *Clim. Change*, **98**, 509–529.

Marsalek J., Maksimović C., Zeman E. and Price R. (eds.). (1998). Hydroinformatic Tools for Planning, Design, Operation and Rehabilitation of Sewer Systems. NATO Science Partnership Subseries 2: Environment, 44, Kluwer Academic Press, p. 552.

Matalas N. C. and Langbein W. B. (1962). Information content on the mean. *J. Geophys. Res.*, **67**(9), 3441–3448.

May B. R. and Hitch T. J. (1989). Periodic fluctuations in extreme hourly rainfalls in the UK. *Met. Mag.*, **118**, 45–50.

May W., Christensen O. B. and Christensen J. H. (2005). Potential future changes in heavy rainfall events in Europe simulated by the HIRHAM regional climate model. Proceedings of the 10th International Conference on Urban Drainage, 21–26 August 2005, Copenhagen, Denmark, p. 7.

McCullagh P. and Nelder J. A. (1989). Generalized Linear Models. Chapman and Hall, London, UK, p. 511.

McGuinness J. L. (1963). Accuracy of estimating watershed mean rainfall. *J. Geoph. Res.*, **68**(16), 4763–4767.

Mearns L. O., Gutowski W., Jones R., Leung R., McGinnis S., Nunes A. and Qian Y. (2009). A regional climate change assessment program for North America. *Eos Trans. AGU*, **90**(36), 311.

Meehl G. A., Arblaster J. M. and Tebaldi C. (2005). Understanding future patterns of increased precipitation intensity in climate model simulations. *Geophys. Res. Lett.*, **32**, L18719.

Menabde M., Harris D., Seed A., Austin G. and Stow D. (1997a). Multiscaling properties of rainfall and bounded random cascades. *Water Resour. Res.*, **33**, 2823–2830.

Menabde M., Seed A., Harris D. and Austin G. (1997b). Self-similar random fields and rainfall simulation. *J. Geophys. Res.*, **102**(D12), 13,509–13,515.

Menabde M., Seed A. and Pegram G. (1999a). A simple scaling model for extreme rainfall. *Water Resour. Res.*, **35**(1), 335–339.

Menabde M., Seed A., Harris D. and Austin G. (1999b). Multiaffine random field model of rainfall. *Water Resour. Res.*, **35**(2), 509–514.

Menéndez C. G., de Castro M., Boulanger J.-P., D'Onofrio A., Sanchez E., Sörensson A. A., Blazquez J., Elizalde A., Jacob D., Le Treut H., Li Z. X., Núñez M. N., Pessacg N., Pfeiffer S., Rojas M., Rolla A., Samuelsson P., Solman S. A. and Teichmann C. (2010). Downscaling extreme month-long anomalies in southern South America. *Clim. Change*, **98**(3–4), 379–403.

Michelson D., Einfalt T., Holleman I., Gjertsen U., Friedrich K., Haase G., Lindskog M. and Jurczyk A. (2005). Weather radar data quality in Europe – quality control and characterization. Review report COST Action 717, Office for Official Publications of the European Communities, Luxembourg, pp. 87.

Middelkoop H., Daamen K., Gellens D., Grabs W., Kwadijk J. C. J., Lang H., Parmet B. W. A. H., Schadler B., Schulla J. and Wilke K. (2001). Impact of climate change on hydrological regimes and water resources in the Rhine basin. *Clim. Change*, **49**, 105–128.

Mielke Jr. P. W. and Berry K. J. (2001). Permutation Methods: A Distance Function Approach. Springer, New York, p. 352.

Mikkelsen P. S., Harremoës P. and Rosbjerg D. (1995). Properties of extreme point rainfall II: Parametric data interpretation and regional uncertainty assessment. *Atmos. Res.*, **37**, 287–304.

Min S.-K., Zhang X., Zwiers F. W. and Hegerl G. C. (2011). Human contribution to more-intense precipitation extremes. *Nature*, **470**, 378–381.

Mishra V., Dominguez F. and Lettenmaier D. P. (2012). Urban precipitation extremes: how reliable are regional climate models? *Geophys. Res. Lett.*, **39**, L03407.

Molnar P. and Burlando P. (2005). Preservation of rainfall properties in stochastic disaggregation by a simple random cascade model. *Atmos. Res.*, **77**, 137–151.

Mora D. and Willems P. (2012). Decadal oscillations in rainfall and air temperature in the Paute River Basin – Southern Andes of Ecuador. *Theoret. Appl. Climatol.*, (1081), 267–282; doi: 0.1007/s00704-011-0527-4.

Moss R. H., Babiker M., Brinkman S., Calvo E., Carter T., Edmonds J., Elgizouli I., Emori S., Erda L., Hibbard K., Jones R., Kainuma M., Kelleher J., Lamarque J.-F., Manning M., Matthews B., Meehl J., Meyer L., Mitchell J., Nakicenovic N., O'Neill B., Pichs R., Riahi K., Rose S., Runci P., Stouffer R., van Vuuren D., Weyant J., Wilbanks T., van Ypersele J.-P. and Zurek M. (2008). Towards New Scenarios for Analysis of Emissions, Climate Change, Impacts, and Response Strategies. Intergovernmental Panel on Climate Change Secretariat, Geneva, Switzerland, p. 132.

Murata F., Hayashi T., Matsumoto J. and Asada H. (2007). Rainfall on the Meghalaya plateau in northeastern India—one of the rainiest places in the world. *Nat. Hazards*, **42**(2), 391–399.

Murphy B. F. and Timbal B. (2008). A review of recent climate variability and climate change in south-eastern Australia. *Int. J. Climatol.*, **28**, 859–879.

Myers B., Beecham S. and van Leeuwen J. (2011). Water quality with storage in permeable pavement basecourse. *Proc. ICE – Water Manage.*, **164**(7), 361–372.

Nakamura M., Kaneda S., Wakazuki Y., Muroi C., Hashimoto A., Kato T., Noda A., Yoshizaki M. and Yasunaga K. (2008). Effects of global warming on heavy rainfall during the Baiu season projected by a cloud-system-resolving model. *J. Disaster Res.*, **3**(1), 15–24.

Nakicenovic N. and Swart R. (eds.). (2000). Special Report on Emissions Scenarios (SRES). A Special Report of the IPCC Working Group III. Cambridge University Press, Cambridge, UK, p. 599.

Nakicenovic N., Alcamo J., Davis G., de Vries B., Fenhann J., Gaffin S., Gregory K. and Grubler A. 2000 Special Report on Emissions Scenarios: A Special Report of Working Group III of the Intergovernmental Panel on Climate Change. Cambridge University Press, Cambridge, UK, p. 599.

Nance L. and Durran D. R. (1997). A modeling study of nonstationary trapped mountain lee waves, Part I: Mean flow variability. *J. Atmos. Sci.*, **54**, 2275–2291.

Nance L. and Durran D. R. (1998). A modeling study of nonstationary trapped mountain lee waves, Part II: Nonlinearity. *J. Atmos. Sci.*, **55**, 1429–1445.

NCEP. (1995). NMC Eta Model Analyses, 3-hourly, 1994 Jul 11–1994 Aug 31. U.S. National Centers for Environmental Prediction Dataset ds068.0 published by the CISL Data Support Section at the National Center for Atmospheric Research, Boulder, USA; Retrieved 31 March 2012, from http://dss.ucar.edu/datasets/ds068.0/.

NERC. (1975). Flood Studies Report. Natural Environmental Research Council, London, p. 1189.

New M., Hewitson B., Stephenson D. B., Tsiga A., Kruger A., Manhique A., Gomez B., Coelho C. A. S., Masisi D. N., Kululanga E., Mbambalala E., Adesina F., Saleh H., Kanyanga J., Adosi J., Bulane L., Fortunata L., Mdoka M. L. and Lajoie R. (2006). Evidence of trends in daily climate extremes over southern and west Africa. *J. Geophys. Res.*, **111**, D14102.

Nguyen V.-T.-V. and Pandey G. R. (1994). Estimation of short-duration rainfall distribution using data measured at longer time scales. *Water Sci. Technol.*, **29**(1–2), 39–45.

Nguyen V.-T.-V., Rousselle J. and McPherson M. B. (1981). Evaluation of areal versus point rainfall with a sparsity of data. *Can. J. Civil Eng.*, **8**(2), 173–178.

Nguyen V.-T.-V., Nguyen T. D. and Ashkar F. (2002). Regional frequency analysis of extreme rainfalls. *Water Sci. Technol.*, **45**(2), 75–81.

Nguyen V.-T.-V., Peyron N. and Rivard G. (2002). Rainfall temporal patterns for urban drainage design in southern Quebec. Proc. Ninth International Conference on Urban Drainage, Portland, Oregon, 8–13 September 2002, p. 16.

Nguyen V.-T.-V., Nguyen T. D. and Gachon P. (2006). On the Linkage of Large-Scale Climate Variability with Local Characteristics of Daily Precipitation and Temperature Extremes: An Evaluation of Statistical Downscaling Methods. Advances in Geosciences, Vol. 4: Hydrological Science. World Scientific Publishing Company, pp. 1–9.

Nguyen V.-T.-V., Nguyen T. -D. and Cung A. (2007). A statistical approach to downscaling of sub-daily extreme rainfall processes for climate-related impacts studies in urban Areas. *Water Science & Technology: Water Supply*, **7**(2), 183–192.

Nguyen V.-T.-V., Desramaut N. and Nguyen T. D. (2008a). Estimation of design storms in consideration of climate variability and change. Proceedings of the 11th International Conference on Urban Drainage, 31 August–5 September 2008, Edinburgh, Scotland, p. 10.

Nguyen V.-T.-V., Desramaut N. and Nguyen T. D. (2008b). Estimation of urban design storms in consideration of GCM-based climate change scenarios. Proceedings International Conference on 'Water & Urban Development Paradigms: Towards an integration of engineering, design and management approaches', Leuven, 15–17 September 2008 J. Feyen, K. Shannon and M. Neville (eds.), CRC Press, Taylor & Francis Group, pp. 347–356.

Nguyen V.-T.-V., Desramaut N. and Nguyen T. D. (2010). Optimal rainfall temporal patterns for urban drainage design in the context of climate change. *Water Sci. Technol.*, **62**(5), 1170–1176.

Nhat L. M., Tachikawa Y., Sayama T. and Takaka K. (2007). A simple scaling characteristic of rainfall and time and space to derive intensity duration frequency relationsips. *Annu. J. Hydraul. Eng. JSCE*, **51**, 73–78.

Nie L., Lindholm O., Lindholm G. and Syversen E. (2009). Impacts of climate change on urban drainage systems – a case study in Fredrikstad, Norway. *Urban Water J.*, **6**(4), 323–332.

Niemczynowicz J. (1982). Areal intensity-duration-frequency curves for short term rainfall events. *Nord. Hydrol.*, **13**(4), 193–204.

Niemczynowicz J. (1988). The rainfall movement – a valuable complement to short-term rainfall data. *J. Hydrol.*, **104** (1–4), 311–326.

Niemczynowicz J. (1989). Impact of the greenhouse effect on sewerage systems – Lund case study. *Hydrol. Sci. J.*, **34**, 651–666.

Ning L. and Qian Y. (2009). Interdecadal change in extreme precipitation over South China and its mechanism. *Adv. Atmos. Sci.*, **26**(1), 109–118.

Noren A. J., Bierman P. R., Steig E. J., Lini A. and Southon J. (2002). Millennial-scale storminess variability in the northeastern Unites States during the Holocene epoch. *Nature*, **419**, 821–824.

NRCC. (1989). Hydrology of floods in canada: a guide to planning and design. In: W. E. Watt (ed.), National Research Council of Canada, Ottawa, Canada, p. 245.

NSSP. (1961). National severe storms project objectives and basic design, NSSP Report No. 1, US Weather Bureau, p. 16.

Ntegeka V. and Willems P. (2008). Trends and multidecadal oscillations in rainfall extremes, based on a more than 100 years time series of 10 minutes rainfall intensities at Uccle, Belgium. *Water Resour. Res.*, **44**, W07402.

Ntegeka V., Willems P., Baguis P. and Roulin E. (2008). Climate change impact on Hydrological Extremes Along Rivers and Urban drainage systems – Phase 1. Development of climate change scenarios for rainfall and ETo. Summary report Phase 1 of CCI–HYDR project by KU Leuven and Royal Meteorological Institute of Belgium, for the Belgian Science Policy Office, Brussels, Belgium, April 2008, p. 56; available 31 March 2012 on http://www.kuleuven.be/hydr/CCI-HYDR.htm.

Nuñez M. N., Solman S. A. and Cabré M. F. (2008). Regional climate change experiments over southern South America. II: climate change scenarios in the late twenty-first century. *Clim. Dyn.*, **32**(7–8), 1081–1095.

Nyeko-Ogiramoi P., Willems P., Ngirane-Katashaya G. and Ntegeka V. (2012). Nonparametric statistical downscaling of precipitation from Global Climate Models. Chapter 6. In: Climate Models, L. Druyan (ed.), pp. 109–136, InTech.

O'Connel P. E. and Todini E. (1996). Modeling of rainfall, flow and mass transport in hydrological systems: an overview. *J. Hydrol.*, **175**, 3–16.

O'Loughlin G., Beecham S., Lees S., Rose L. and Nicholas D. (1995). On-site stormwater detention systems in Sydney. *Water Sci. Technol.*, **32**(1), 69–175.

Olsson J. (1998). Evaluation of a scaling cascade model for temporal rainfall disaggregation. *Hydrol. Earth Syst. Sci.*, **2**, 19–30.

Olsson J., Niemczynowicz J. and Berndtsson R. (1993). Fractal analysis of high-resolution rainfall time series. *J. Geophys. Res.*, **98**(D12), 23265–23274.

Olsson J. and Niemczynowicz J. (1996). Multifractal analysis of daily spatial rainfall distributions. *J. Hydrol.*, **187**, 29–43.

Olsson J., Berggren K., Olofsson M. and Viklander M. (2009). Applying climate model precipitation scenarios for urban hydrological assessment: a case study in Kalmar City, Sweden. *Atmos. Res.*, **92**, 364–375.

Olsson J., Dahné J., German J., Westergren B., von Scherling M., Kjellson L., Ohls F. and Olsson A. (2010). A study of future discharge load on stockholm's main sewer system, SMHI Reports climatology, No. 3, (in Swedish).

Olsson J., Gidhagen L., Gamerith V., Gruber G., Hoppe H. and Kutschera P. (2012b). Downscaling of short-term precipitation from regional climate models for sustainable urban planning. *Sustainability*, **4**, 866–887.

Olsson J., Uvo C. B., Jinno K., Kawamura A., Nishiyama K., Koreeda N., Nakashima T. and Morita O. (2004). Neural networks for rainfall forecasting by atmospheric downscaling. *J. Hydrol. Eng.*, **9**(1–12).

Olsson J., Willén U. and Kawamura A. (2012a). Downscaling extreme regional climate model (RCM) precipitation for urban hydrological applications. *Hydrol. Res.*, **43**, 341–351.

Onof C. and Arnbjerg-Nielsen K. (2009). Quantification of anticipated future changes in high resolution design rainfall for urban areas. *Atmos. Res.*, **92**(3), 350–363.

Onof C., Chandler R. E., Kakou A., Northrop P. J., Wheather H. S. and Isham V. (2000). Rainfall modeling using Poisson-cluster processes. *Stoch. Environ. Res. Risk Assess.*, **14**, 384–411.

Onof C., Townend J. and Kee R. (2005). Comparison of two hourly to 5-min rainfall disaggregators. *Atmos. Res.*, **77**, 176–187.

Osborn T. J., Hulme M., Jones P. D. and Basnett T. A. (2000). Observed trends in the daily intensity of United Kingdom precipitation. *Int. J. Climatol.*, **20**, 347–364.

Over T. M. and Gupta V. K . (1996). A space–time theory of mesoscale rainfall using random cascades. *J. Geophys. Res.*, **101**(D21), 26319–26331.

Overeem A., Buishand T. A. and Holleman I. (2009). Extreme rainfall analysis and estimation of depth-duration-frequency curves using weather radar. *Water Resour. Res.*, **45**, W10424.

Pagliara S., Viti C., Gozzini B., Meneguzzo F. and Crisci A. (1998). Uncertainties and trends in extreme rainfall series in Tuscany, Italy: effects on urban drainage networks design. *Water Sci. Technol.*, **37**, 195–202.

Pahl-Wostl C. (2008). Requirements for adaptive water management. In: Adaptive and Integrated Water Management. Coping with Complexity and Uncertainty. Springer; ISBN 978-3-540-75940-9, 1–22.

Palmer T. N. and Räisänen J. (2002). Quantifying the risk of extreme seasonal precipitation events in a changing climate. *Nature*, **415**, 512–514.

Pao-shan Y. and Lin C. S. (2004). Regional rainfall intensity formulas based on scaling property of rainfall. *J. Hydrol.*, **295**(1–4), 108–123.

Parey S., Malek F., Laurent C. and Dacunha-Castelle D. (2007). Trends and climate evolution: statistical approach for very high temperatures in France. *Clim. Change*, **81**(3–4), 331–352.

Pathirana A. (2008). Global dimming – a cure for climate change or an agent for making its impacts worse? Proceedings of the 11th International Conference on Urban Drainage, 31 August–5 September 2008, Edinburgh, Scotland, p. 8.

Pathirana A.,, Herath S. and Yamada T. (2005). Simulating orographic rainfall with a limited-area, non-hydrostatic atmospheric model under idealized forcing. *Atmos. Chem. Phys.*, **5**, 215–226.

Pathirana A., Herath S., Yamada T. and Swain D. (2007). Impacts of absorbing aerosols on South Asian rainfall. *Clim. Change*, **85**(1), 103–118.

Pathirana A., Veerbeek W., Denekew H., Banda A. and Zevenburgen C. (2012). Impact of urban growth-driven landuse change on microclimate and extreme precipitation – a sensitivity study. *Atmos. Sci.*, submitted.

Pathirana A., Yamaguchi M. and Yamada T. (2003). Idealized simulation of airflow over a mountain ridge using a mesoscale atmospheric model. *Annu. J. Hydraul. Eng. (JSCE)*, **47**, 31–36.

Parkinson J. (2003). Assessing responses – drainage and stormwater management strategies for low-income urban communities. *Environ. Urban.*, **15**(2), 115–126.

Parkinson J., Schutze M. and Butler D. (2005). Modelling the impacts of domestic water conservation on the sustainability of urban sewerage design. *J. Chart. Inst. Water Environ. Manage.*, **19**(1), 49–56.

Pavan V., Tomozeiu R., Cacciamani C. and Di Lorenzo M. (2008). Daily precipitation observations over Emilia-Romagna: mean values and extremes. *Int. J. Climatol.*, **28**, 2065–2079.

Perica S. and Foufoula-Georgiou E. (1996). A Model for multiscale disaggregation of spatial rainfall based on coupling meteorological and scaling descriptions. *J. Geophys. Res.*, **101**(D21), **26**, 347–361.

Peterson T. C., Zhang X., Brunet-India M. and Vázquez-Aguirre J. L. (2008). Changes in North American extremes derived from daily weather data. *J. Geophys. Res.*, **113**, D07113.

Peters-Lidard C. D. and Wood E. F. (1994). Estimating storm areal average rainfall intensity in field experiments. *Water Resour. Res.*, **30**(7), 2119–2131.

Pettitt A. N. (1979). A non-parametric approach to the change-point problem. *Appl. Stat.* **28**(2), 126–135.

Peyron N., Nguyen V.-T.-V. and Rivard G. (2005). An Optimal Design Storm for the Design of Urban Drainage Systems (in French). Annales du bâtiment et des travaux publics, France, **3**, pp. 35–42.

Pezzaniti D., Beecham S. and Kandasamy J. (2009). Influence of clogging on the effective life of permeable pavements. *J. Water Manage., Inst. Civ. Eng. UK*, **162**(3), 211–220.

Philander S. G. (1990). El Niño, La Niña, and the Southern Oscillation. Academic Press, San Diego, CA, p. 293.

Piani C., Haerter J. O. and Coppola E. (2010). Statistical bias correction for daily precipitation in regional climate models over Europe. *Theoret. Appl. Climatol.*, **99**, 187–192.

Pickands J. (1975). Statistical inference using extreme order statistics. *Ann. Stat.*, **3**, 119–131.

Pielke R. A. (2002). Mesoscale Meteorological Modeling. 2nd edn, Academic Press, San Diego, USA, p. 676.

Pielke R. A., Cotton W. R., Walko R. L., Tremback C. J., Lyons W. A., Grasso L. D., Nicholls M. E., Moran M. D., Wesley D. A., Lee T. J. and Copelan J. H. (1992). A comprehensive meteorological modeling system RAMS. *Meteorol. Atmos. Phys.*, **49**, 69–91.

Pilgrim D. H. (ed.) (1998). Australian Rainfall and Runoff: A Guide to Flood Estimation. Volumes I and II, Reprinted edition, Institution of Engineers, Canberra, Australia.

Pilgrim D. H. and Cordery I. (1975). Rainfall temporal pattern for design floods. *ASCE J. Hydr. Div.*, **101**(HY1), 81–95.

Plosz B. G., Liltved H. and Ratnaweera H. (2009). Climate change impacts on activated sludge wastewater treatment: a case study from Norway. *Water Sci. Technol.*, **60**(2), 533–541.

Pouget L., Russo B., Escaler I., Redaño Á., Ribalaygua J. and Theias H. (2011). Climate change impacts on extreme rainfalls and design storms for flood risk assessment in urban areas. Application to Barcelona city. Proceedings of the 12th International Conference on Urban Drainage, 10–16 September 2011, Porto Alegre, Brazil, p. 6.

Pryor S. C., Howe J. A. and Kunkel K. E. (2009). How spatially coherent and statistically robust are temporal changes in extreme precipitation in the contiguous USA? *Int. J. Climatol.*, **29**(1), 31–45.

Quirmbach M., Papadakis I., Einfalt T., Langstadtler G. and Mehlig B. (2009). Analysis of precipitation data from climate model calculations for North Rhine-Westphalia. CD-ROM Proceedings of the 8th International Workshop on Precipitation in Urban Areas, 10–13 December 2009, St. Moritz, Switzerland, pp. 134–138.

Räisänen J. (2007). How reliable are climate models? *Tellus*, **59A**, 2–29.

Räisänen J., Hansson U., Ullerstig A., Döscher R., Graham L. P., Jones C., Meier H. E. M., Samuelsson P. and Willén U. (2004). European climate in the late twenty-first century: regional simulations with two driving global models and two forcing scenarios. *Clim. Dyn.*, **22**, 13–31.

Rajagopalan B. and Lall U. (1999). A k-nearest-neighbor simulator for daily precipitation and other variables. *Water Resour. Res.*, **35**, 3089–3101.

Rajeevan M., Bhate J. and Jaswal A. K. (2008). Analysis of variability and trends of extreme rainfall events over India using 104 years of gridded daily rainfall data. *Geophys. Res. Lett.*, **35**, L18707.

Rascko P., Szeidl L. and Semenov M. (1991). A serial approach to local stochastic weather models. *Ecol. Model.*, **57**, 27–41.

Rauch W. and de Toffol S. (2006). On the issue of trend and noise in the estimation of extreme rainfall properties. *Water Sci. Technol.*, **54**, 17–24.

Rauch W., Bertrand-Krajewski J.-L., Krebs P., Mark O., Schilling W., Schütze M. and Vanrolleghem P. A. (2002). Deterministic modelling of integrated urban drainage systems. *Water Sci. Technol.*, **45**(3), 81–94.

Refsgaard J. C., Arnbjerg-Nielsen K., Drews M., Halsnæs K., Jeppesen E., Madsen H., Markandya A., DHI, Olesen J. E., Porter J. R. and Christensen J. H. (2012). Climate change adaptation strategies: water management options under high uncertainty – a danish example. *Mitig. Adapt. Strateg. Glob. Change*, doi: 10.1007/s11027-012-9366-6, Open Access.

Renard B. (2006). Détection et prise en compte d'éventuels impacts du changement climatique sur les extrêmes hydrologiques en France. Doctoral dissertation, INPG, Cemagref, p. 360.

Renard B. and Lang M. (2007). Use of a Gaussian copula for multivariate extreme value analysis: some case studies in hydrology. *Adv. Water Resour.*, **30**, 897–912.

Renard B., Lang M. and Bois P. (2006). Statistical analysis of extreme events in a non-stationary context via a Bayesian framework: case study with peak-over-threshold data. *Stoch. Environ. Res. Risk Assess.*, **21**(2), 97–112.

Renard B., Lang M., Bois P., Dupeyrat A., Mestre O., Niel H., Sauquet E., Prudhomme C., Parey S., Paquet E., Neppel L. and Gailhard J. (2008). Regional methods for trend detection: assessing field significance and regional consistency. *Water Resour. Res.*, **44**, W08419.

Restrepo-Posada P. J. and Eagleson P. S. (1982). Identification of independent rainstorms. *J. Hydrol.*, **55**, 303–319.

Randall D. A. *et al.* (2007). Climate models and their evaluation. In: The Physical Science Basis. Contribution of Working Group I to the Fourth Assessment Report of the Intergovernmental Panel on Climate Change, S. Solomon *et al.* (eds.), Cambridge University Press, Cambridge, UK, and New York, USA, pp. 589–662.

Rauch W., Bertrand-Krajewski J.-L., Krebs P., Mark O., Schilling W., Schutze M. and Vanrolleghem P. A. (2002). Deterministic modeling of integrated urban drainage systems. *Water Sci. Technol.*, **45**(3), 81–94.

Richardson C. W. (1981). Stochastic simulation of daily precipitation, temperature, and solar radiation. *Water Resour. Res.*, **17**(1), 182–190.

Richardson G. R. A. (2010). Adapting to Climate Change: An Introduction for Canadian Municipalities. Natural Resources Canada, Ottawa, Ontario, p. 40.

Risbey J. and Kandlikar M. (2007). Expressions of likelihood and confidence in the IPCC uncertainty assessment process. *Clim. Change*, **85**(1–2), 19–31.

Rodriquez-Iturbe I. and Mejia J. M. (1974). On the transformation of point rainfall to areal rainfall. *Water Resour. Res.*, **10**(4), 729–735.

Rodriguez-Iturbe I., Cox D. R. and Isham V. (1987a). Some models for rainfall based on stochastic point processes. *Proc. R. Soc. Lond. A*, **410**, 269–288.

Rodriguez-Iturbe I., Febres de Power B. and Valdes J. B. (1987b). Rectangular pulses point process models for rainfall: analysis of empirical data. *J. Geophys. Res.*, **92**, 9645–9656.

Rosbjerg D., Madsen H. and Rasmussen P. F. (1992). Prediction in partial duration series with generalized paretodistributed exceedances. *Water Resour. Res.*, **28**(11), 3001–3010.

Rosbjerg D. and Madsen H. (2004). Advanced approaches in PDS/POT modelling of extreme hydrological events. In: Hydrology: Science & Practice for the 21st Century, Volume 1. British Hydrological Society, pp. 217–220.

Rosenberg E. A., Keys P. W., Booth D. B., Hartley D., Burkey J., Steinemann A. C. and Lettenmaier D. P. (2010). Precipitation extremes and the impacts of climate change on stormwater infrastructure in Washington State. *Clim. Change*, **102**(1–2), 319–349.

Rosenzweig C. and Abramopoulos F. (1997). Land-surface model development for the GISS GCM. *J. Clim.*, **10**, 2040–2056.

Rummukainen M. (2010). State-of-the-art with regional climate models. *Wiley Interdisc Rev.: Clim. Change*, **1**, 82–96.

Rupp D. E., Keim R. F., Ossiander M., Brugnach M. and Selker J. S. (2009). Time scale and intensity dependency in multiplicative cascades for temporal rainfall disaggregation. *Water Resour. Res.*, **45**, W07409.

Sadri S., Madsen H., Mikkelsen P. S. and Burn D. H. (2009). Analysis of extreme rainfall trends in Denmark, 33rd IAHR Congress: Water engineering for a sustainable environment. International Association of Hydraulic Engineering & Research (IAHR), ISBN: 978–94–90365-01–1, pp. 1731–1738.

Saji N. H., Goswami B. N., Vinayachandran P. N. and Yamagata T. (1999). A dipole mode in the tropical Indian Ocean. *Nature*, **401**, 360–363.

Schaefer M. G. (1990). Regional analyses of precipitation annual maxima in Washington State. *Water Resour. Res.*, **26** (1), 119–131.

Schertzer D. and Lovejoy S. (1987). Physical modeling and analysis of rain and clouds by anisotropic scaling multiplicative processes. *J. Geophys. Res.*, **92**, 9693–9714.

Schertzer D. and Lovejoy S. (1991). Scaling nonlinear variability in geodynamics: multiple singularities, observables, universality classes. In: Non-linear Variability in Geophysics, Scaling and Fractals, D. Schertzer and S. Lovejoy (eds.), Kluwer Academic, Dordrecht, pp. 41–82.

Schilling W. and Fuchs L. (1986). Errors in stormwater modelling – a quantitative assessment. *J. Hydraul. Eng.*, **112**(2), 111–123.

Schimek G., Fleming P. and Hartley D. (2008). Preparing for an uncertain future: Seattle Public Utility's climate change and Urban Drainage adaptation strategy. Proceedings of the 11th International Conference on Urban Drainage, 31 August–5 September, Edinburgh, Scotland, p. 10.

Schmidli J. and Frei C. (2005). Trends of heavy precipitation and wet and dry spells in Switzerland during the 20th century. *Int. J. Climatol.*, **25**, 753–771.

Schmidt G. A., Ruedy R.,, Hansen J. E., Aleinov I., Bell N., Bauer M., Bauer S., Cairns B., Canuto V., Cheng Y., Del Genio A., Faluvegi G., Friend A. D., Hall T. M., Hu Y., Kelley M., Kiang N. Y., Koch D., Lacis A. A., Lerner J., Lo K. K., Miller R. L., Nazarenko L., Oinas V., Perlwitz J. P., Perlwitz Ju., Rind D., Romanou A., Russell G. L., Sato Mki., Shindell D. T., Stone P. H., Sun S., Tausnev N., Thresher D. and Yao M.-S. (2006). Present day atmospheric simulations using GISS Model E: comparison to *in-situ*, satellite and reanalysis data. *J. Clim.*, **19**, 153–192.

Schreck C. J. and Semazzi F. H. M. (2004). Variability of the recent climate of Eastern Africa. *Int. J. Climatol.*, **24**, 681–701.

Schreider S., Smith D. I. and Jakeman A. J. (2000). Climate change impacts on urban flooding. *Clim. Change*, **47**, 91–115.

Schüpp M. (1968). Kalender der wetter- und witterungslagen von 1955 bis 1967 im zentralen alpengebiet. Veroeffentlichung der SMA, Zuerich, **11**.

Seed A. W., Srikanthan R. and Menabde M. (1999). A space and time model for design storm rainfall. *J. Geophys. Res.*, **104**(D24), 31,623–31,630.

Segond M.-L., Neokleous N., Makropoulos C., Onof C. and Maksimovic C. (2007). Simulation and spatial-temporal disaggregation of multi-site rainfall data for urban drainage applications. *Hydrol. Sci. J.*, **52**(5), 917–935.

Semadeni-Davies A., Hernebring C., Svensson G. and Gustafsson L.-G. (2008). The impacts of climate change and urbanisation on drainage in Helsingborg, Sweden: combined sewer system. *J. Hydrol.*, **350**, 100–113.

Semenov M. A. and Barrow E. M. (1997). Use of a stochastic weather generator in the development of climate change scenarios. *Clim. Change*, **35**(4), 397–414.

Semenov M. A. and Stratonovitch P. (2010). Use of multi-model ensembles from global climate models for assessment of climate change impacts. *Clim. Res.*, **41**, 1–14.

Semenov M. A., Brooks R. J., Barrow E. M. and Richardson C. W. (1998). Comparison of the WGEN and LARS-WG stochastic weather generators for diverse climates. *Clim. Res.*, **10**, 95–107.

Sen Roy S. and Balling R. C. (2004). Trends in extreme daily precipitation indices in India. *Int. J. Climatol.*, **24**, 457–466.

Sevruk B. (1989). Reliability of precipitation measurements. Proceedings of the International Workshop on Precipitation Measurement. World Meteorological Organisation, Geneva, pp. 13–19.

Seyoum S. D., Vojinovic Z., Price R. K. and Weesakul S. (2012). Coupled 1D and noninertia 2D flood inundation model for simulation of urban flooding. *J. Hydraul. Eng.*, **138**(1), 23–34.

Shang H., Yan J., Gebremichael M. and Ayalew S. M. (2011). Trend analysis of extreme precipitation in the Northwestern Highlands of Ethiopia with a case study of Debre Markos. *Hydrol. Earth Syst. Sci.*, **15**, 1937–1944.

Shem W. and Shepherd J. M. (2008). On the impact of urbanization on summertime thunderstorms in Atlanta: two numerical model case studies. *Atmos. Res.*, **92**(2), 172–189.

Shepherd J. M., Carter M., Manyin M., Messen D. and Burian S. (2010). The impact of urbanization on current and future coastal precipitation: a case study for Houston. *Environ. Plann. B: Plann. Des.*, **37**(2), 284–304.

Sharif M. and Burn D. H. (2007). Improved K-nearest neighbor weather generating model. *J. Hydrol. Eng.*, **12**(1), 42–51.

Sharif M., Burn D. H. and Wey K. M. (2007). Daily and hourly weather generation using a K-nearest neighbor approach. 18th CSCE Hydrotechnical Conference, 22–24 August 2007, Winnipeg, Manitoba, Canada, p. 10.

Sharma M., Coulibaly P. and Dibike Y. (2011). Assessing the need for downscaling RCM data for hydrologic impact study. *J. Hydrol. Eng.*, **16**, 534–539.

Sharma D., Das Gupta A. and Babel M. (2007). Spatial disaggregation of bias-corrected GCM precipitation for improved hydrologic simulation: Ping River Basin, Thailand. *Hydrol. Earth System Sci.*, **11**, 1373–1390.

Shea D. J., Worley S. J., Stern I. A. and Hoar T. J. (1994). An Introduction to Atmospheric and Oceanographic Data. NCAR Technical Note NCAR/TN-404 + IA. National Center for Atmospheric Research, Boulder, Colorado, USA, p. 132.

Siegel S. and Castellan N. J. (1988). Non Parametric Statistics for the Behavioural Sciences. 2nd edn., McGraw–Hill, New York, p. 399.

Sieker F. (1998). On-site stormwater management as an alternative to conventional sewer systems: a new concept spreading in Germany. *Water Sci. Technol.*, **38**(10), 65–71.

Sifalda V. (1973). Entwicklung eines berechnungsregens fűr die bemessung von kanalnetzen. *GWF-Wasser/Abwasser*, **114**, 435–440.

Silva V. B. S., Kousky V. E., Shi W. and Higgins R. W. (2007). An improved historical daily precipitation analysis for Brazil. *J. Hydrometeor.*, **8**, 847–861.

Simmonds I. and Hope P. (1997). Persistence characteristics of Australian rainfall anomalies. *Int. J. Climatol.*, **17**, 597–613.

Simões N., Ochoa S., Leitão J. P., Pina R., Sá Marques A. and Maksimović C. (2011). Urban drainage models for flood forecasting: 1D/1D, 1D/2D and hybrid models. 12th International Conference on Urban Drainage, Porto Alegre/Brazil, 11–16 September 2011.

Singleton A. T. and Reason C. J. C. (2006). Numerical simulations of a severe rainfall event over the Eastern Cape coast of South Africa: sensitivity to sea surface temperature and topography. *Tellus Ser. A Dyn. Meteorol. Oceanogr.*, **58**(3), 355–367.

Sivakumar B., Sorooshian S., Gupta V. J. and Gao X. (2001). A chaotic approach to rainfall disaggregation. *Water Resour. Res.*, **37**, 61–72.

Sivapalan M. and Blöschl G. (1998). Transformation of point rainfall to areal rainfall: Intensity-duration-frequency curves. *J. Hydrol.*, **204**, 150–167.

Skamarock W. C., Klemp J. B., Dudhia J., Gill D. O., Barker D. M., Wang W. and Powers J. G. (2005). A Description of the Advanced Research WRF Version 2. NCAR Technical Note NCAR/TN–468 + STR. National Center for Atmospheric Research, Mesoscale and Microscale Meteorology Division, Boulder, Colorado, USA, p. 88.

Smith I. (2004). An assessment of recent trends in Australian rainfall. *Aust. Meteorol. Mag.*, **53**, 163–173.

Smith I., Mcintose P., Ansell T. J., Reason C. J. C. and Mcinnes K. (2000). Southwest western Australian winter rainfall and its association with Indian ocean climate variability. *Int. J. Climatol.*, **20**, 1913–1930.

Sokol Z. (2003). Utilization of regression models for rainfall estimates using radar-derived rainfall data and rain gauge data. *J. Hydrol.*, **278**, 144–152.

Srikanthan R. and McMahon T. A. (2002). Stochastic generation of annual, monthly and daily climate data: a review. *Hydrol. Earth Syst. Sci.*, **5**, 653–670.

Staes J., Willems P., Marbaix P., Vrebos D., Bal K. and Meire P. (2011). Impact of climate change on river hydrology and ecology: a case study for interdisciplinary policy oriented research. Final Report SUDEM–CLI project Research Programme Science for a Sustainable Development, Belgian Science Policy Office, Brussels, p. 100.

Stahre P. (2006). Sustainability in Urban Storm Drainage. Planning and Examples. Svenskt Vatten, Stockholm, ISBN 91–85159-20-4.

Staufer P., Siekmann M., Roder S. and Pinnekamp J. (2008). Sustainable development of regional water resources management confronting climate trends and extreme weather. Proceedings of the 11th International Conference on Urban Drainage, 31 August–5 September, Edinburgh, Scotland, p. 10.

Stedinger R. J., Vogel R. M. and Foufoula-Georgiou E. (1993). Frequency analysis of extreme events. In: Handbook of Hydrology, D. R. Maidment (ed.), McGraw–Hill, New York, p. 66.

Stern N. (2007). The Economics of Climate Change: The Stern Review. Cambridge University Press, Cambridge, UK, and New York, USA, p. 712.

Stern R. D. and Coe R. (1984). A model fitting analysis of daily rainfall data. *J. R. Stat. Soc. A*, **147**, 1–34.

Stone D. A., Weaver A. J. and Zwiers F. W. (2000). Trends in Canadian precipitation intensity. *Atmos.-Ocean*, **38**(2), 321–347.

Strupczewski W. G. and Kaczmarek Z. (2001). Non-stationary approach to at-site flood frequency modelling II. Weighted least squares estimation. *J. Hydrol.*, **248**(1–4), 143–151.

Strupczewski W. G., Singh V. P. and Feluch W. (2001). Non-stationary approach to at-site flood frequency modelling I. Maximum likelihood estimation. *J. Hydrol.*, **248**(1–4), 123–142.

Su B. D., Jiang T. and Jin W. B. (2006). Recent trends in observed temperature and precipitation extremes in the Yangtze River basin, China. *Theor. Appl. Climatol.*, **83**(1–4), 139–151.

Sugahara S., da Rocha R. P. and Silveira R. (2009). Non-stationary frequency analysis of extreme daily rainfall in Sao Paulo, Brazil. *Int. J. Climatol.*, **29**(9), 1339–1349.

Sunyer M. A. and Madsen H. (2009). A comparison of three weather generators for extreme rainfall simulation in climate change impact studies. CD-ROM Proceedings of the 8th International Workshop on Precipitation in Urban Areas, 10–13 December 2009, St. Moritz, Switzerland, pp. 109–113.

Sunyer M. A., Madsen H. and Ang P. H. (2012). A comparison of different regional climate models and statistical downscaling methods for extreme rainfall estimation under climate change. *Atmos. Res.*, **103**, 119–128.

Svenskt Vatten (2011). Precipitation data for design and analysis of sewage systems. Report P104, Svenskt Vatten AB, Stockholm, Sweden, ISSN 1651-4947, p. 110. (in Swedish).

Svenskt Vatten (2007). The Impact of Climate Change on Common Sewage Systems. Svenskt Vatten AB, Stockholm, Sweden, ISSN 1651–6893, p. 44. (in Swedish).

Tait S. J., Ashley R. M., Cashman A., Blanksby J. and Saul A. J. (2008). Sewer system operation into the 21st century, study of selected responses from a UK perspective. *Urban Water J.*, **5**(1), 77–86.

Taye M. T. and Willems P. (2011). Influence of climate variability on representative QDF predictions of the upper Blue Nile Basin. *J. Hydrol.*, **411**, 355–365.

Taye M. T. and Willems P. (2012). Temporal variability of hydro-climatic extremes in the Blue Nile basin. *Water Resour. Res.*, in press; doi: 10.1029/2011WR011466.

Tchiguirinskaia I., Bonell M. and Hubert P. (eds.) (2004). Scales in hydrology and water management – echelles en hydrologie et gestion de l'eau. Kovacs Colloquium, Paris, September 2004, IAHS Publication 287, p. 170.

Tebaldi C. and Knutti R. (2007). The use of the multi-model ensemble in probabilistic climate projections. *Phil. Trans. R. Soc. A-Math. Phys. Eng. Sci.*, **365**(1857), 2053–2075.

Tebaldi C. and Sanso B. (2009). Joint projections of temperature and precipitation change from multiple climate models: a hierarchical Bayesian approach. *J. Roy. Stat. Soc. A Stat. Soc.*, **172**, 83–106.

Tebaldi C., Hayhoe K., Arblaster J. M. and Meehl G. A. (2006). Going to the extremes. An intercomparison of model-simulated historical and future changes in extreme events. *Clim. Change*, **79**(3–4), 185–211.

Tebaldi C., Smith R., Nychka D. and Mearns L. (2005). Quantifying uncertainty in projections of regional climate change: a Bayesian approach to the analysis of multi-model ensembles. *J. Clim.*, **18**, 1524–1540.

Tessier Y., Lovejoy S. and Schertzer D. (1993). Universal multifractals: theory and observations for rain and clouds. *J. Appl. Meteorol.*, **32**, 223–250.

Thapliyal V. and Kulshrestha M. (1992). Which is the rainiest place in the world? *Mausam*, **43**, 331–332.

Themeßl M., Gobiet A. and Leuprecht A. (2010). Empirical-statistical downscaling and error correction of daily precipitation from regional climate models. *Int. J. Climatol.*, **31**, 1531–1544.

Thielen J. (2000). The possible influence of urban surfaces on rainfall development: a sensitivity study in 2D in the meso-scale. *Atmos. Res.*, **54**(1), 15–39.

Thompson D. W. J. and Solomon S. (2002). Interpretation of recent Southern Hemisphere climate change. *Science*, **296**, 895–899.

Thorndahl S. and Willems P. (2008). Probabilistic modeling of overflow, surcharge, and flooding in urban drainage using the first order reliability method and parameterization of local rain series. *Water Res.*, **42**, 455–466.

Thorndahl S. and Rasmussen M. R. (2011). Marine X-band weather radar data calibration. *Atmos. Res.*, **103**, 33–44.

Thorndahl S., Beven K., Jensen J. and Schaarup-Jensen K. (2008). Event based uncertainty assessment in urban drainage modelling, applying the GLUE methodology. *J. Hydrol.*, **357**(3–4), 421–437.

Timbal B. (2004). Southwest Australia past and future rainfall trends. *Clim. Res.*, **26**, 233–249.

Timbal B. and Jones D. (2008). Future projections of winter rainfall in southeast Australia using a statistical downscaling technique. *Clim. Change*, **86**, 165–187.

Trenberth K. E. (1984). Signal versus noise in the Southern Oscillation. *Mon. Weather Rev.*, **112**, 326–332.

Trenberth K. E. and Solomon A. (1994). The global heat balance: heat transports in the atmosphere and ocean. *Clim. Dyn.*, **10**, 107–134.

Trenberth K. E., Dai A., Rasmussen R. M. and Parsons D. B. (2003). The changing character of precipitation. *Bull. Am. Meteorol. Soc.*, **84**, 1205–1217.

Trenberth K. E., Fasullo J. and Smith L. (2005). Trends and variability in column-integrated atmospheric water vapor. *Clim. Dyn.*, **24**, 741–758.

Tsakalias G. and Koutsoyiannis D. (1999). A comprehensive system for the exploration and analysis of hydrological data. *Water Resour. Manage.*, **13**, 269–302.

Smith D. M., Scaife A. A. and Kirtman B. P. (2012). What is the current state of scientific knowledge with regard to seasonal and decadal forecasting? *Environ. Res. Lett.*, **7**(1), 015602.

Smith R. L., Tebaldi C., Nychka D. and Mearns L. O. (2009). Bayesian modelling of uncertainty in ensembles of climate models. *J. Am. Stat. Assoc.*, **104**, 97–116.

Svensson C., Olsson J. and Berndtsson R. (1996). Multifractal properties of daily rainfall in two different climates. *Water Resour. Res.*, **32**, 2463–2472.

Trenberth K. E. and Caron J. M. (2000). The Southern Oscillation revisited: sea level pressures, surface temperatures and precipitation. *J. Clim.*, **13**, 4358–4365.

UK Institute of Hydrology (2008). Flood Estimation Handbook. CEH, Wallingford, UK.

UKWIR (2003). Climate Change and the Hydraulic Design of Sewerage Systems, Report no. 03/CL/10. Vol. I–III, Water Industry Research Ltd, UK.

UN (2012). International Strategy for Disaster Risk Reduction. Secretariat of the United Nations International Strategy for Disaster Risk Reduction, Geneva, Switzerland; Retrieved 31 March 2012, from www.unisdr.org.

Uppala S. M., Kållberg P. W., Simmons A. J., Andrae U., da Costa Bechtold V., Fiorino M., Gibson J. K., Haseler J., Hernandez A., Kelly G. A., Li X., Onogi K., Saarinen S., Sokka N., Allan R. P., Andersson E., Arpe K., Balmaseda M. A., Beljaars A. C. M., van de Berg L., Bidlot J., Bormann N., Caires S., Chevallier F., Dethof A., Dragosavac M., Fisher M., Fuentes M., Hagemann S., Hólm E., Hoskins B. J., Isaksen L., Janssen P. A. E. M., Jenne R., McNally A. P., Mahfouf J.-F., Morcrette J.-J., Rayner N. A., Saunders R. W., Simon P., Sterl A., Trenberth K. E., Untch A., Vasiljevic D., Viterbo P. and Woollen J. (2005). The ERA-40 re-analysis. *Q.J.R. Meteorol. Soc.*, **131**, 2961–3012.

US Army Corps of Engineer (1982). HEC-1 Flood Hydrograph Package. Davis, California.

Vaes G., Willems P. and Berlamont J. (2001). Rainfall input requirements for hydrological calculations. *Urban Water J.*, **3**(1–2), 107–112.

Vaes G., Willems P. and Berlamont J. (2005). Areal rainfall correction coefficients for small urban catchments. *Atmos. Res.*, **77**, 48–59.

van den Berg M. J., Vandenberghe S., De Baets B. and Verhoest N. E. C. (2011). Copula-based downscaling of spatial rainfall: a proof of concept. *Hydrol. Earth Syst. Sci.*, **15**, 1445–1457.

van der Linden P. and Mitchell J. F. B. (eds.) (2009). ENSEMBLES: Climate Change and its Impacts: Summary of research and results from the ENSEMBLES project. Met Office Hadley Centre, FitzRoy Road, Exeter EX1 3PB, UK. p. 160.

van Luijtelaar H., Gastkemper H. G. and Beenen A. S. (2008). Heavier rainfall due to climate change: how to deal with effects in urban areas. Proceedings of the 11th International Conference on Urban Drainage, 31 August – 5 September, Edinburgh, Scotland, p. 10.

van Luijtelaar H., Stapel W., Moens M. and Dirzwager A. (2005). Climatic change and urban drainage: strategies. Proceedings of the 10th International Conference on Urban Drainage, 21–26 August 2005, Copenhagen, Denmark, p. 8.

van Meijgaard E., van Ulft L., van de Berg W., Bosveld F., van den Hurk B., Lenderink G. and Siebesma A . (2008). The KNMI Regional Atmospheric Climate Model RACMO, Version 2.1. Technical report no. 302, Royal Netherlands Meteorological Institute, De Bilt, The Netherlands.

Van Steenbergen N. and Willems P. (2012). Method for testing the accuracy of rainfall-runoff models in predicting peak flow changes due to rainfall changes, in a climate changing context. *J. Hydrol.*, **414–415**, 425–434.

Vandenberghe S., Verhoest N. E. C., Buyse E. and De Baets B. (2010). A stochastic design rainfall generator based on copulas and mass curves. *Hydrol. Earth Syst. Sci.*, **14**, 2429–2442.

Vandenberghe S., Verhoest N. E. C., Onof C. and De Baets B. (2011). A comparative copula-based bivariate frequency analysis of observed and simulated storm events: a case study on Bartlett-Lewis modeled rainfall. *Water Resour. Res.*, **47**, W07529.

Vanhaute W. J., Vandenberghe S., Scheerlinck K., De Baets B. and Verhoest N. E. C. (2012). Calibration of the modified Bartlett-Lewis model using global optimization techniques and alternative objective functions. *Hydrol. Earth Syst. Sci.*, **16**, 873–891.

Veneziano D. and Iacobellis V. (2002). Multiscaling pulse representation of temporal rainfall. *Water Resour. Res.*, **38**(8), 1138.

Ventura V., Paciorek C. J. and Risbey J. S. (2004). Controlling the proportion of falsely-rejected hypotheses when conducting multiple tests with climatological data. *J. Clim.*, **17**, 4343–4356.

Venugopal V., Foufoula-Georgiou E. and Sapozhnikov V. (1999). Evidence of dynamic scaling in space-time rainfall. *J. Geophys. Res.*, **104**(D24), 31599–31610.

Verhoest N., Troch P. A. and De Troch F. P. (1997). On the applicability of Bartlett-Lewis rectangular pulses models for calculating design storms at a point. *J. Hydrol.*, **202**(1–4), 108–120.

Verhoest N. E. C., Vandenberghe S., Cabus P., Onof C., Meca-Figueras T. and Jameleddine S. (2010). Are stochastic point rainfall models able to preserve extreme flood statistics? *Hydrol. Process.*, **24**, 3439–3445.

Verworn H.-R., Krämer S., Becker M. and Pfister A . (2008). The impact of climate change on rainfall runoff statistics in the Emscher-Lippe region. Proceedings of the 11th International Conference on Urban Drainage, 31 August–5 September, Edinburgh, Scotland, p. 10.

Villarini G., Smith J. A., Baeck M. L., Vitolo R., Stephenson D. B. and Krajewski W. F. (2011). On the frequency of heavy rainfall for the Midwest of the United States. *J. Hydrol.*, **400**(1–2), 103–120.

Vincent L. A. and Mekis E. (2006). Changes in daily and extreme temperature and precipitation indices for Canada over the twentieth century. *Atmos.-Ocean*, **44**(2), 177–193.

von Storch V. H. (1995). Misuses of statistical analysis in climate research. In: Analysis of Climate Variability: Applications of Statistical Techniques, H. von Storch and A. Navarra (eds.), Springer–Verlag, Berlin, pp. 11–26.

Voss R., May W. and Roeckner E. (2002). Enhanced resolution modelling study on anthropogenic climate change: changes in extremes of the hydrological cycle. *Int. J. Climatol.*, **22**, 755–777.

Vrac M. and Naveau P. (2007). Stochastic downscaling of precipitation: from dry events to heavy rainfalls. *Water Resour. Res.*, **43**, W07402.

Vrac M., Marbaix P., Peillard D. and Naveau P. (2007b). Non-linear statistical downscaling of present and LGM precipitation and temperatures over Europe. *Clim. Past*, **3**, 669–682.

Vrac M., Stein M., Hayhoe K. and Liang X.-Z. (2007a). A general method for validating statistical downscaling methods under future climate change. *Geophys. Res. Lett.*, **34**, L18701.

Vrac M., Stein M. and Hayhoe K. (2007c). Statistical downscaling of precipitation through nonhomogeneous stochastic weather typing. *Clim. Res.*, **34**, 169–184.

Walker W. E., Harremoës P., Rotmans J., van der sluijs J. P., van Asselt M. B. A., Janssen P. and Krayer von Krauss M. P. (2003). Defining uncertainty. a conceptual basis for uncertainty management in model-based decision support. *Integr. Assess.*, **4**(1), 5–17.

Wallis J. R., Schaefer M. G., Barker B. L. and Taylor G. H. (2007). Regional precipitation-frequency analysis and spatial mapping for 24-h and 2-h durations for Washington State. *Hydrol. Earth Syst. Sci.*, **11**, 415–442.

Wan Zin W. Z., Jamaludin S., Deni S. M. and Jemain A. A. (2010). Recent changes in extreme rainfall events in Peninsular Malaysia: 1971–2005. *Theor. Appl. Climatol.*, **99**(3–4), 303–314.

Wang X. L. L. and Swail V. R. (2002). Trends of Atlantic wave extremes as simulated in a 40-year wave hindcast using kinematically reanalyzed wind fields. *J. Clim.*, **15**, 1020–1035.

Wang Y. Q., Leung L. R., McGregor J. L., Lee D. K., Wang W. C., Ding Y. H. and Kimura F. (2004). Regional climate modeling: progress, challenges, and prospects. *J. Meteorol. Soc. Japan*, **82**(6), 1599–1628.

Wang L., Onof C. and Maksimovic C. (2010). Reconstruction of sub-daily rainfall sequences using multinomial multiplicative cascades. *Hydrol. Earth Syst. Sci. Discuss.*, **7**, 5267–5297.

Ward J. D., Werner A. D., Nel W. P. and Beecham S. (2011). The influence of constrained fossil fuel emissions scenarios on climate and water resource projections. *Hydrol. Earth Syst. Sci.*, **15**(6), 879–1893.

Waters D., Watt W. E., Marsalek J. and Andreson B. C. (2003). Adaptation of a storm drainage system to accommodate increased rainfall resulting from climate change. *J. Environ. Plann. Manag.*, **46**, 755–770.

Watt W. E., Waters D. and McLean R. (2003). Climate variability and urban stormwater infrastructure in Canada: context and case studies. Toronto-Niagara region study report and working paper series, Report 2003–1. Meteorological Service of Canada, Waterloo, Ontario, p. 27.

Wenzel H. G. (1982). Rainfall for urban stormwater design. In: Urban Storm Water Hydrology. Water Resour. Monogr. Ser., 7, D. F. Kibler (ed.), American Geophysical Union, Washington, DC, pp. 35–67.

Wheater H. S., Chandler R. E., Onof C. J., Isham V. S., Bellone E., Yang C., Lekkas D., Lourmas G. and Segond M.-L. (2005). Spatial-temporal rainfall modelling for flood risk estimation. *Environ. Res. Risk Assess.* **19**, 403–416.

Wilby R. L. and Harris I. (2006). A framework for assessing uncertainties in climate change impacts: low-flow scenarios for the River Thames, UK. *Water Resour. Res.*, **42**(2), W02419.

Wilby R. L., Wigley T. M. L., Conway D., Jones P. D., Hewitson B. C., Main J. and Wilks D. S. (1998). Statistical downscaling of general circulation model output: a comparison of methods. *Water Resour. Res.*, **34**, 2995–3008.

Wilby R. L., Dawson C. W. and Barrow E. M. (2002). SDSM – a decision support tool for the assessment of regional climate change impacts. *Environ. Modell. Softw.*, **17**, 147–159.

Wilks D. S. (1993). Comparison of the three-parameter probability distributions for representing annual extreme and partial duration precipitation series. *Water Resour. Res.*, **29**(10), 3543–3549.

Wilks D. S. (1998). Multisite generalization of a daily stochastic precipitation generation model. *J. Hydrol.*, **210**(1–4), 178–191.

Wilks D. S. and Wilby R. L. (1999). The weather generation game: a review of stochastic weather models. *Prog. Phys. Geogr.*, **23**, 329–357.

Willems P. (2000). Compound IDF-relationships of extreme precipitation for two seasons and two storm types. *J. Hydrol.*, **233**, 189–205.

Willems P. (2001). Stochastic description of the rainfall input errors in lumped hydrological models. *Stoch. Environ. Res. Risk Assess.*, **15**, 132–152.

Willems P. (2006). Random number generator or sewer water quality model? *Water Sci. Technol.*, **54**(6–7), 387–394.

Willems P. (2008). Quantification and relative comparison of different types of uncertainties in sewer water quality modelling. *Water Res.*, **42**, 3539–3551.

Willems P. (2009). A time series tool to support the multi-criteria performance evaluation of rainfall-runoff models. *Environ. Modell. Softw.*, **24**(3), 311–321.

Willems P. (2011). Revision of urban drainage design rules based on extrapolation of design rainfall statistics. 12nd International Conference on Urban Drainage, Porto Alegre/Brazil, 10–15 September 2011, p. 8.

Willems P. (2012). Model uncertainty analysis by variance decomposition. *Phys. Chem. Earth*, **42–44**, 21–30.

Willems P. and Berlamont J. (2002a). Accounting for the spatial rainfall variability in urban modelling applications. *Water Sci. Technol.*, **45**(2), 105–112.

Willems P. and Berlamont J. (2002b). Probabilistic emission and immission modelling: case-study of the combined sewer – WWTP – receiving water system at Dessel (Belgium). *Water Sci. Technol.*, **45**(3), 117–124.

Willems P. and Yiou P. (2010). Multidecadal oscillations in rainfall extremes. *Geoph. Res. Abstr.*, **12**, EGU2010–EGU10270: EGU General Assembly 2010, 2–7 May 2010, Vienna.

Willems P. and Vrac M. (2011). Statistical precipitation downscaling for small-scale hydrological impact investigations of climate change. *J. Hydrol.*, **402**, 193–205.

Willems P., Guillou A. and Beirlant J. (2007). Bias correction in hydrologic GPD based extreme value analysis by means of a slowly varying function. *J. Hydrol.*, **338**, 221–236.

Willems P., Arnbjerg-Nielsen K., Olsson J. and Nguyen V.-T.-V. (2012). Climate change impact assessment on urban rainfall extremes and urban drainage: methods and shortcomings. *Atmos. Res.*, **103**, 106–118.

Williams B. K., Szaro R. C. and Shapiro C. D. (2007). Adaptive Management: The U.S. Department of the Interior Technical Guide. Adaptive Management Working Group, U.S. Department of the Interior, Washington, DC; ISBN 978-0-7277-3449-5, 491–504.

Wilson C. B., Valdes J. B. and Rodriguez-Iturbe I. (1979). On the influence of the spatial distribution of rainfall on storm runoff. *Water Resour. Res.*, **15**(2), 321–328.

WMO (2008a). Guide to Hydrological Practices. Volume I: Hydrology – From Measurement to Hydrological Information. WMO Report No.168, 6th edn, World Meteorological Organization, Geneva, Switzerland, p. 296.

WMO (2009a). Guidelines on Analysis of Extremes in a Changing Climate in Support of Informed Decisions for Adaptation. WMO-TD no. 1500, WCDMP no.72, World Meteorological Organization, Geneva, Switzerland, p. 52.

WMO (2009b). Manual on Estimation of Probable Maximum Precipitation (PMP). WMO Series No. 1045, 3rd edn, World Meteorological Organization, Geneva, Switzerland, p. 259.

WMO (2009c). Guide to Hydrological Practices. Volume II: Management of Water Resources and Application of Hydrological Practices. WMO Report No.168, 6th edn, World Meteorological Organization, Geneva, Switzerland, p. 302.

Wong T. H. F. (2005). An overview of water sensitive design practices in Australia. Proceedings of the 10th International Conference on Urban Drainage, 21–26 September 2005, Copenhagen, Denmark.

Woolhiser D. A. (1992). Modeling daily precipitation – progress and problems. In: Statistics in the Environmental and Earth Sciences, A. T. Walden and P. Guttorp (eds.), Halsted Press, pp. 71–89.

Worsley K. J. (1979). On the likelihood ratio test for a shift in location of normal populations. *J. Am. Stat. Assoc.*, **74**, 365–376.

Yang C., Chandler R. E., Isham V. and Wheater H. S. (2005). Spatial-temporal rainfall simulation using generalized linear models. *Water Resour. Res.*, **41**, W11415.

Yang W., Andréasson J., Graham L. P., Olsson J., Rosberg J. and Wetterhall F. (2010) Improved use of RCM simulations in hydrological climate change impact studies. *Hydrol. Res.*, **41**, 211–229.

Yao C., Yang S., Qian W., Lin Z. and Wen M. (2008). Regional summer precipitation events in Asia and their changes in the past decades. *J. Geophys. Res.*, **113**, D17107.

Yarnal B. (1984). A procedure for the classification of synoptic weather maps from gridded atmospheric pressure data. *Comput. Geosci.*, **10**, 397–410.

Yatagai A., Arakawa O., Kamiguchi K., Kawamoto H., Nodzu M. I. and Hamada A. (2009). A 44-year daily gridded precipitation dataset for Asia based on a dense network of rain gauges. *SOLA*, **5**, 137–140.

Seleshi Y. and Camberlin P. (2006). Recent changes in dry spell and extreme rainfall events in Ethiopia. *Theor. Appl. Climatol.*, **83**(1–4), 181–191.

Yen B. C. and Chow V. T. (1980). Design hyetograph for small drainage structures. *J. Hyd. Div. Am. Soc. Civ. Eng.*, **106** (HY6), 1055–1976.

Yiou P., Goubanova K., Li Z. X. and Nogaj M. (2008). Weather regime dependence of extreme value statistics for summer temperature and precipitation. *Nonlinear Process. Geophys.*, **15**, 365–378.

Yoo C. and Ha E. (2002). Basin average rainfall and its sampling error. *Water Resour. Res.*, **38**(11), 1259.

You Q., Kang S., Aguilar E., Pepin N., Flügel W.-A., Yan Y., Xu Y., Zhang Y. and Huang J. (2010). Changes in daily climate extremes in China and their connection to the large scale atmospheric circulation during 1961–2003. *Clim. Dyn.*, **36**(11–12), 2399–2417.

You Q., Kang S., Aguilar E. and Yan Y. (2008). Changes in daily climate extremes in the eastern and central Tibetan Plateau during 1961–2005. *J. Geophys. Res.*, **113**, D07101.

Young K. C. (1994). A multivariate chain model for simulating climatic parameters from daily data. *J. Appl. Meteorol.*, **33**, 661–671.

Yu P.-S., Yang T.-C. and Lin C.-S. (2004). Regional rainfall intensity formulas based on scaling property of rainfall. *J. Hydrol.*, **295**(1–4), 108–123.

Yue S. and Wang C. Y. (2002). Regional streamflow trend detection with consideration of both temporal and spatial correlation. *Int. J. Climatol.*, **22**(8), 933–946.

Yue S., Pilon P., Phinney B. and Cavadias G. (2002). The influence of autocorrelation on the ability to detect trend in hydrological series. *Hydrol. Process.*, **16**, 1807–1829.

Zalina M. D., Desa M. N. M., Nguyen V.-T.-V. and Amir K. (2002). Statistical analysis of extreme rainfall processes in Malaysia. *Water Sci. Technol.*, **45**(2), 63–68.

Zhai P., Zhang X., Wan H. and Pan X. (2005). Trends in total precipitation and frequency of daily precipitation extremes over China. *J. Clim.*, **18**, 1096–1108.

Zhang J. (2007). Likelihood moment estimation for the generalized pareto distribution. *Aust. N. Z. J.Stat.*, **49**(1), 69–77.

Zhang J. and Stephens M. A. (2009). A new and efficient estimation method for the generalized pareto distribution. *Technometrics*, **51**(3), 316–325.

Zhang X., Vincent L. A., Hogg W. D. and Niitsoo A. (2000). Temperature and precipitation trends in Canada during the 20th century. *Atmos.–Ocean*, **38**(3), 395–429.

Zhou Q., Mikkelsen P. S., Halsnæs K. and Arnbjerg-Nielsen K. (2012). Framework for economic pluvial flood risk assessment considering climate change effects and adaptation benefits. *J. Hydrol.*, **414–415**, 539–549.

Zorita E. and von-Storch H. (1998). The analog method as a simple statistical downscaling technique: comparison with more complicated methods. *J. Clim.*, **12**, 2474–2489.

Appendices

These appendices provide training material on some state-of-the-art methods described in the technical boxes of the main text. They are accompanied with an electronic supplement provided on the following IWA Water Wiki website: http://www.iwawaterwiki.org/xwiki/bin/view/Articles/ICCREUDS

Some training material is uploaded on restricted pages. To access those pages, you need to create first your own IWA Water Wiki user account. Once this is done, you can email your Wiki username together with the password "CCIGUR" to the Water Wiki Community Manager (WaterWiki@iwap.co.uk), who will grant you access to be restricted pages.

The electronic supplement contains electronic training material in the form of scripts in R, Matlab and Python. It also contains example datasets that can be used by scientists, students, teachers, urban water engineers, urban planners or other technical experts to train themselves or to provide training on the methods.

The electronic supplement is designed to be dynamic, such that more practical examples and training material can/will be provided over time. Readers therefore are invited to add their own examples.

Appendix A

Use of open source software R for statistical downscaling and rainfall extreme value analysis

A.1 INTRODUCTION

The statistical language R is open source software and freely available from www.r-project.org. Numerous extra packages have been developed by the users, each with a collection of functions developed specially for specific statistical techniques for example extreme value analysis. Some packages look alike with only small difference in the included functions and hence it can be difficult to create an overview. In the following section, a short (and hence incomplete) list of functions is given, which are relevant in relation to downscaling and extreme value statistics. Furthermore, a few examples of application are provided for selected packages. Note that most basic statistical functions are a part of the base functionality of R, meaning that no additional package is required for their use.

A.2 R PACKAGES

- ismev: A package developed to match the content of Coles (2001), and which therefore covers different aspects of extreme value analysis
- POT: A package specifically for Peak over Threshold analysis
- lmom: L-moments for a high variety of distributions
- lmomrfa: L-moments for regional frequency analysis
- mcmc: A package for basic Monte Carlo Markov Chain
- nlme: A package for extended linear modelling
- geepack: Generalized estimation equations for GLMs
- vgam: Tools for fitting vector GLMs and additive models
- quantreg: Quantile regression
- Kendall: Various versions of the non-parametric Mann-Kendall test for trend
- anm: The analogue method
- clim.pact: Climate analysis and empirical-statistical downscaling
- netcdf: Read and write netCDF files

For some packages a technical documentation is included and sometimes published in journals for statistical software. Help on specific functions (e.g. the function for linear regression `lm`) is obtained from:

```
?lm
```

The code behind the function can be obtained from the command:

```
lm
```

Selecting "Help → html help → Packages" gives you an overview of all installed packages and which functions they include. Furthermore, there is a high variety of forums for R related questions, among others: https://stat.ethz.ch/mailman/listinfo/r-help.

A.3 EXTREME VALUE ANALYSIS (POT)

Available for demonstrations we have a PDS of extreme exceedances. These are point rainfall extremes for a duration of 3 hours [μm/s] measured in the central part of Jutland, Denmark. Measures have been taken to ensure that the extremes are independent. Furthermore, a threshold level of 1.1 μm/s has been selected. The series represent 29.6 years of observations. We load the data in R and call the POT package (which first should be installed manually by the user):

```
library(POT)
data <- read.table('22421.txt', header=T, sep=',')
lambda <- length(data$extremes)/29.6 #the average number of events pr year
```

The data set contains two columns, one with the extreme intensities and one with the year of their occurrence. A GPD can be fitted to the extreme intensities using only one line of code; by the ML method (for details see Box 2.4) or by the method of L-moments (for details see Box 2.3). The latter is equivalent to unbiased probability weighted moments, which explain the notation in the example:

```
model.mle <- fitgpd(data$extremes, 1.1, 'mle') #maximum likelihood
model.pwmu <- fitgpd(data$extremes, 1.1, 'pwmu') #unbiased probability
weighted moments
```

If we look at the created object `model.mle` R is giving the following information:

```
model.mle
Estimator: MLE
Deviance: 14.83922
AIC: 18.83922

Varying Threshold: FALSE

Threshold Call: 1.1
Number Above: 80
Proportion Above: 1

Estimates
scale  shape
0.3037 0.2846

Standard Error Type: observed
```

```
Standard Errors
scale    shape
0.05064  0.12774

Asymptotic Variance Covariance
scale    shape
scale  0.002564 -0.003568
shape -0.003568  0.016318

Optimization Information
Convergence: successful
Function Evaluations: 32
Gradient Evaluations: 11
```

The output gives information on: the GPD parameters estimates ($\hat{\beta}$, $\hat{\gamma}$) together with their covariance and standard error (from the observed information matrix), the Akaike information criterion (AIC) goodness-of-fit measure of the model, and the progress of the numerical optimisation procedure. An almost identical output is obtained from `model.pwmu`:

```
model.pwmu
Estimator: PWMU

Varying Threshold: FALSE

Threshold Call: 1.1
Number Above: 80
Proportion Above: 1

Estimates
scale  shape
0.3015 0.3039

Standard Error Type:

Standard Errors
scale    shape
0.05491  0.15330

Asymptotic Variance Covariance
scale    shape
scale  0.003015 -0.005226
shape -0.005226  0.023501

Correlation
scale  shape
scale  1.0000 -0.6208
shape -0.6208  1.0000

Optimization Information
Convergence: NA
Function Evaluations: NA
```

Note that the optimization information is unavailable as the L-moment equations are solved analytically. From the two methods we obtain parameter estimations that are very much alike. Larger differences may occur, for example when outliers are present in the data set. Note that several other estimation methods are supported by the POT package. The return level plots (Figure A1) are obtained from:

```
retlev(model.mle,npy=lambda)
mtext(text='Maximum Likelihood', side = 3, line = 0.4, cex = 1.3)
retlev(model.pwmu,npy=lambda)
mtext(text='L-moments', side = 3, line = 0.4, cex =1.3)
```

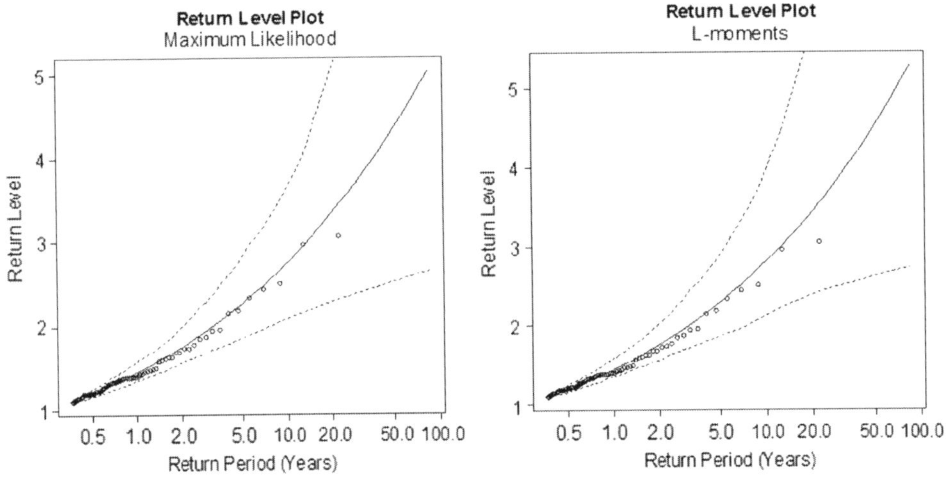

Figure A1 Return level plots obtained from the POT package.

The POT package also contains functions which can be helpful in relation to threshold selection.

A.4 NON-STATIONARY GPD PARAMETER ESTIMATION (ISMEV)

We now fit a non-stationary extreme value distribution with time-depended parameters, $\beta(t)$ and $\gamma(t)$ using the ML method. The time dependency can be described by a variety of functions and many authors only look at variation in β, due to the high uncertainty on the estimate of γ. In this example we compare these two models:

$$\beta(t) = \beta$$
$$\beta(t) = b_0 + b_1 t$$

The first corresponds to a stationary situation reviewed in Box 2.4 and in the example above. The latter leads to the following likelihood function:

$$L(b_0, b_1, \gamma) = \prod_{i=1}^{n} \frac{1}{b_0 + b_1 t_i} \left(1 + \gamma \frac{x_t - x_t}{b_0 + b_1 t_i}\right)^{-1/\gamma - 1}$$

A GPD with covariate depended parameters solved by the ML method is possible using the `ismev` package:

```
library(ismev)
data <- read.table('22421.txt', header=T, sep=',')
lambda <- length(data$extremes)/29.6
data$time <- data$year-1978 # the covariate "years after 1978"
model.nonstat <- gpd.fit(data$extremes, 1.1, lambda, ydat = as.matrix(data$time),
sigl=1)
#where ydat = matrix of covariates
#sigl = 1 means that the scale parameters depends on the variable in the first column of
#the covariate matrix
#The function furthermore needs information on the threshold (1.1) and the average
number of events pr year (lambda)
```

It is possible to use many other covariates than time. One could for example link variations in the scale parameter to variations in climatic indices (e.g. NAO). Note that function documentation recommends that the covariates are centred and scaled, to obtain a variable with mean zero and variance one. This will reduce the uncertainty on the estimate of b_0 in the cases where the covariate values are large. However, this transformation makes the direct interpretation of b_1 more difficult. In the non-transformed case b_1 corresponds to the annual rate of change.

Note that there is a similar function for fitting a non-stationary GEV distribution, which can be applied for annual maxima.

The parameters and their uncertainty are:

```
model.nonstat$mle # maximum likelihood estimates (b0, b1 and gamma)

0.472801636 -0.009998748 0.265088291

model.nonstat$se # the standard errors of the estimates (b0, b1 and gamma)

0.109000765 0.004514471 0.114322542
```

To compare the goodness of this model with a model with stationary parameters we can use the AIC. This must partly be computed by hand as the `gpd.fit` function only gives us the negative logarithm of the likelihood evaluated at the ML estimates:

```
2*model.nonstat$nllh + 2*3 #AIC note the model has 3 estimated parameters

16.32728
```

All information is collected in the table below, with parameter estimates, their uncertainty in parenthesis, and the AIC of the model. The parameters of the stationary model are computed in the example of Section A3. Based on this information we would conclude that the model with a linear decrease in the scale parameter describes the data best. However, the 95% confidence interval for b_1 cover values which are very close to zero.

Model	β	b_0	b_1	γ	**AIC**
stationary	0.30 (0.05)	–	–	0.28 (0.13)	18.84
$\beta(t) = b_0 + b_1 t$	–	0.47 (0.11)	−0.01 (0.005)	0.27 (0.11)	16.33

Appendix B

Use of Matlab for statistical downscaling and bias correction of RCM precipitation by quantile-quantile mapping

B.1 INTRODUCTION

In this appendix, an example of Matlab code for bias correction of daily RCM precipitation using a quantile mapping version called Distribution-Based Scaling (DBS; Yang *et al.* 2010) is given. DBS assumes that the frequency distribution of intensities can be accurately approximated using two gamma distributions, one for low and intermediate precipitation intensities (up to the 95% quantile) and one for extreme intensities (above the 95% quantile). Before distribution fitting, the number of wet days in the RCM are adjusted by using an intensity threshold and setting all values below it to zero.

B.2 STEP-BY-STEP PROCEDURE

Let X_{OBS} be a time series of observed non-zero precipitation in a reference period, X_{SIM_REF} the corresponding original (raw) data from an RCM simulation in the same period and X_{SIM_FUT} RCM simulated data for a future period. Let `x_obs` be the Matlab vector with the X_{OBS} values. In the same way, let the variables `x_sim_ref` and `x_sim_fut` be the X_{SIM_REF} and X_{SIM_FUT} values.

(1) The first step is to adjust the number of wet days in the simulation period. A wet day is defined as a day with precipitation. First, the percentage of wet days in the observation period is calculated – call this value p:

```
p=100*sum(x_obs==0)/numel(x_obs);
```

(2) Calculate the p'th percentile of X_{SIM_REF}, and call this value y:

```
y=prctile(x_sim_ref,p);
```

(3) Set all values in X_{SIM_REF} that are less than or equal to y to 0. Now the number of wet days in both time series is equal.

```
x_sim_ref(x_sim_ref<=y)=0;
```

(4) Remove any zeros from the observed time series and split the series in two at the 95% quantile, with the purpose of separating between extreme intensities (the highest 5%) and low-to-intermediate (normal) intensities. Apply the upper index N to the normal data and the upper index E to the extreme data. With this terminology, our two new time series are X^N_{OBS} and X^E_{OBS}.

Remove zeros:

```
x_obs(x_obs==0)=[];
```

Identify the 95% quantile:

```
p95obs=prctile(x_obs,95);
```

Split the observed time series:

```
x_obs_n=x_sim(x_obs < p95sim);
x_obs_e=x_sim(x_obs > p95sim);
```

(5) Repeat step 4 for the RCM simulated time series in the reference period, generating $X^N_{SIM\_REF}$ and $X^E_{SIM\_REF}$.

Remove zeros:

```
x_sim_ref(x_sim_ref==0)=[];
```

Identify the 95% quantile:

```
p95sim_ref=prctile(x_sim_ref,95);
```

Split the RCM simulated time series:

```
x_sim_ref_n=x_sim(x_sim_ref < p95sim_ref);
x_sim_ref_e=x_sim(x_sim_ref > p95sim_ref);
```

(6) Fit gamma distributions to the series representing normal intensities in the observations, X^N_{OBS}, with the ML method to get the parameter estimates α^N_{OBS} and β^N_{OBS}. In the same way, fit a gamma distribution to X^N_{SIM} with the ML method to get the parameter estimates α^N_{SIM} and β^N_{SIM}:

```
[alpha_obs_n,beta_obs_n]=gamfit(x_obs_n);
[alpha_sim_n,beta_sim_n]=gamfit(x_sim_ref_n);
```

(7) Similarly, fit gamma distributions to the series representing extreme intensities in the observations, X^E_{OBS}, with the ML method to get the parameter estimates α^E_{OBS} and β^E_{OBS}. In the same way, fit a gamma distribution to X^E_{SIM} with the MLE-method to get the parameter estimates α^E_{SIM} and β^E_{SIM}:

```
[alpha_obs_e,beta_obs_e]=gamfit(x_obs_e);
[alpha_sim_e,beta_sim_e]=gamfit(x_sim_ref_e);
```

(8) Define the DBS mapping function, denoted D, as:

$$D(x, \alpha_{SIM}, \beta_{SIM}, \alpha_{OBS}, \beta_{OBS}) := F^{-1}[F(x, \alpha_{SIM}, \beta_{SIM}), \alpha_{OBS}, \beta_{OBS}]$$

The DBS scaling is basically a quantile-quantile match between the observed data and the simulated data. For each simulated value x, we calculate its percentile value from by putting it into $F(x, \alpha_{SIM}, \beta_{SIM})$. Then we pick out this percentile p from the inverse CDF $F^{-1}(p, \alpha_{OBS}, \beta_{OBS})$, ending up with a precipitation value. The greater the difference between the simulated and observed parameters, the greater the difference in the CDF functions, and the greater the scaling.

(9) Apply the DBS scaling to all elements in X_{SIM_REF} by putting them through the DBS function using their respective ML estimation parameters.
If $x \in X_{SIM}^{N}$ then scale x according to $x = D(x, \alpha_{SIM}^{N}, \beta_{SIM}^{N}, \alpha_{OBS}^{N}, \beta_{OBS}^{N})$.
If $x \in X_{SIM}^{E}$ then scale x according to $x = D(x, \alpha_{SIM}^{E}, \beta_{SIM}^{E}, \alpha_{OBS}^{E}, \beta_{OBS}^{E})$.

Pre-allocate a vector X_{COR_REF} for bias corrected data:

```
n=numel(x_sim_ref);
x_cor_ref=zeros(n,1); %pre-allocate x_sim_dbs

for i=1:n
if x_sim(i) < p95sim
                x1=gamcdf(x_sim(i),alpha_sim_n,beta_sim_n);
                x_cor_ref(i)=gaminv(x1,alpha_obs_n,alpha_obs_e);
else
                x1=gamcdf(x_sim(i),alpha_sim_e,beta_sim_e)
                x_cor_ref(i)=gaminv(x1,alpha_obs_e,alpha_obs_e);
end
end
```

In the code above, the variable `x1` is the inner part of the DBS function, that is $F(x, \alpha_{SIM}, \beta_{SIM})$ part (see step 8).

(10) To bias correct the future scenario and generate X_{COR_FUT}, repeat steps 3 and 9 using as input X_{SIM_FUT}.

B.3 FINAL REMARKS

The code uses built-in Matlab functions that have been confirmed with version 7.12.0.

Other probability distributions than the gamma distribution are possible, both for the normal data and the extreme data. For example, a Gumbel distribution may work well for extreme data, since it is an extreme value distribution and we want it to fit the highest 5% data.

One could also split the data at another quantile than the 95% quantile. Different quantiles will give different number of elements to our time series (a lower quantile value will assign more elements to the extreme data), and will thus give different gamma distributions and different scaling.

Appendix C

Running Weather Research Forecast (WRF) Limited Area Atmospheric Model (LAM) on PC

This is a hands-on tutorial on the basic use of the Weather Research and Forecasting (WRF) Model on a typical personal computer (PC). At the onset, it should be noted that WRF is a well-documented and supported model and therefore the user should always refer to the standard model documentation for detailed information on the use of the model. The objective of this tutorial is to provide easy to follow instructions to get started on performing very basic simulations on a PC. At the end of this tutorial we provide a list of resources to find comprehensive information on how to take your study further.

C.1 LEARNING OBJECTIVES

After successfully finishing this tutorial the reader should be able to a) set-up a nested modelling domain for a region of interest and initialize it with standard geographical data, b) use a historical or operational global atmospheric dataset to prepare initial and boundary conditions for the domain, for a specific period, c) run the model and obtain the results, and d) perform elementary post-processing on the output to prepare basic graphs including precipitation.

C.2 STRUCTURE OF THIS TUTORIAL

In the next section we present a sort of background on this tutorial. What was it like to run LAMs in the past (about 10 years ago)? What are the challenges that discouraged people using these? What made us want to write a "yet-another" tutorial on running LAMs? We try to answer these types of questions. We also describe the tools and techniques contributed in making the "simulation jig" we use in this tutorial.

If you are not interested in answers to such questions at least initially, feel free to skip the next section. Skipping it will not affect your understanding of the WRF model or effectively using it.

C.3 BACKGROUND

The traditional challenge in using LAMs

Atmospheric models are demanding computer applications that were traditionally run on dedicated supercomputers. For example when MM5 was released in 1994, it was targeted largely for Cray supercomputers (like Cray-3 of NCAR). While operating systems like Windows and Macintosh are popular among personal computer users, these super computers often run some variant UNIX® operating system. However, over the years the situation changed and many users started using them on personal computer systems, employing Linux: a UNIX-like operating system running on personal computers. Atmospheric models depend on the UNIX operating systems for many advanced services, making it difficult to port them to Windows. Even today atmospheric models are exclusively used on UNIX (or Linux) operating systems.

Atmospheric models were designed to run on a number of different hardware platforms (e.g. Cray supercomputers, IBM supercomputers, personal computers running Linux). Due to this reason, it is common to provide these models in the form of source-code (typically written in Fortran and C), which needs to be compiled in the target computer system. This process can be challenging for inexperienced users and poses the second barrier for widespread use of these models. It was not unusual for individual users to spend weeks if not months trying to compile a LAM system.

STRC Environmental Model System (EMS)

The National Weather Service – Science Operations Officers – Science and Training Resource Center (NWS-SOO-STRC) started providing pre-compiled binaries of WRF and its support programs several years ago (STRC-EMS system). This relieved the user from the burden of having to compile the source-code before running the WRF system – it is provided in already complied form.

Yet-another WRF tutorial?

Linux operating system has become a truly user friendly and easy to use one during the last decade. However, many computer users are unfamiliar with the system. This is perceived as one of the primary challenges in using LAMs by new users. To overcome this challenge, in this tutorial we offer a solution that can run the WRF model within a window (virtual machine) of a Windows PC.

Basically what we have done is:

(1) Create a virtual Linux computer ("computer" running within a window).
(2) Install STRC-EMS system in this virtual computer.
(3) Add Chimplot – an easy to use plotting program to view output of the model.

The result is a system that allows you to get familiar with the WRF model (its elementary use) in an afternoon without unduly bothering about technical issues.

C.4 INSTALLING WRF-LIVE SYSTEM ON YOUR PC

System requirements

We have tested this system on Windows 7 and XP computers. Atmospheric models are demanding applications in terms of computer processing power and memory. The system presented in this tutorial will need the following resources:

	Minimum	Recommended
Memory	2 GB	3 GB
Hard drive space	20 GB	30 GB

The WRF model can efficiently utilize multiple-core processors. If your computer has dual-core or a quad-core processor, you can make use of it to run your simulations faster. However this is not a necessity to complete this tutorial. A single core processor will work.

If your computer's hard drive does not have enough space, you can use an external hard drive. But make sure that what you use is relatively fast. For example, a typical USB externam memory, even if it has adequate space, may not be fast enough to run this system.

Install VMware player

VMware player is a proprietary but free software product that allows you to run an operating system within other operating systems. For example with VMware you can run Windows XP (guest) within a computer running Windows 7 (host). You will see Windows XP working within an application window of host (Windows 7) system. Figure C1 shows a screenshot of a Linux guest running on Windows (host).

Figure C1 A Linux guest running on Windows 7 host using VMware.

With the permission of VMware Inc., we have included a version of VMware player with WRF-live. WRF-live system will run fine with this included version. However, the software is often updated and by the time you are reading this, there may be a newer version of it online. The latest version can be downloaded from: http://www.vmware.com/products/player/.

Installation of VMware player is easy. Just double click on the file VMware-player-X.X.Xxxxxx.exe (Either the version we have provided with WRF-live or the latest version you have downloaded). Follow the instructions on the screen (default choices are fine). You may have to restart your computer. After this, you can go ahead and install WRF-live system.

Install and start WRF-live system

Select a disk drive that has plenty of space: this is important! Unless your drive has at least 20GB of space free, probably you will not succeed in running the tutorial. Simply extract WRF-live.rar file to this drive. It will create a directory called `WRF-live`.

Running WRF-live is easy. Just double click on the file `RUN.bat` in the top level of WRF-live directory (Figure C2). First time when you do this VMware player may complain, asking you whether you copied or moved the virtual machine (Figure C3). Just click “I copied it”.

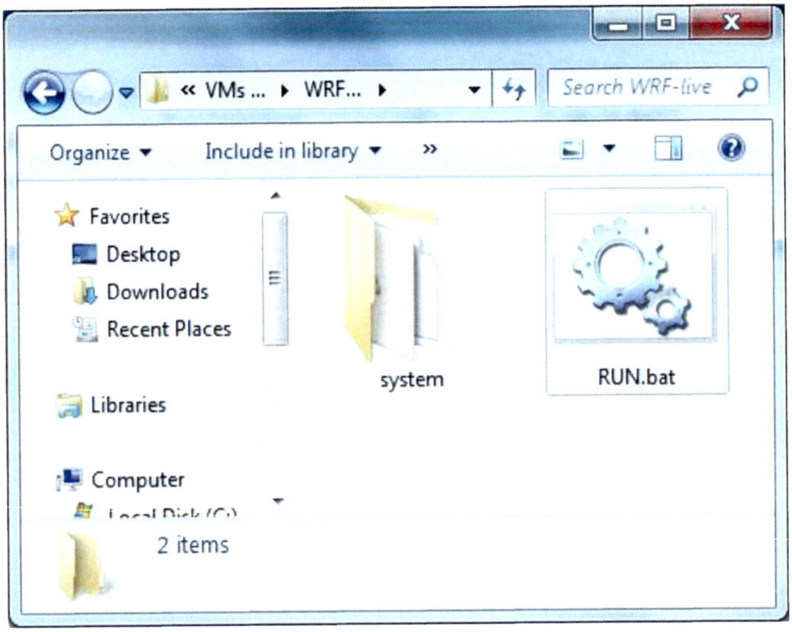

Figure C2 RUN.bat file. Click on this to start WRF-live.

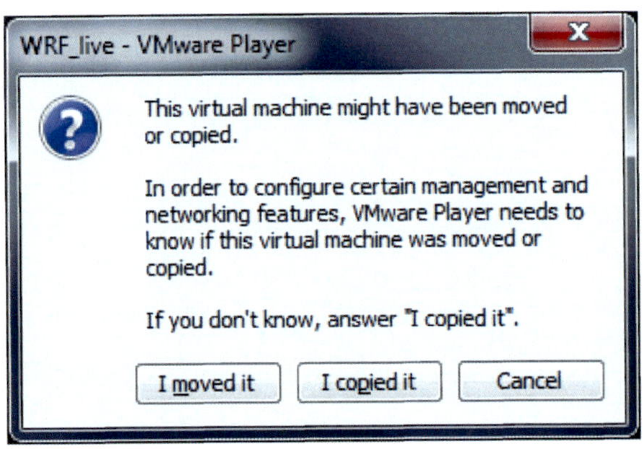

Figure C3 Did you copy/move it. Click “I...”.

In a minute or two the WRF-live will just look like another computer desktop (Figure C4). Login using the username "wrf" and password "wrf" (Figure C5).

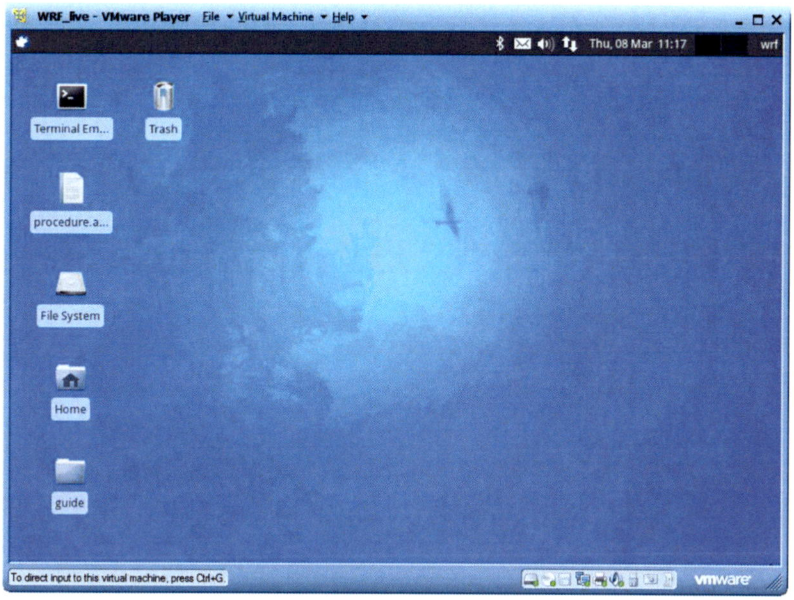

Figure C4 WRF-live desktop. Your desktop might be slightly different in appearance.

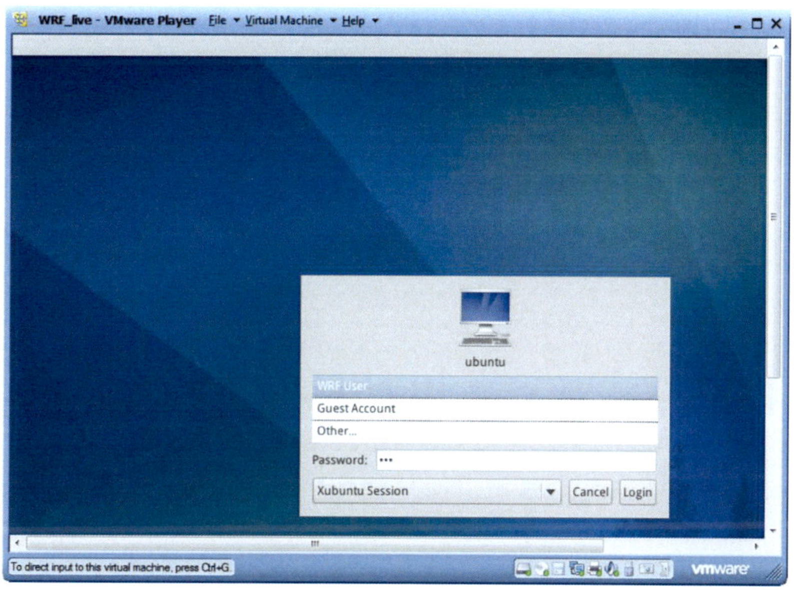

Figure C5 Login to the virtual machine. Username: wrf, password: wrf.

At this point maximize your WRF-live window so that you have plenty of space to work on.

After this point, all the instructions describe what you should do *within* this window, unless we say otherwise

The terminal

Double-click the "terminal" icon on the desktop (Figure C6). This should open the terminal program (Figure C7).

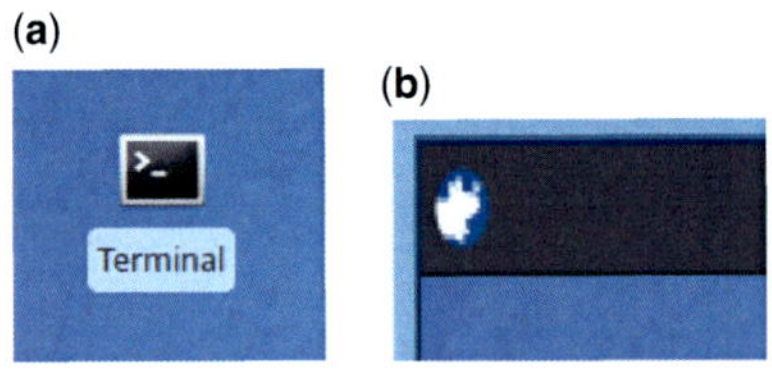

Figure C6 The terminal icon (a). You can also find this on the start (little mouse) > accessories menu. Start menu button is on the top left corner of the screen (b).

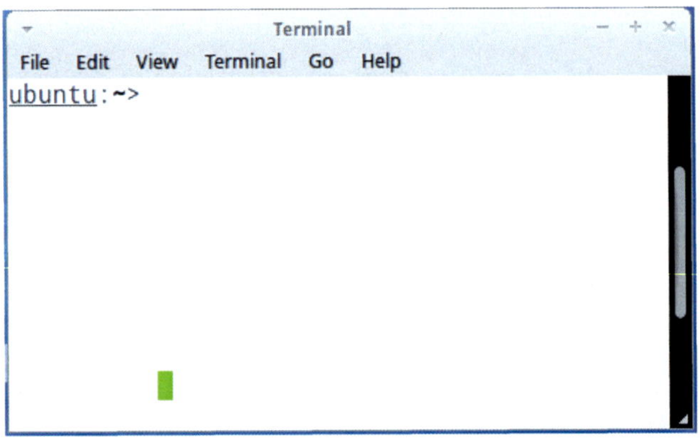

Figure C7 Terminal program. It looks very similar to the "dos" prompt in windows.

This is a good time to take a break and be familiar with some basic UNIX commands. An excellent tutorial can be found at http://www.ee.surrey.ac.uk/Teaching/Unix/. Just the basics like how to change to a directory, list contents of a directory, create a directory, copy/move a file are enough. We recommend you spend around an hour or two learning this.

A look around your WRF-Live system

Like in windows computers, WRF-Live (which is based on Linux) can also use a graphical file manager program to move around in the file system. File manager can be started by clicking the "File System" icon on the WRF-Linux desktop or using the "Start-menu" (accessories -> File Manager). Use the file manager to explore `/opt` directory. All the programs necessary to do this tutorial are under that directory. For example, in a later section we create a new simulation domain under: `/opt/wrf_ems/wrfems/runs`.

It is also possible to navigate the file system using the terminal (Figure C8). The command for changing directory is *cd* and listing the files in a directory is `ls`. For example, `cd /opt/wrf_ems/wrfems/runs` will change to `/opt/wrf_ems/wrfems/runs` directory. After that, if you use the command `ls` you will get a list of files in the `/opt/wrf_ems/wrfems/runs` directory.

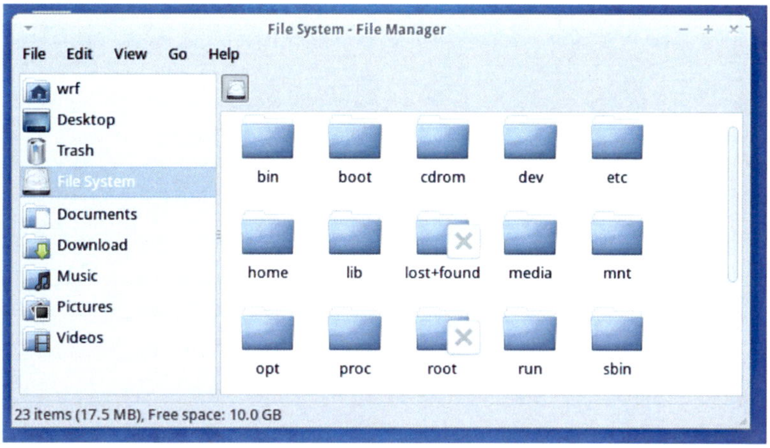

Figure C8 Navigate the file system using the terminal.

C.5 MUMBAI CASE STUDY

Background

On 26 July 2005, an unexpected heavy rainfall event occurred over Mumbai city and surrounding areas. The heavy rainfall started around 06GMT, and several regions within a 100 km radius recorded extremely high rainfall over the next 24-hour period. The Indian Meteorology Department raingauge at Santacruz recorded 944 mm during this period. The event caused nearly 500 deaths and is classified as a billion USD natural disaster (Lei *et al.* 2008).

A number of studies on this event has reported that it is a difficult event to capture in terms of storm position and accuracy (Lei *et al.* 2008; Deb *et al.* 2008; among others).

In this tutorial, we conduct an elementary simulation of this event.

Setting up the domain

Our objective here is to learn to use the WRF model. Keeping this in mind, we shall keep our domain resolutions and number of cells within certain limits, so as not to create a time-consuming simulation. There are many reported studies, including the ones referred to above, that will provide more suitable domain parameters.

Open a terminal window and type `dwiz`. This will open the WRF Domain Wizard, a graphical program that can be used to create domains easily. Select "Create a new Domain" option (Figures C9 and C10).

The next screen (Figure C11) shows a map of the world (The map may be centred on North America, use the scrollbars to move around). Move the map to show India. Then draw a square on the map to approximately indicate your modelling domain. Then specify the longitude and latitude of the centre-point by entering the correct numbers (Table C1) and select the projection to Mercator. After doing this click the large button (not shown in Figure C11) labelled "update map". The domain wizard will clip the world map to match your selection and display it (Figure C12). Now, in the "grid options" area specify the horizontal grid dimensions (50, 50) and grid spacing (30km). Set the geographic data resolution to 5m (Figure C12).

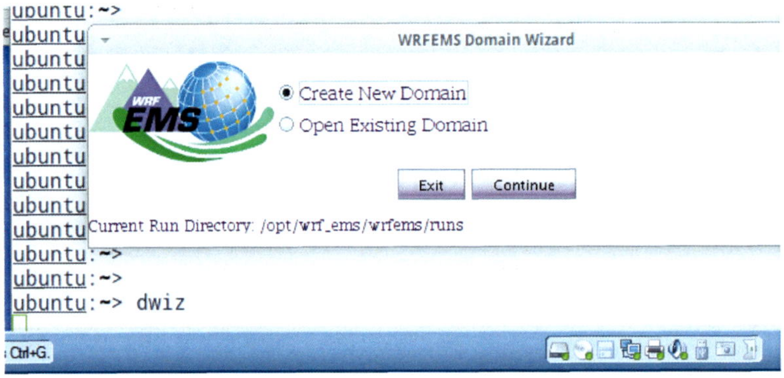

Figure C9 Start screen of WRF Domain Wizard.

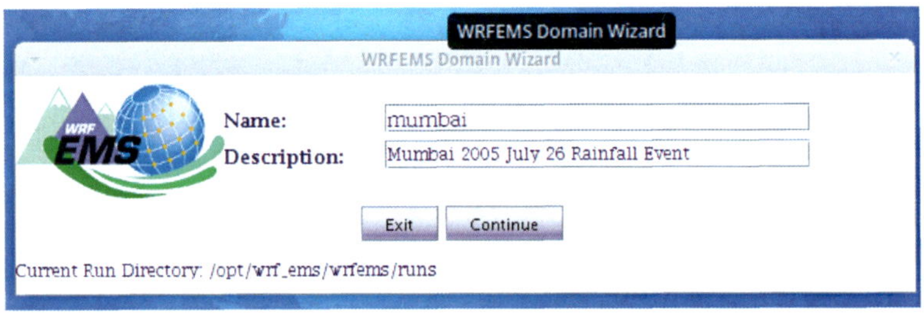

Figure C10 Give the name "mumbai" and some meaningful description.

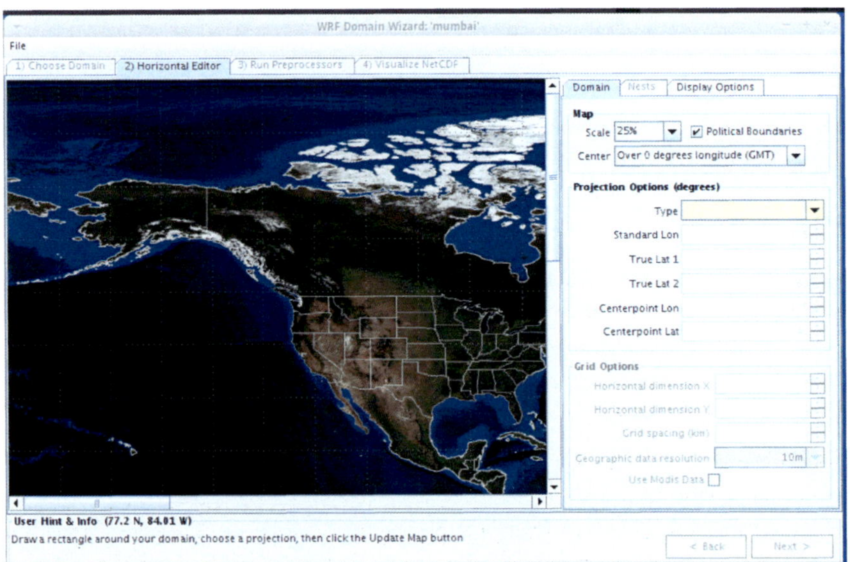

Figure C11 The screen for specifying domains.

Table C1 Domain parameters.

Domain centre	19.080 N
	72.855 E
Number of nesting levels	2
Domain resolutions:	
Parent	30 km
Child	10 km
Grid dimensions:	
Parent	50 × 50
Child	30 × 30

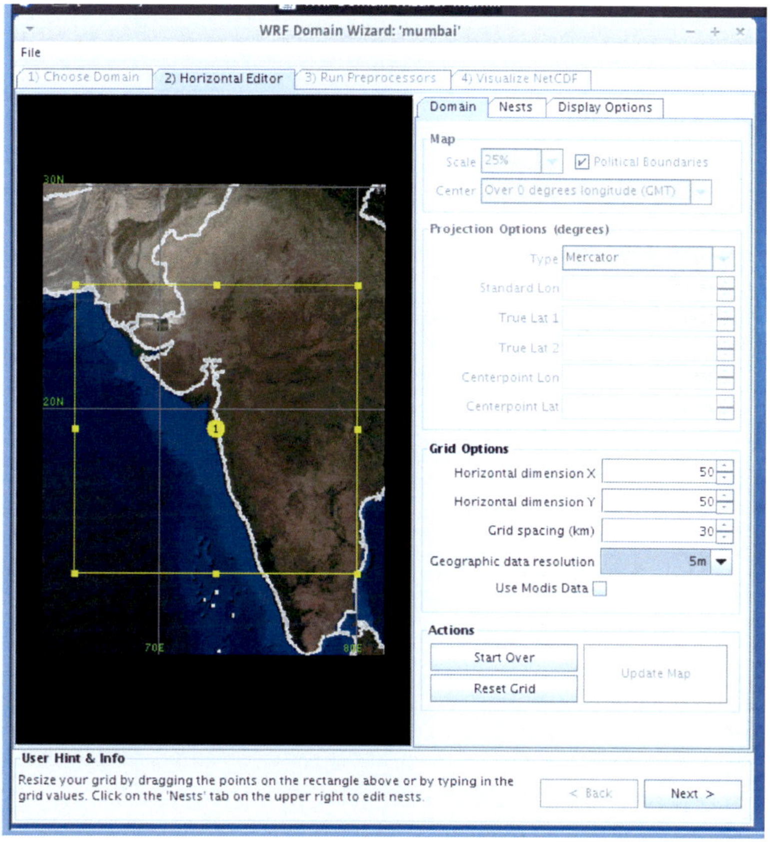

Figure C12 Parent domain ready.

Now, go to the tab "nests" and click on the button "new" to create the child (nested) domain (Figure C13). Change the geographic data resolution to 2m. As shown in Figure 11, make sure that the "nest coordinates" values are 19, 32, 32, 19, respectively. The grid spacing ratio should be 3. Clicking "OK" will create the child domain in the centre of the parent domain (Figure C14). The resulting domain (with the grid coordinates and grid spacing ratio mentioned above) is 40x40.

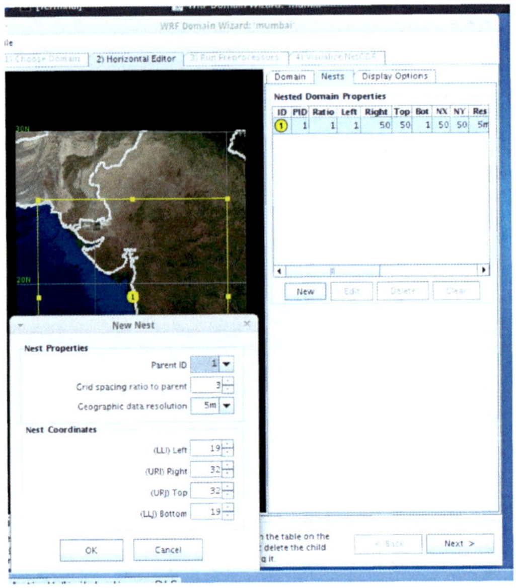

Figure C13 Create child domain.

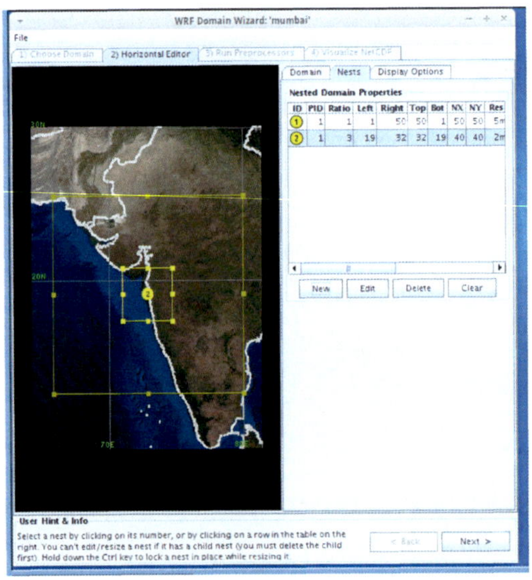

Figure C14 Child domain ready.

After making sure that the size (NX, NY) of the parent and child domains are 50x50, and 40x40 grid-cells respectively, click "Next". In the next screen, click the button "Localize domains". This will start the process of generating the modelling domains using geographical data and may take a few minutes to complete. If everything is OK, you should see a message:

```
***Successful completion of program geogrid.exe***
```

near the end of the long output (Figure C15).

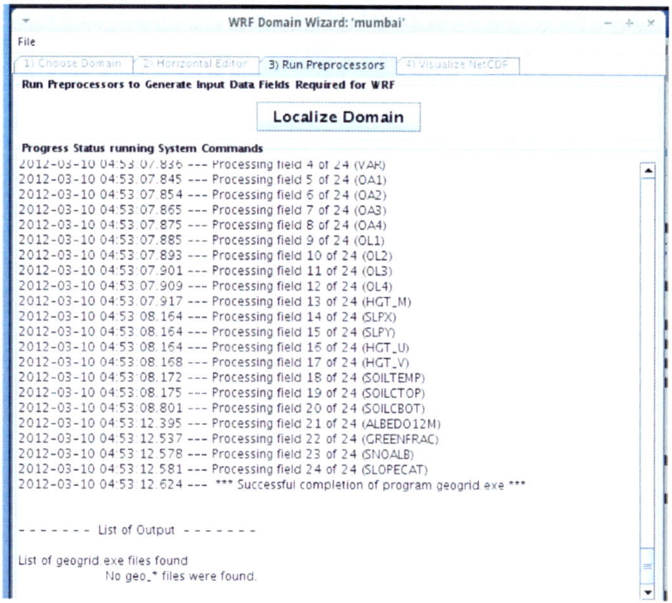

Figure C15 Successful completion of program geogrid.exe.

Click "Next" to go to the tab "Visualize NetCDF". Here the two domains that have been created should be listed (`geo_em.d02.nc`, `geo_em.d03.nc`). You are able to examine these domains using the viewer called Panoply. Select the `geo_em.d01.nc` and click "View in Panoply".

Click on one of the datasets and variables (Figure C16).

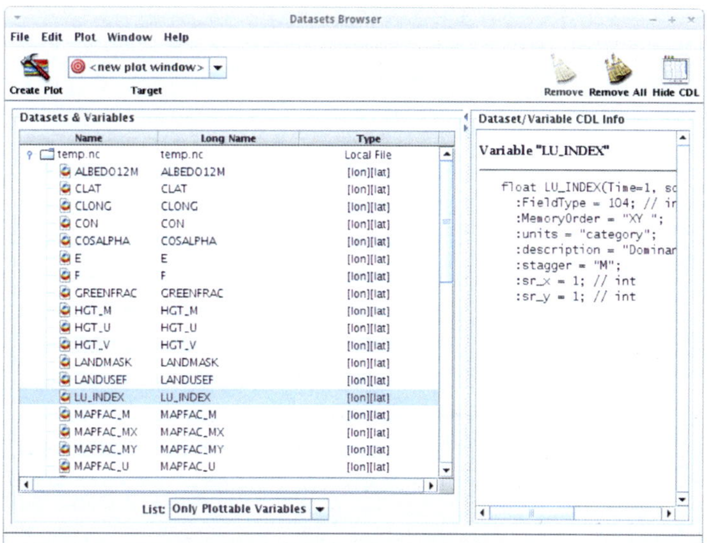

Figure C16 Panoply showing the data in domain 1 (geo_em.d01.nc).

Click on the Create Plot icon. Take some time to explore the domains using panoply. Interesting variables to explore could be the terrain height, landuse, vegetation fraction (seasonal), and so on (Figure C17). After this you can close both Panoply and WRF Domain Wizard programs.

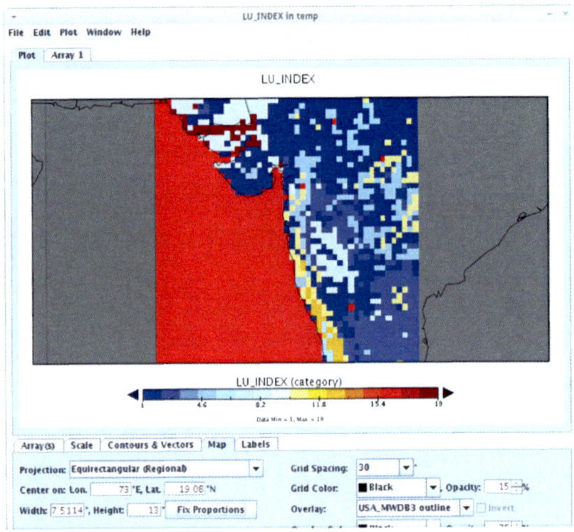

Figure C17 Plot of landuse index of domain 1.

Initial and boundary condition

Now that we have created the model domains, it is time to download some global data for boundary and initial conditions. This step (and the rest of our project) is done using the terminal.

Start a terminal window and change the directory to the newly created domain:

```
cd /opt/wrf_ems/wrfems/runs/mumbai
```

Your prompt should now look like below:

```
ubuntu:/opt/wrf_ems/wrfems/runs/mumbai >
```

We use the command `ems_prep` program to download atmospheric data and setup the boundary and initial conditions for our domains. Go ahead and type command `ems_prep` and press "Enter". Running the command like this (without any command arguments) will give a screen full of text explaining how to use the command. Now let us examine the capabilities of this program:

```
ems_prep -dslist
```

will give a list of data sources available (on the internet) to be used. We are going to use the Climate Forecast System (cfs) historical reanalysis data (0.5 deg resolution). The full command to do this is:

```
ems_prep -domains 1,2 -date 20050726 -cycle 00 -length 30 -dset cfsrptile
```

Option	Explanation
-domain 1,2	Process for both domain 1 and 2
-date 20050726	Start date of initialization data
-cycle 00	Start hour of initialization data
-length 30	Prepare data for 30 hours (since 20050726 00:00)
-dset cfsrptile	User cfs data

This command may take some time to complete – anything from a few minutes to a few tens – depending on your internet connection speed. Make sure that it does not produce any error messages before moving to the next section.

Run the simulation

Running the simulation with default model parameters is surprisingly easy. However, to do a useful simulation, it is always necessary to change the model parameters from default values. This is explained later in this tutorial. For the moment let us be content with the default parameters. The command `ems_run –domains 1,2` will start your simulation.

Be preared to wait. It is usual for the simulation to take more than an hour. Do not close the terminal window until the model has completed running. Following is how you can monitor the progress of your simulation during this time.

Open a new terminal window, and do the following to monitor the progress of your model run:

```
cd /opt/wrf_ems/wrfems/runs/Mumbai
tail -f rsl.out.0000
```

This will give you a running commentary of the model progress:

```
Timing for main: time 2005-07-26_21:29:10 on domain  2:  0.49510 elapsed seconds.
Timing for main: time 2005-07-26_21:30:00 on domain  2:  0.95303 elapsed seconds.
Timing for main: time 2005-07-26_21:30:00 on domain  1:  3.71809 elapsed seconds.
Timing for Writing wrfout_d02_2005-07-26_21:30:00 for domain  2:  0.13994 elapsed
seconds.
Timing for main: time 2005-07-26_21:30:50 on domain  2:  0.61399 elapsed seconds.
Timing for main: time 2005-07-26_21:31:40 on domain  2:  0.47157 elapsed seconds.
Timing for main: time 2005-07-26_21:32:30 on domain  2:  0.47762 elapsed seconds.
Timing for main: time 2005-07-26_21:32:30 on domain  1:  2.56080 elapsed seconds.
Timing for main: time 2005-07-26_21:33:20 on domain  2:  0.49284 elapsed seconds.
Timing for main: time 2005-07-26_21:34:10 on domain  2:  0.48890 elapsed seconds.
....
....
... wrf: SUCCESS COMPLETE WRF
```

Once you see the last line (SUCCESS COMPLETE WRF), you can rest assured that the simulation is finished. If you want now close the terminal in which you ran the `ems_run` command.

The output files of the model are now stored in the `wrfprd` sub-directory. The command:

```
cd /opt/wrf_ems/wrfems/runs/mumbai/wrfprd
ls
```

will give the following output:

```
wrfout_d01_2005-07-26_00:00:00 wrfout_d02_2005-07-26_00:00:00
```

Post processing the results

There are numerous ways of post-processing WRF model output. However, again in this tutorial we will cover one of the easiest. Issue the following command in a terminal: `chimp mumbai`.

The program will ask you for the choice of domain you want to open. Type the number and press Enter. This will open the selected output file in the chimplot program (Figure C18).

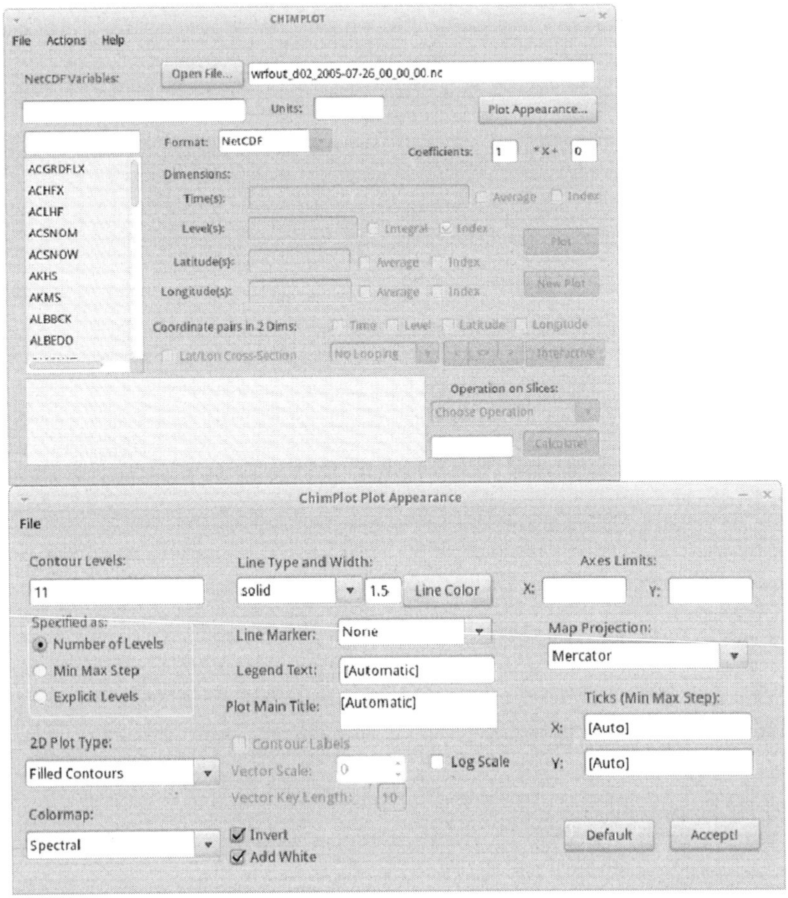

Figure C18 Chimplot main window (top). The windows for customizing plots (bottom).

Chimplot can be used to produce variety of plots of WRF output including contour plots of 2d variables (e.g. rainfall), cross-sections of 3d variables (e.g. relative humidity) and time series plots (e.g. rainfall time series for a given grid-point).

Plot rainfall

WRF simulation computes two types of rainfall: RAINC (rainfall output from cumulus parameterization) and RAINNC (explicit rainfall output) – both accumulated values since the start of the simulation. In our inner domain cumulus parameterization is switched off.

You can confirm this by checking the file:

```
/opt/wrf_ems/wrfems/runs/mumbai/conf/ems_run/run_physics.conf
```

for the parameter `CU_PHYSICS`. It says `CU_PHYSICS=1,0` meaning cumulus parameterization is switched on for the first (parent) domain and off for the second (child) domain. Therefore, to obtain the total rainfall we can just plot RAINNC. Double click the RAINNC on the list of variables (Figure C19) so that it will be displayed in the text box above. Select “Time” for looping variable. Then click the Plot button. In a few seconds the button’s text will change to “Show”.

Click on “Show” button to view the plot. The time plot shown can be changed by pressing the “<” and “>” signs on the CHIMPLOT window.

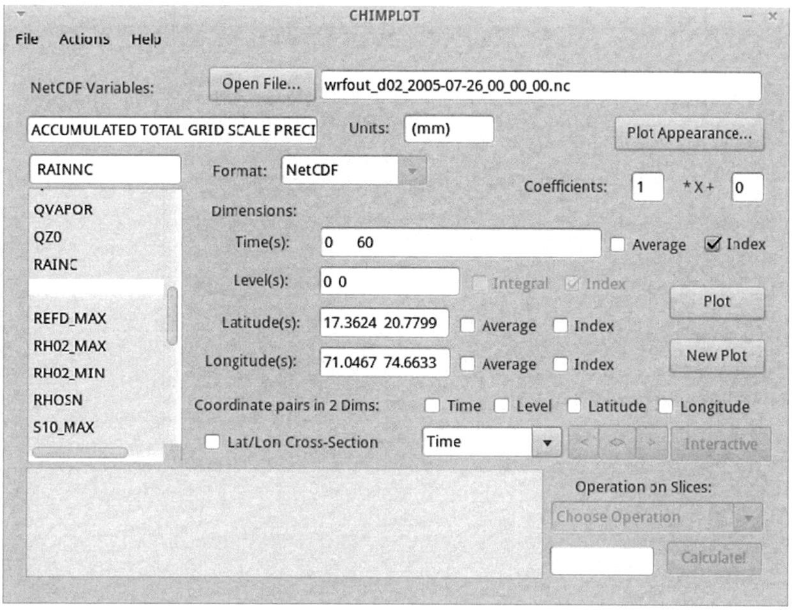

Figure C19 Annotated view of the chimplot main window: (a) Variable list, (b) Selected variable, (c) Variable for looping, (d) Controls for looping, (e) Plot button.

A better looking plot can be obtained by changing plot parameters (Figure C20 bottom).

Figure C21 shows a cross section plot of the cloud water mixing ratio (QCLOUD). This is obtained by selecting the QCLOUD variable (which is 4d, meaning x,y,z and t), setting latitude to 17.81 (instead of the range) and selecting Time as loop variable.

Figure C20 shows the accumulated rainfall time series at (74.2 E, 17.81 N). It can be plotted by selecting the RAINNC variable and fixing latitude and longitude values (instead of range).

Another important feature of Chimplot is its ability to export data. After making a plot, just select `File->Save ASCII` on the main menu, to do this.

Finishing off

To save your computers resources (e.g. Memory) its best not to leave it running, when you are not working with the WRF-live system.

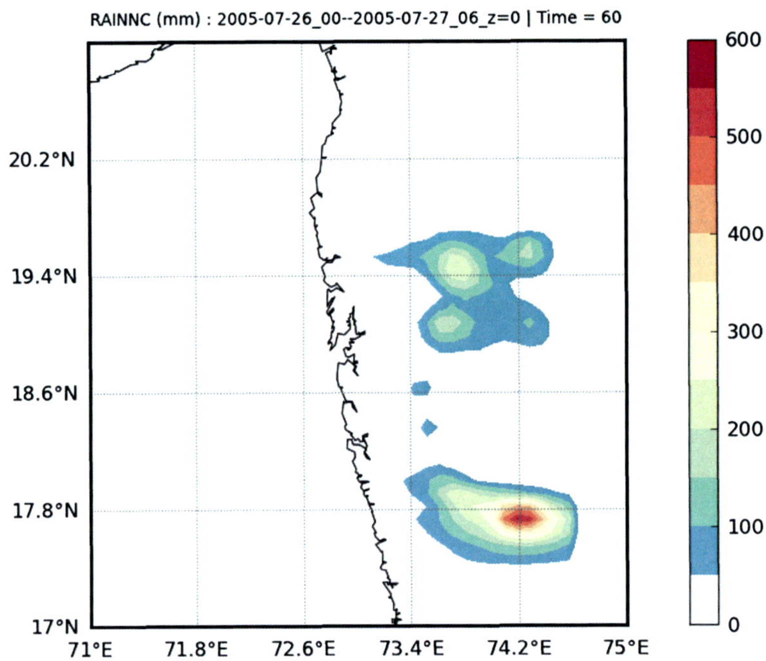

Note: your plot may show slightly different.

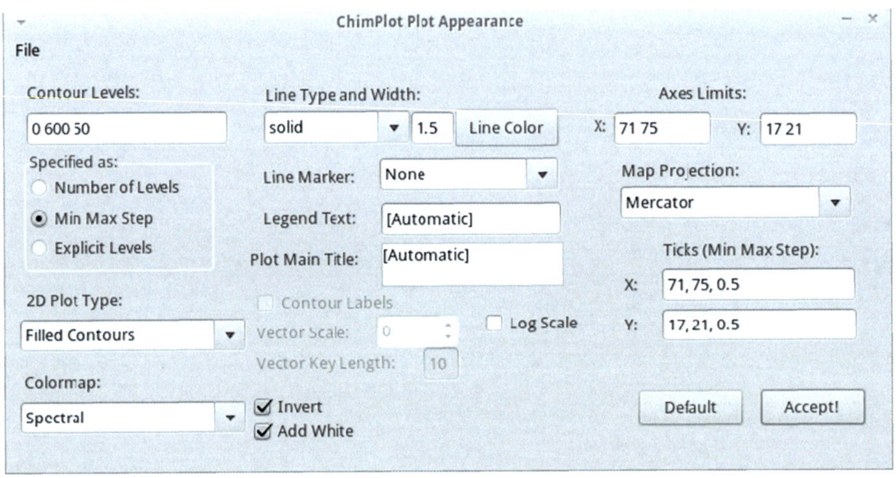

Figure C20 Accumulated rainfall (RAINNC) at the end of the simulation (top). The parameters used to obtain this plot (bottom).

Just close the VMware window.VMware will ask to confirm whether you would like to "suspend the Vistual Machine and exit". Just say yes and this will save the status of WRF-live system in your computer's hard drive. (This will take a minute or so, so please be patient after closing the window). After letting this process complete, you can safely shutdown your (parent or host) computer.

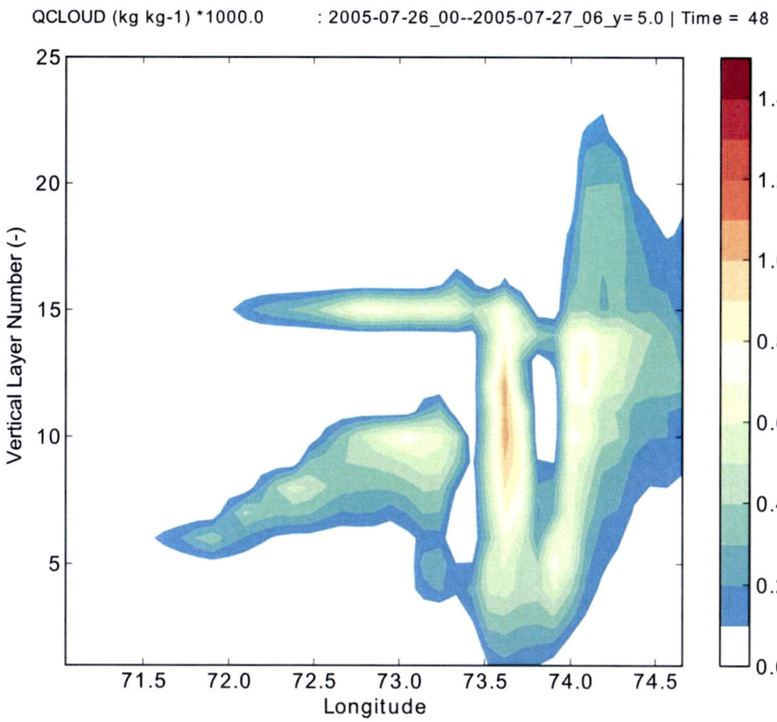

Note: your plot may show slightly different.

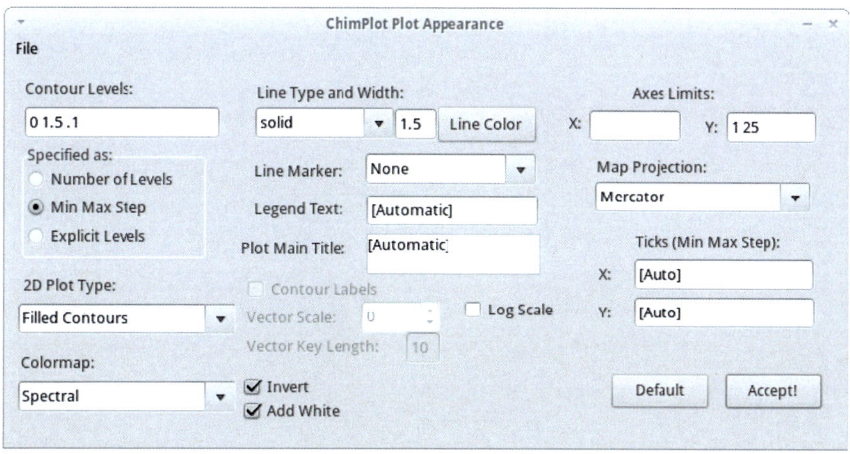

Figure C21 Cloud water mixing ratio (g/kg) along a longitudinal cross-section (at 17.81 N) after 24 hours of start of the simulation (top). The parameters used to obtain this plot (bottom).

Getting files out

For all practical purposes, WRF-live system is a separate "virtual" computer, having its own file system. There are situations in which you would like to export some files (say simulation results or an image file). This can be done by setting up a shared directory between the WRF-live system and the host computer.

Select "Virtual Machine -> Virtual Machine settings" on the VMware window. A new window will pop-up Select the "options" tab and choose "Shared folders" option. Set "Folder Sharing" to enabled. Click "Add" button to add a directory on the host computer to the share list. Say you select "Desktop" on your host computer, then, it will appear as: `/mnt/hgfs/Desktop` on WRF-live system. Any file you copy to `/mnt/hgfs/Desktop` within the WRF-live system will appear automatically on your desktop.

C.6 WHERE TO GO FROM HERE

We have taken a whirlwind tour of the WRF-live system to learn the elementary use of WRF model. As stated at the onset, our objective for this task was to learn how to operate WRF model in a PC. We did not cover many topics that are extremely important in order to use WRF effectively for research or operational purposes. There are many excellent resources out there that can be used to learn the proper use of the model. We will make some recommendations later in this section.

Utilizing multiple processor cores

Many modern computers, even laptops, have more than one processor core. WRF model can very efficiently utilize these to speed-up simulations. Here's how to do this in WRF-live system:

First, find out how many processor cores your computer has. For typical consumer computers, this is either one, two or four. Usually this is indicated in the computer's documentation (printed or online). If you have only one core, skip reading this section.

There are two steps: First you have to get Vmware to use multiple processor cores to run WRF-live system. Then you have to tell the WRF model to use those four cores when running a simulation.

Step 1:
First, shutdown the WRF-live system: Open a terminal and type the following:

```
sudo /sbin/shutdown -h
```

This will ask your password (remember the original password for WRF-live is "wrf"). The command will shutdown the WRF-live system.

Then go to the WRF-live directory (Where you copied WRF-live files before staring the tutorial) and edit the file RUN.bat with notepad (Write click on the file and select "Edit").

If you have four processor cores change as follows:

RUN.bat file for a single processor	RUN.bat file for four processor cores

`.\system\WRF_live.vmx`
`rem .\system\WRF_live-2CPU.vmx`
`rem .\system\WRF_live-4CPU.vmx`
`exit`
	`rem .\system\WRF_live.vmx`
	`rem .\system\WRF_live-2CPU.vmx`
	`.\system\WRF_live-4CPU.vmx`
	`exit`

Now when you start WRF-live by clicking on RUN.bat file, the VMware will utilize all four cores.

Step 2:
Use the domain wizard to create the simulation domain (say "mumbai"). Then in WRF-live system, use the file manager to navigate to the directory: `/opt/wrf_ems/wrfems/runs/mumbai/conf/ems_run` and edit the file `run_ncpus.conf` (Double clicking on the file will open it for editing.) Make the following changes in the middle of the file (from line 105 to 110):

Original run_ncpus.conf file	Run_ncpus.conf using four processor cores
`# go down in flames. And flames is never good unless you're grill'n something.`	`# go down in flames. And flames is never good unless you're grill'n something.`
`#`	`#`
`REAL_NODECPUS=local:1`	`REAL_NODECPUS=local:1`
`WRFM_NODECPUS=local:1`	`WRFM_NODECPUS=local:4`
`# 2. DECOMP (Possible values 0, 1, or 2; Note changes from V3.0)`	`# 2. DECOMP (Possible values 0, 1, or 2; Note changes from V3.0)`
`#`	`#`

Then save the file. Now the "mumbai" simulation will use four processor cores to run.

Do we need VMware?

The purpose of VMware in this tutorial was to run Linux system on top of a windows system. While this made our lives much easier, but not having to fiddle with the computer too much, it is not the most efficient way of running WRF model. When you use VMware, two operating systems (Windows and Linux) share valuable resources like memory and the CPUs. This is fine if you just wanted to complete this tutorial, or want to run a few small WRF simulations. However, if you want to use the model for research, you may end up using it more often. In this scenario, our suggestion is to install Linux without using VMware. While this is more complicated, it is not too difficult to do. Install an easy to use, modern Linux distribution like Mint-Linux, or Ubuntu. Then copy the `/opt` directory with all its content to the new system. Also copy the file `/home/wrf/.cshrc` to your "home directory" in the new system. If needed get the help of someone who knows a bit of Linux to do this. Linux these days can be safely installed alongside Windows.

Chimplot and others

Chimplot (http://www.lmd.polytechnique.fr/chimplot/), as you have already seen is a really user-friendly plotting system for WRF. However, the tool is still at its infancy. Check the website of the tool to see whether the author has released a new version that may include new features.

However, there are many other capable plotting programs available for WRF. Vapour is a 3D visualization system (http://www.vapor.ucar.edu/) that can be used to produce beautiful 3D animations from WRF output. It can be run on windows too. You can export the output of wrf model from wrfprd directory to windows and use Vapor to visualize it. Another tool is RIP (http://www.mmm.ucar.edu/wrf/users/docs/ripug.htm) which is already installed in WRF-live system. Yet another is NCAR command language (NCL - http://www.ncl.ucar.edu/) that is a scripting language to produce plots from

WRF output. There are many example scripts on WRF web site. Unidata's Integrated Data Viewer (IDV) is a freely available, multiplatform visualization and analysis tool that can view WRF output. Finally, general purpose mapping and plotting tool like GMT (http://gmt.soest.hawaii.edu/) can be used to produce maps with WRF output (http://assela.pathirana.net/Plotting_WRF_Model_output_with_ GMT_%28Generic_Mapping_Tools%29).

More information

As stated before there are some excellent online sources to help you to learn WRF model. Our first suggestion is the "The National Weather Service Science Operations Officers Science and Training Resource Center (NWS-SOO-STRC)" (http://strc.comet.ucar.edu/software/newrems/). In WRF-live system we use the WRF binaries and other tools (collectively known as STRC-EMS) provided by NWS-SOO-STRC as is. STRC-EMS users guide describes in detail, various EMS commands (ems_prep, ems_run, etc.) and all the parameter files that can be changed to customize a simulation. For example, `run_physics.conf` file lists all the WRF physics options. (e.g. Cumulus parameterization, cloud microphysics and planetary boundary layer parameterization choice.) For any serious use of WRF model, changing such parameters is routinely needed.

The next source for information is the WRF official documentation available at Mesoscale and Miscroscale Meteorology Section of University Cooperation for Atmospheric Research (UCAR) (http://www.mmm.ucar.edu/wrf/users/). Particularly useful are the "Technical description of ARW", WRF Dynamics and Physics Document. But, be careful when reading WRF User's guide available in this site, for it describes a different (classical) way of using WRF model than that of STRC-EMS system that we have adopted for this tutorial. While simulation results will be identical, technical details of running the model components are quite different in the two systems.

Index